THE YEAR IN DYSLIPIDAEMIA 2003

THE YEAR IN DYSLIPIDAEMIA

2003

PAUL DURRINGTON, MICHAEL MACKNESS
PAUL MILLER, ALAN REES

CLINICAL PUBLISHING SERVICES
OXFORD

Distributed worldwide by
CRC Press
Boca Raton London New York Washington, D.C.

Clinical Publishing Services Ltd

Oxford Centre for Innovation
Mill Street, Oxford OX2 0JX, UK

Tel: +44 1865 811116
Fax: +44 1865 251550
Web: www.clinicalpublishing.co.uk

Distributed by:

CRC Press LLC
2000 NW Corporate Blvd
Boca Raton, FL 33431, USA
E-mail: orders@crcpress.com

CRC Press UK
23–25 Blades Court
Deodar Road
London SW15 2NU, UK
E-mail: crcpress@itps.co.uk

First published 2003

A catalogue record for this book is available from the British Library

ISBN 1 904392 07 5

Commissioning Editor: Jonathan Gregory
Project Manager: Rosemary Osmond
Typeset by Footnote Graphics Limited, Warminster, Wiltshire
Printed in Spain by TG Hostench SA, Barcelona

Contents

PART IV

Lipid-lowering trials

Paul Durrington

Contributors

Paul N Durrington, BSc, MBChB, MD, FRCPath, FRCP, Clinical Research Division II, University Medical Unit, Manchester Royal Infirmary, Oxford Road, Manchester

Marc Evans, Mb BCh, MRCP, MD, Department of Medicine, University Hospital of Wales, Heath Park, Cardiff, Wales

Michael Mackness, BSc (Hons), PhD, Lecturer in Medicine, Paraoxonase Research, Department of Medicine, Manchester Royal Infirmary, Oxford Road, Manchester

J Paul Miller, BM, BCh, MSc, DPhil, FRCP, Department of Medicine and Gastroenterology, South Manchester University Hospitals, Wythenshawe Hospital, Manchester

J Alan E Rees, BSc, MD, FRCP, Department of Medicine, University Hospital of Wales, Heath Park, Cardiff, Wales

Part I

Lipid disorders

1

Primary hypercholesterolaemia

The cholesterol quartet

Most people with primary hypercholesterolaemia have no easily discernible genetic abnormality. Their hyperlipidaemia is often described as 'common' or 'polygenic' and is probably the result of multiple minor genetic influences compounded by lifestyle factors, particularly the interrelated effects of diet, lack of exercise and obesity.

There are however four single gene abnormalities of cholesterol metabolism which lead to relatively severe hypercholesterolaemia and premature coronary heart disease (CHD). Goldstein and Brown have called these the cholesterol quartet |**1**| and they are discussed below. The commonest, the best known and the most severe of these is familial hypercholesterolaemia (FH) |**2**|, described clinically more than 60 years ago |**3**| and shown in 1973 by Brown and Goldstein to be the result of mutations in the low-density lipoprotein (LDL)-receptor, of which about 700 have now been described. Familial defective apolipoprotein B-100 (FDB) results from a single amino acid substitution in apoB-100, usually at position 3500, which also interferes with LDL removal by the LDL-receptor as a result of impaired binding |**4**|.

The rare autosomal recessive hypercholesterolaemia (ARH) can cause hypercholesterolaemia almost as severe as that in homozygous FH, but is distinguished by the fact that the parents, who are obligate heterozygotes, have normal cholesterol levels. This contrasts with the situation in FH, a dominant disorder with close to full penetrance and a gene dosage effect, in which the heterozygotes have hypercholesterolaemia and the homozygotes are even more severely afflicted.

In sitosterolaemia, perhaps better called phytosterolaemia, plant sterols together with cholesterol accumulate in plasma. Sitosterol is the predominant phytosterol found together with lesser amounts of campesterol and stigmasterol |**5**|.

The molecular defects in both ARH and phytosterolaemia have recently been described and are discussed below.

Familial hypercholesterolaemia

Detection and diagnosis

Population and even high-risk screening strategies are an inefficient way of detecting new cases of FH and there is widespread agreement that a case-finding approach, which investigates the relatives of known cases, is the most effective. As will be seen from the studies reviewed there have been different approaches to the diagnosis of FH. Genetic testing for mutations in the LDL-receptor is regarded as the gold standard. It is not uncommon however to fail to find a receptor mutation in more than half of FH cases diagnosed by clinical and biochemical criteria. A case-finding strategy based solely on genetic testing will therefore exclude many of these kindreds from the screening programme. Not all will have FH but many are likely to have high-risk disorders that would benefit from earlier detection.

Increasingly we are faced with a semantic problem in relation to the definition of FH. Strictly the term should be reserved for those patients with autosomal dominant hypercholesterolaemia due to mutations in the LDL-receptor. In practice this is not feasible for most cases who are diagnosed on clinical grounds |**6**| are not currently subject to genetic testing, which in any case often gives a low yield. Those patients who have the FH phenotype but possess other mutations, such as FDB and FH3 (see below), will usually be regarded as having FH until proved otherwise by DNA testing.

Information about mutations in the LDL-receptor giving rise to FH can be found at the FH website (http://www.ucl.ac.uk/fh), based at University College London and recently described by Heath *et al.* |**7**|. At the time of writing their report 683 mutations had been recorded.

A molecular genetic service for diagnosing individuals with familial hypercholesterolaemia (FH) in the United Kingdom.

Heath KE, Humphries SE, Middleton-Price H, Boxer M. *Eur J Hum Genet* 2001; **9**: 244–52.

BACKGROUND. A genetic diagnostic service for FH has been established over the last four years in the Clinical Molecular Genetics Laboratory at Great Ormond Street Hospital for Children NHS Trust (GOSH), London. In total there have been 368 referrals; 227 probands and 141 family members, which have come from a number of lipid clinics and from general practitioners. FH is caused by mutations in the low-density lipoprotein receptor gene (LDLR) and these are analyzed by single strand conformational polymorphism (SSCP), DNA sequencing and direct assays. The clinically indistinguishable disorder, FDB, is caused by one of three mutations in the apolipoprotein B-100 gene (apoB) which are analyzed by direct assays.

INTERPRETATION. Mutations predicted to be pathogenic were found in 76 probands, 67 in LDLR (23 previously undescribed) and nine in apoB. The mutation detection rate was 53% in paediatric probands, 32% in adults with a 'definite' FH diagnosis (tendon xanthoma-positive) and 14% in adults with a 'possible' FH diagnosis (tendon xanthoma-negative). The predicted loss of sensitivity that would result from reducing the number of exons tested has been assessed, and a molecular screening strategy suitable for UK patients is proposed. A similar strategy may be useful for other countries where genetic heterogeneity results in a wide mutation spectrum for FH.

Comment

The UK molecular genetic service for the detection of mutations in patients with FH or FDB described here has proved valuable to UK lipidologists not only for research, but also in individual cases, where it is unclear whether or not the patient actually has FH. There are for example families without tendon xanthomata but who from their lipid levels and pattern of inheritance are suspected of having FH (or FDB, which is clinically indistinguishable). In other instances there are individuals with modest hyperlipidaemia who are members of families with known FH, but it is unclear if they too have the disease or some other milder disorder. These patients may be reluctant to take drug treatment unless their risk is perceived to be particularly high, which it may not be in young patients or those without other risk factors. FH patients however have a higher risk than would be predicted simply from the ratio of their total to high-density lipoprotein (HDL) cholesterol, probably because their LDL levels have been high since birth, justifying earlier treatment and making them unsuitable for the usual risk assessment methods [8].

The clinical criteria used for the diagnosis of 'definite' or 'possible' FH are those of The Simon Broome FH Register [6]. It is notable that, even in 'definite' FH, mutations are found in less than half of adult cases. Reasons for this include the possibility that they do not in fact have the disorder or that they have a mutation which is not one of the approximately 700 so far described, and which is not identified. Moroever the technique used (SSCP) is useful for screening large numbers of samples but has a sensitivity of only 75–85%.

Recent work suggests that there is also a group of patients (designated FH3) with autosomal dominant hypercholesterolaemia, who have mutations neither in the LDL-receptor nor in the apoB gene, but who may have tendon xanthomata [9–11]. The function of the defective gene(s) and its frequency amongst clinically defined FH populations are unknown. Other approaches to the diagnosis of FH include assay of LDL-receptors in fibroblasts or leucocytes, but this is time-consuming, expensive and not infrequently yields equivocal results. Imaging of Achilles tendons can be helpful (see Descamps *et al.* below), but there is no single method at present that will give an unequivocal answer in all cases.

Review of first 5 years of screening for familial hypercholesterolaemia in The Netherlands.

Umans-Eckenhausen MA, Defesche JC, Sijbrands EJ, Scheerder RL, Kastelein JJ. *Lancet* 2001; **357**: 165–8.

BACKGROUND. FH is a common lipid disorder that predisposes for premature cardiovascular disease (CVD). A screening programme was set up in The Netherlands

in 1994 to: establish the feasibility of active family screening supported by DNA diagnostics; assess whether or not active identification of these patients with FH would lead to more cholesterol-lowering treatment; and compare diagnosis by DNA analysis with that by cholesterol measurement. Both DNA analysis and measurement of cholesterol concentrations were used to screen families in which a functional mutation in the LDL-receptor gene had been detected. In the first 5 years, 5442 relatives of 237 people with FH were screened; 2039 individuals were identified as heterozygous by LDL-receptor gene mutation analysis. At the time of examination, 667 of these adults with FH (39%) received some form of lipid-lowering treatment; 1 year later, this percentage had increased to 93%. In addition, laboratory analysis showed that for carriers as well as non-carriers 18% would have been misdiagnosed by cholesterol measurement alone, with sex-specific and age-specific 90th percentiles of the general Dutch population as diagnostic criteria.

INTERPRETATION. Targeted family screening with DNA analysis proved to be highly effective in identifying patients with hypercholesterolaemia. Most of the identified patients sought treatment and were successfully started on cholesterol-lowering treatment to lower the risk of premature CVD. Our findings could have wider relevance for the screening of other prevalent genetic disorders in the population at large.

Comment

About 500 subjects will need to be tested to identify one new FH case in random population screening. A high-risk screening strategy will often miss cases since not all have a notably adverse family history particularly where the gene has passed down the female line. FH subjects commonly have normal blood pressure and are often not obese so they are not particularly likely to be identified because of the presence of multiple risk factors. It is known that many cases go unrecognized |**12**|. A more effective approach to their identification is likely to be through the screening of relatives of known cases.

In this study the 237 index cases of FH were initially diagnosed clinically and the diagnosis confirmed by the identification of a mutation in the LDL-receptor. Where an affected first-degree relative was found then their relatives were also tested. Second-degree relatives were tested only if the relevant first-degree relative was not available. The finding of an average of 8.6 cases of FH for each index case is impressive. Only half of these knew that they had elevated serum cholesterol and only 39% of those under the age of 60 years were on treatment with statins. In most of the remainder, treatment was initiated within a year illustrating the power of this approach.

The most favourable total cholesterol concentration to distinguish between carriers and non-carriers amongst the relatives was the age- and sex-specific 90th centile for the Dutch population. Using this there was still about an 18% false positive and false negative rate. Accordingly the authors emphasize the need for genetic testing in parallel with clinical diagnosis and undoubtedly this will give added precision. The disadvantage of this approach is that only kindreds of patients with a known LDL mutation can be screened making it impracticable in areas where genetic testing is not readily available. Moreover there is usually a substantial proportion of cases in whom FH is diagnosed clinically but in whom no mutation can be found (see Heath *et al.* above).

Bhatnager *et al.* [13] used clinical criteria alone for the diagnosis of FH and demonstrated that screening 200 first-degree relatives of FH patients yielded 121 new cases. About 60 000 blood tests would be needed to identify a similar number of cases by random population screening. Given the findings of the Dutch group in the current work one might predict some lack of sensitivity from a clinical approach without concomitant DNA testing. This seems unlikely to have been great given that 121 cases were identified where 100 would have been predicted. There may of course have been some false positive diagnoses, though some of these are likely to have been in patients with FDB or other high-risk disorders.

Low-density lipoprotein receptor (LDLR) gene mutations in Canadian subjects with familial hypercholesterolemia, but not of French descent.

Wang J, Huff E, Janecka L, Hegele RA. *Hum Mutat* 2001; **18**: 359.

BACKGROUND. Heterozygous FH is a relatively common autosomal dominant disorder, which is characterized by elevated plasma concentrations of LDL cholesterol and early CHD. FH results from mutations in the gene encoding the LDL-receptor (LDLR). In Canada, there is a founder effect for LDLR mutations in FH among individuals of French descent, most of whom reside in the province of Quebec. However, the spectrum of mutations in other regions, specifically in the populous and predominantly English-speaking province of Ontario, has not been studied. In this study, the coding regions, promoter and intron-exon boundaries of the LDLR gene were sequenced in 60 Ontario FH subjects from a variety of ethnic backgrounds other than French Canadian. We found 25 LDLR mutations in 34 subjects. Eleven LDLR mutations were novel, including two in-frame deletions of a single amino acid (one each in exons 2 and 4), two larger deletions that shifted the reading frame (one each in exons 4 and 10), five missense mutations (C42R, A370T, T413M, L561P and E760D) and two splice acceptor mutations (one each in introns 3 and 8).

INTERPRETATION. The results indicate that FH is more genetically diverse in Ontario than in Quebec. The results are also consistent with findings from investigations of the LDLR in FH conducted in other countries, in which polymerase chain reaction (PCR)-based, exon-by-exon sequencing uncovers small mutations in about half of the subjects screened. The gap in molecular diagnosis suggests that lesions not found by this sequencing strategy, such as larger scale LDLR mutations that cannot be amplified, may underlie a substantial number of cases of FH. Alternatively, there might be genetic heterogeneity underlying the FH phenotype, with contributions from other single or multiple genes.

Comment

The full text of this paper is published only on-line. FH results from mutations in the LDLR of which about 700 have been described to date. In many countries, including the UK, the FH population is very heterogeneous with multiple mutations responsible for the disorder. In a few countries or communities there is a strong founder effect with a small number of mutations being responsible for most cases. These include Lebanese Christians, Afrikaners in South Africa and French Canadians who reside predominantly in Quebec. Five mutations are responsible for two-thirds of cases of FH in Quebec. This work demonstrates that elsewhere, in the English-speaking population of Ontario, the situation is similar to that in Europe with many mutations of the LDLR underlying the manifestations of FH. These are thought to represent the ethnic origin of immigrants to Canada since FH has not been identified in Canadian aboriginees.

Clinical *versus* molecular diagnosis of heterozygous familial hypercholesterolaemia in the diverse South African population.

Vergotine J, Thiart R, Kotze MJ. *S Afr Med J* 2001; **91**: 1053–9.

BACKGROUND. FH is a common genetic disease characterized by strikingly elevated plasma cholesterol concentration, which can lead to premature coronary death if left untreated. In this study DNA diagnosis of FH, which allows detection before onset of clinical symptoms, was evaluated against biochemical parameters routinely used to identify subjects with FH. A population-based strategy was used to identify LDLR gene defects in South Africans with clinical signs of FH, followed by a family-based DNA screening approach for presymptomatic diagnosis of FH. DNA screening of 790 at-risk relatives for the FH-related mutations identified in 379 index cases, allowed accurate disease diagnosis in an additional 338 relatives and exclusion of the relevant mutation in 452 individuals. The sensitivity and specificity of the diagnosis, based on total cholesterol values measured in family members of FH heterozygous index cases with one of the three founder-related mutations, D154N, D206E and V408M, were 89.3% and 81.9%, respectively.

INTERPRETATION. The predominance of ten LDLR gene mutations in the local population justifies population-directed DNA diagnosis of FH in South Africa on a routine basis, particularly since expression of the defective gene measured in biochemical tests does not allow accurate diagnosis of FH in all cases. DNA testing provides a definitive tool for family tracing aimed at pre-clinical diagnosis and preventive treatment of FH.

Comment

The prevalence of FH in the Afrikaner population of South Africa is about 1 in 70 compared to an average of around 1 in 500 in many other societies. This is believed to be the result of a founder effect and about 90% of cases result from one of three mutations. These mutations are also common in the South African coloured population of mixed ancestry in which they may account for about 20% of cases (actually 46% in this study). A different mutation is also common in South African Jews and Indians.

This study reports only on the screening of relatives of FH index cases in whom an LDLR mutation had been identified. The finding of 338 affected relatives amongst 790 at-risk relatives is close to expectations. The use of DNA testing is particularly appropriate in communities where a large proportion of FH cases result from a relatively small number of defined mutations. This should not however, even in these communities, detract from family screening, by clinical and biochemical criteria, of the relatives of clinically diagnosed FH cases in whom no mutation can be identified.

Pre-natal diagnosis of familial hypercholesterolemia: importance of DNA analysis in the high-risk South African population.

Vergotine J, Thiart R, Langenhoven E, Hillermann R, De Jong G, Kotze MJ. *Genet Couns* 2001; **12**: 121–7.

Background. In this report on the outcome of the first prenatal diagnosis performed for FH in a South African family, the aim was to demonstrate the value of a population-directed screening strategy to identify FH patients in populations with an enrichment for certain LDLR gene mutations. Prenatal diagnosis was offered to an Afrikaner couple, both partners being heterozygous for the FH mutation D206E, whose first child was diagnosed with heterozygous FH and the second with homozygous FH. Genomic DNA isolated from parental peripheral blood and subsequently amniotic fluid was amplified by the PCR and subjected to mutation analysis. Heterozygosity for mutation D206E was confirmed in both parents, whilst this mutation was not detected in DNA directly amplified from amniotic fluid. To exclude the possibility of a false negative result due to the limited number of cells in the uncultured amniotic fluid sample, cells were also cultured *in vitro*, and the DNA extracted and subjected to a second round of analysis. This confirmed the absence of mutation D206E in the fetus.

Interpretation. This case illustrates the application of a DNA-based mutation detection technique as a simple and rapid diagnostic aid that can be carried out at a relatively early gestational stage. Prenatal diagnosis of FH, aimed at the detection of homozygous cases, is particularly feasible in populations and families with molecularly defined LDLR gene mutations.

Comment

The advent of the statins has made the treatment of heterozygous FH relatively straightforward though there remain difficult and poorly responsive cases. The risk of heterozygous FH would very rarely be an indication for prenatal diagnosis with a view to termination of pregnancy. Homozygous FH however remains a very high-risk disorder with impaired prognosis despite heroic treatments which include frequent and essentially lifelong LDL-apheresis or plasma exchange, or even liver transplantation. Thus prenatal diagnosis is something that parents who are both carriers of an LDLR mutation might well wish to explore according to their religious or ethical principles.

The use of Achilles tendon ultrasonography for the diagnosis of familial hypercholesterolemia.

Descamps OS, Leysen X, Van Leuven F, Heller FR. *Atherosclerosis* 2001; **157**: 514–8.

BACKGROUND. Differentiating FH from other causes of hypercholesterolaemia has important clinical and therapeutic implications but is often not possible by standard clinical criteria. As accumulation of cholesterol in tendon is generally considered as pathognomonic of FH, the sensitivity and specificity of clinical and ultrasonographic tendon characteristics were evaluated using the data of 127 genetically ascertained FH subjects and 160 controls with various lipid profiles. Upon clinical examination, none of the controls and 29% of FH individuals (17% FH women and 38% FH men) presented with xanthomata in Achilles tendons (AT), but no female and only 6% of male FH patients also showed xanthomata in the extensor tendon of the hand. Amongst all possible quantitative parameters (thickness, breadth, section and roundness) of AT measured by ultrasonography, the thickness presented the best receiver operating curves. AT thickness above 5.8 mm was the most useful threshold for diagnosis of FH, procuring sensitivity of 75% and specificity of 85%. Analysis of variation of AT thickness with age and sex indicated that this clinical criterion performed better in females older than 45 and in males under 45. In patients carrying the APOB-R3500Q mutation, AT-thickness appeared significantly less important compared with those carrying LDLR mutations.

INTERPRETATION. In conclusion, this study recommends identification of possible FH individuals amongst hypercholesterolaemic patients using a criterion of AT-thickness over 5.8 mm eventually associated with a specific genetic test for APOB-R3500Q mutation.

Comment

Tendon xanthomata are often regarded by clinicians as synonymous with a diagnosis of FH. They are however by no means specific and may also be found in FDB, type III hyperlipoproteinaemia, phytosterolaemia and cerebrotendinous xanthomatosis. The latter two conditions are exceedingly rare and in the vast majority of patients with a type IIa or IIb phenotype the diagnosis will be FH, a small proportion having FDB. As a result, quantitative imaging techniques have been used for many years in an attempt to improve the sensitivity and precision of diagnosis in FH |**14**|.

Ultrasonography has been used by several groups over the last 15 years. The present study has the advantage of examining the sensitivity and specificity of ultrasound as diagnostic tool in a group of patients with proven LDLR mutations. It is a useful additional approach but there is currently no method that will give an unequivocal answer in all patients.

What determines risk in FH?

There is significant concordance in the age at which clinical atherosclerosis first becomes apparent within families with FH |**15**|. This may be partly explained by LDL concentration and the nature of the LDLR mutation |**2,16**|. Other factors are also likely to be involved which presumably include other conventional risk factors that are independent of the LDLR defect. Some risk factors may however be influenced by the receptor defect. One such is lipoprotein (a) (Lp(a)) concentration, which is increased in the presence of LDLR defects |**17**|, though there has been some controversy about the role of Lp(a) in explaining variability of risk in FH.

Mortality over two centuries in large pedigree with familial hypercholesterolaemia: family tree mortality study.

Sijbrands EJ, Westendorp RG, Defesche JC, de Meier PH, Smelt AH, Kastelein JJ. *Br Med J* 2001; **322**: 1019–23.

BACKGROUND. The aim of this study was to estimate all cause mortality from untreated FH free from selection for coronary artery disease. A family tree mortality study used a large pedigree in The Netherlands traced back to a single pair of ancestors in the 19th century. The subjects were all members of the pedigree aged over 20 years with 0.5 probability of carrying a mutation for FH. The main outcome measure was all cause mortality. A total of 70 deaths took place among 250 people analyzed for 6950 person years. Mortality was not increased in carriers of the mutation during the 19th and early 20th century; it rose after 1915, reached its maximum between 1935 and 1964 (standardized mortality ratio 1.78, 95% confidence interval [CI] 1.13 to 2.76; *P* = 0.003), and fell thereafter. Mortality differed significantly between two branches of the pedigree (relative risk 3.26, 95% CI 1.74 to 6.11; *P* = 0.001).

INTERPRETATION. Risk of death varies significantly among patients with FH. This large variability over time and between branches of the pedigree points to a strong interaction with environmental factors. Future research is required to identify patients with familial hypercholesterolaemia who are at extreme risk and need early and vigorous preventive measures.

Comment

The subjects in this unusual and intriguing study were part of a large kindred with the so-called Afrikaner-2 mutation. The authors were able to study standardized mortality ratios directly in those from whom DNA was available and who were known to have FH. The rest of the pedigree had a 0.5 chance of being affected and were assumed to exhibit 50% of the excess mortality due to carriage of the FH mutation.

There is a suggestion that carriage of FH may have conferred a survival advantage during the 19th century, perhaps from increased resistance to infection. Standardized mortality rose and peaked in the mid 20th century, with significant differences between different arms of the family despite the fact that they all carried the same mutation. Many affected subjects had a normal life-span and this has been observed in FH previously |**18,19**|. This study provides an elegant demonstration of the interaction of a particular mutation with environmental and other, unidentified, genetic factors.

Importance of HDL cholesterol levels and the total/HDL cholesterol ratio as a risk factor for coronary heart disease in molecularly defined heterozygous familial hypercholesterolaemia.

Real JT, Chaves FJ, Martinez-Uso I, Garcia-Garcia AB, Ascaso JF, Carmena R. *Eur Heart J* 2001; **22**: 465–71.

BACKGROUND. The aim of this study was to assess the relationship of the lipid profile to CHD in a group of heterozygous FH subjects with similar age, sex, body mass index (BMI), prevalence of angiotensin converting enzyme DD genotype and type of LDLR mutation. A total of 66 molecularly defined heterozygous FH subjects, 33 of whom had CHD, was studied. Clinical features, cardiovascular risk factors and lipid parameters were compared in both groups. FH patients with CHD showed significantly lower values of mean plasma HDL cholesterol and a higher total/HDL cholesterol ratio as compared with FH subjects free of CHD. Total and LDL cholesterol concentrations were higher in patients with CHD, without reaching statistical significance. No differences in plasma lipoprotein(a) levels on absolute and log-transformed values were observed between the two groups. In the whole FH group, plasma HDL cholesterol levels were related to plasma triglyceride (TG) values and to LDLR gene 'null mutations'.

INTERPRETATION. In FH subjects of similar age, gender, BMI, systolic and diastolic blood pressure, and genetic factors that could influence CHD risk, plasma HDL cholesterol values and total/HDL cholesterol ratios are two important coronary risk factors. Hence, treatment of FH should focus not only on lowering total and LDL cholesterol levels, but also on increasing HDL cholesterol values for CHD prevention. More prospective and intervention trials should be conducted to establish the relationship of HDL cholesterol levels and CHD in FH.

Comment

The 33 heterozygotes with CHD in this study were well matched by the CHD-free controls though they had a much higher prevalence of xanthomata (72 *vs* 27%). Their total and LDL cholesterol (LDL-C) were numerically but insignificantly higher and apoB concentrations did not differ between the groups. HDL cholesterol (HDL-C) was the only lipid variable measured to show a significant difference.

Association between the TaqIB polymorphism in the cholesteryl ester transfer protein gene locus and plasma lipoprotein levels in familial hypercholesterolemia.

Carmena-Ramon R, Ascaso JF, Real JT, Najera G, Ordovas JM, Carmena R.
Metabolism 2001; **50**: 651–6.

BACKGROUND. Cholesteryl ester transfer protein (CETP) facilitates the exchange of triglycerides (TG) and cholesteryl ester between lipoprotein particles. Subjects with FH have been reported to have higher CETP activities, which could contribute to the lower HDL-C levels and increased cardiovascular risk observed in some of these patients. Several polymorphisms have been reported in the CETP locus; the common TaqIB polymorphism is associated, in normolipidaemic subjects, with decreased CETP activity and levels and with increased HDL-C levels. No data is available on the influence of this polymorphism in FH subjects. This investigation examined the TaqIB polymorphism in a group of 101 FH heterozygotes from Valencia, Spain. A frequency of 0.43 was observed for the B2 allele, similar to those reported in the general population. Based on analysis of variance (ANOVA), significant associations were found between the presence of the B2 allele and increased plasma HDL-C (*P* <0.04) and apolipoprotein A-I (apoA-I) levels (*P* <0.01). An opposite association was observed for LDL-C levels, with the B2/B2 subjects having lower levels than B1/B1 and B1/B2 subjects. The plasma apoB levels followed the same trend as those for LDL-C. In addition, the response to a National Cholesterol Education Program (NCEP)-I diet was studied in 77 of these subjects. The TaqIB polymorphism did not have a significant effect over the individual dietary response for any of the variables examined, as demonstrated by the lack of significant gene by diet interactions.

INTERPRETATION. In summary, the CETP TaqIB polymorphism is associated with a less atherogenic lipid profile, consisting of lower LDL-C, higher HDL-C levels, and a lower LDL-C/HDL-C ratio in heterozygous FH subjects. Moreover, the B2 allele was associated with a lower appearance of arcus cornealis, xanthomata, and clinical arteriosclerotic disease in these subjects.

Comment

Some of the variability in HDL-C and any related variability in vascular disease risk seen in patients with FH (see Real *et al.* above) may be mediated through differences in CETP activity. Although subjects homozygous for the B2 allele of the TaqIB polymorphism of the CETP gene locus are said to have lower CETP activities and, as in this study, higher HDL-C, there was no obvious dose-response in that heterozygotes (B1/B2) had HDL-C levels indistinguishable from the B1/B1 subjects. The finding of lower LDL-C in the B2 homozygotes makes it difficult to determine to what extent the differences in HDL-C explain the difference in CHD status of the two groups. The conclusions of this study in FH are consonant with those in a more heterogeneous population from Framingham [**20**].

Another modality which might be influenced by the metabolic defect in FH is triglyceride-rich lipoprotein remnant metabolism. Remnants, which are widely believed to be atherogenic [**21**], are thought to be cleared by virtue of their apoE component, through the LDLR and the LRP, though other receptors have also been implicated [**22**]. Thus their clearance could be impaired in FH patients, though again that would not necessarily imply that variability in risk in FH was dependent upon remnant metabolism. The literature to date contains conflicting results with regard to remnant metabolism in FH [**23**].

Elevated apolipoprotein B-48 and remnant-like particle-cholesterol in heterozygous familial hypercholesterolaemia.

Dane-Stewart CA, Watts GF, Mamo JC, Dimmitt SB, Barrett PH, Redgrave TG. *Eur J Clin Invest* 2001; **31**: 113–17.

BACKGROUND. Apolipoprotein B-48 (apoB-48) is a marker of TG-rich lipoprotein (TRL) remnants of intestinal origin. Chylomicron remnants are causally related to atherosclerosis. We have shown previously that fasting plasma apoB-48 may predict post-prandial lipaemia. Remnant-like particle-cholesterol (RLP-C) may also reflect TRL remnants. This study aimed to determine whether subjects with heterozygous FH had an accumulation of remnants of intestinal origin, as reflected by fasting plasma apoB-48 and RLP-C levels. The fasting plasma concentrations of apoB-48 and RLP-C were measured in 15 subjects with heterozygous FH and 15 age- and sex-matched, normolipidaemic subjects. ApoB-48 was determined using SDS-PAGE and a Western blotting/enhanced chemi-luminescence technique. RLP-C was measured using an immuno-separation assay. Serum apolipoprotein B-100 (apoB-100) levels were measured using immunonephelometry; lipids were assayed enzymatically. Compared with controls, FH subjects had significantly elevated plasma concentrations of apoB-48 (29.3 median, 16.7–45.1 mg/l range *vs* 12.8, 7.3–28.6; *P* = 0.001) and RLP-C (16.2, 1.5–114.3 mg/dl *vs* 8.5, 5.0–13.5; *P* = 0.003), as well as serum total apoB-100 (1.9, 1.3–2.6 g/l *vs* 1.0, 0.3–1.3; *P* <0.001), LDL- (8.1, 4.6–10.4 mmol/l *vs* 3.5, 2.4–4.4; *P* <0.001) and TG (1.5, 0.6–5.6 mmol/l *vs* 1.0, 0.4–1.8; *P* = 0.018). There was no significant difference in HDLC.

INTERPRETATION. The findings suggest that patients with heterozygous FH have elevated plasma concentrations of TRL remnants, including those of intestinal origin. This may be a consequence of decreased clearance of these particles by the LDLR.

Comment

Patients with FH can have either the IIa (LDL raised) or IIb (very low-density lipoprotein [VLDL] and LDL raised) phenotype. The FH and control patients in this study were not matched for serum TG levels. All the controls were normotriglyceridaemic whereas the highest TG level in the FH group was 5.6 mmol/l.

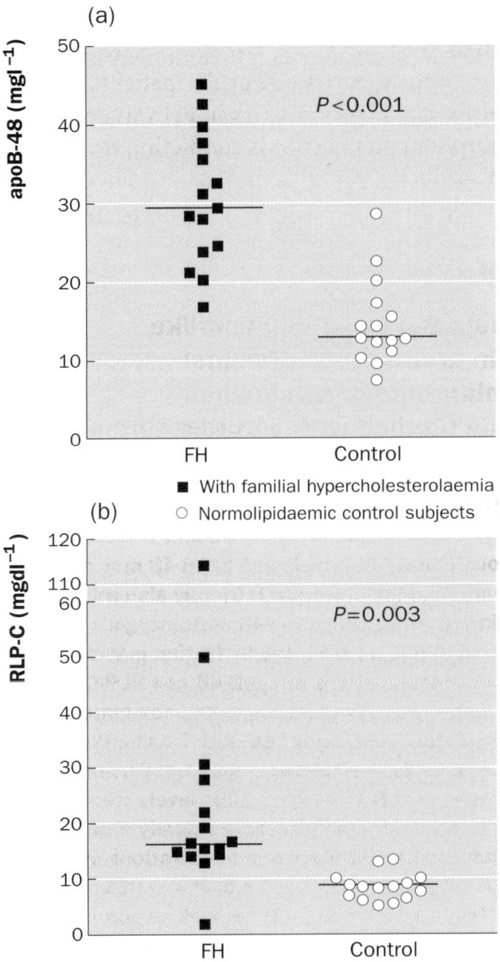

Fig. 1.1 Fasting plasma apoB-48 (a) and RLP-C (b) concentrations in patients with familial hypercholesterolaemia and normolipidaemic control subjects. Horizontal bars indicate median values. Source: Dane-Stewart *et al.* (2001).

It might be asked if different remnant concentrations would be observed between FH patients and controls matched for plasma TG concentrations. There was however no correlation within the FH group between apoB-48 and TG levels. Moreover there was no relationship between apoB-48 and RLP-C concentrations presumably because the former comprise chylomicron remnants and the latter also contain a contribution from remnants of hepatic VLDL.

The obvious explanation for the accumulation of remnants in FH would be defective clearance by the LDLR, and the authors postulate this as a possibility, though it does not accord with other work by some of the same authors, using different methodology, and described below in the next paper. Previous work on remnant metabolism in FH has also given conflicting results |**24–26**|.

The present work seems to leave little doubt however that remnant concentrations are increased at least in some patients with FH – there was some overlap with the control group – whatever the mechanism. The notion that some of the variability in prognosis in FH depends on differences in remnant metabolism is intriguing and deserves further study.

Chylomicron remnant metabolism in familial hypercholesterolaemia studied with a stable isotope breath test.

Watts GF, Barrett PH, Marais AD, *et al. Atherosclerosis* 2001; **157**: 519–23.

BACKGROUND. Chylomicron remnant metabolism was studied using a stable isotope breath test in 25 patients with FH (10 homozygotes, 15 heterozygotes), and in 15 normolipidaemic controls. A lipid emulsion mimicking the composition of chylomicron remnants and labelled with cholesteryl ^{13}C-oleate was injected intravenously; $^{13}CO_2$ was measured subsequently in breath using isotope-ratio mass spectrometry. The fractional catabolic rate (pools/h) of the emulsion, derived from a compartmental model, did not differ significantly among the groups: homozygous FH mean 0.20 (SEM 0.05), heterozygous FH 0.12 (0.02), controls 0.16 (0.03).

INTERPRETATION. The authors suggest that the catabolism of chylomicron remnants from plasma is not impaired in FH and that the hepatic uptake of these particles is not dependent on functional LDLRs.

Comment

A variety of methods has been used to assess post-prandial chylomicron remnant metabolism including the measurement of apoB-48 and retinyl esters, but none is entirely satisfactory. The authors have recently validated a stable isotope breath test for this purpose (see Chapter 2 – Hypertriglyceridaemia and mixed hyperlipidaemia) |**27**|. Using this technique the authors found no evidence of impaired

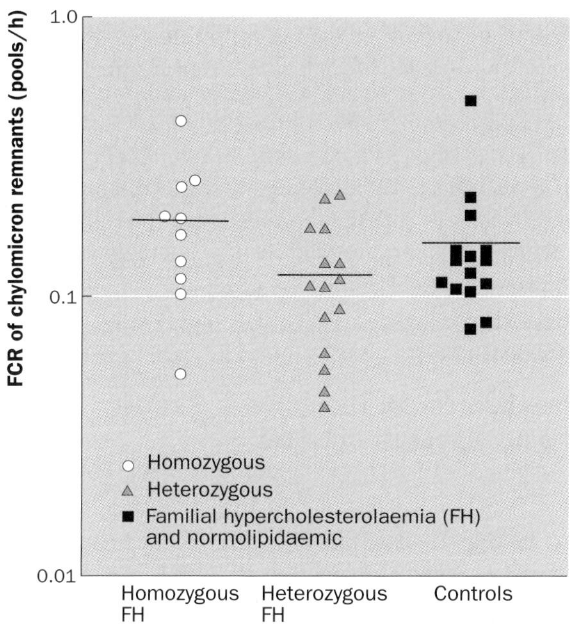

Fig. 1.2 Fractional catabolic rate (FCR) of chylomicron remnant-like emulsion (logarithmic scale) in homozygous and heterozygous patients with FH, and in the normolipidaemic control subjects. Geometric mean shown. $P = 0.118$ for difference by ANOVA. Source: Watts *et al.* (2001).

remnant metabolism in patients with homozygous or heterozygous FH. This accords with previous work from South Africa using retinyl palmitate |**24**|. Moreover the clearance of the synthetic remnants was not influenced by the type of LDLR mutation. The conclusion is that the remnants are likely to be cleared by virtue of their apoE component, via the LRP, and independently of the LDLR. Other reports |**25,26,28**| however have demonstrated accumulation of remnants in FH and are discussed further.

Familial defective apolipoprotein B-100 (FDB)

FDB resembles FH in many ways. It is a single gene co-dominant disorder which gives rise to raised LDL concentrations and premature CHD. Tendon xanthomata can be a feature. With a frequency of 1 in 500–700 it has a similar prevalence in affected societies. Clinically it is indistinguishable from FH. Unlike FH nearly all cases occur in individuals of Caucasian origin. Most cases result from the substitution of glutamine for arginine at position 3500. Very rarely tryptophan may replace arginine at the same site or cysteine may substitute for arginine at position 3531. The mutations are outside the LDLR binding site, which is found in the region of residues 3359–3369 demonstrating the importance of the conformation of apoB in binding. Only four mutations in apoB-100 have been found that lead to defective binding and hypercholesterolaemia, despite an extensive search. This contrasts with the situation in FH where approaching 700 mutations of the LDLR have been identified [7]. In general the manifestations of FDB are milder than those in FH, so that despite similar gene frequencies in some populations considerably less than 10% of phenotypic FH patients prove to have FDB.

The molecular mechanism for the genetic disorder familial defective apolipoprotein B100.

Boren J, Ekstrom U, Agren B, Nilsson-Ehle P, Innerarity TL. *J Biol Chem* 2001; **276**: 9214–8.

BACKGROUND. FDB is a genetic disorder in which LDLs bind defectively to the LDLR, resulting in hypercholesterolemia and premature atherosclerosis. FDB is caused by a mutation (R3500Q) that changes the conformation of apoB-100 near the receptor-binding site. The authors previously showed that arginine, not simply a positive charge, at residue 3500 is essential for normal receptor binding and that the carboxyl terminus of apoB-100 is necessary for mutations affecting arginine 3500 to disrupt LDLR binding. Thus, normal receptor binding involves an interaction between arginine 3500 and tryptophan 4369 in the carboxyl tail of apoB-100. W4369Y LDL and R3500Q LDL isolated from transgenic mice had identically defective LDL binding and a higher affinity for the monoclonal antibody MB47, which has an epitope flanking residue 3500.

INTERPRETATION. The authors conclude that arginine 3500 interacts with tryptophan 4369 and facilitates the conformation of apoB-100 required for normal receptor binding of LDL. Based on their findings, they developed a model that explains how the carboxyl terminus of apoB-100 interacts with the backbone of apoB-100 that enwraps the LDL particle. The model also explains how all known ligand-defective mutations in apoB-100, including a newly discovered R3480W mutation in apoB-100, cause defective receptor binding.

Comment

ApoB-100 is the ligand within LDL that determines binding to the LDLR. The presence of a normal apoB-100 molecule within a lipoprotein particle is not in itself sufficient to permit binding. If it were, newly secreted VLDL would be cleared by the receptor before its TG component had undergone lipolysis and delivered its fatty acids to peripheral tissues. Conformation is also important. Previous work suggests that most of the apoB-100 molecule wraps around the LDL particle like a belt with the carboxyl terminus folding round and crossing over the belt not far from amino acid 3500. This enables the interaction, demonstrated in this paper, between tryptophan in the carboxyl tail of apoB and arginine at position 3500. It is proposed that modulation of this interaction during lipolysis prevents premature clearance of VLDL but allows binding and removal of delipidated LDL via the LDLR.

Compound heterozygous familial hypercholesterolemia and familial defective apolipoprotein B-100 produce exaggerated hypercholesterolemia.

Tai ES, Koay ES, Chan E, *et al.* *Clin Chem* 2001; **47**: 438–43.

Background. FH and FDB represent ligand-receptor disorders that are complementary. Individuals with both FH and FDB are unusual. This study reports a family with both disorders and the impact of the mutations on the phenotypes of the family members. SSCP and denaturing gradient gel electrophoresis (DGGE) were used for genetic analysis of all 18 exons and the promoter region of the LDLR and DGGE for genetic analysis of the apolipoprotein B-100 (apoB-100) gene. The functional significance of the apoB-100 mutation was studied using a U937 cell proliferation assay. Fasting serum lipid profiles were determined for the index case and seven first-degree relatives. One of the patients' sisters had a missense mutation (Asp(407) →Lys) in exon 9 of the LDLR and a serum LDL-C concentration of 4.07 mmol/l. Four other first-degree relatives had hyperlipidaemia but no LDLR mutation. However, these subjects had a mutation of the apoB-100 gene (Arg(3500) →Trp). The cell proliferation rate of U937 cells fed with LDL from other subjects with the same mutation was fourfold less than that of controls. The index case had both FH- and FDB-related mutations. Her serum LDL-C (9.47 mmol/l) was higher than that of all other relatives tested.

Interpretation. Existence of both FH and FDB should be considered in families with LDLR mutations in some but not all individuals with hypercholesterolaemia or when some individuals in families with FH exhibit exaggerated hypercholesterolaemia.

Comment

The presence of an FDB allele exaggerates the metabolic defect in heterozygous FH. This is predictable in that the normal LDLR population will be unable to clear the LDL population which posseses the abnormal apoB.

By contrast the coexistence of type III hyperlipoproteinaemia in an FH heterozygote attenuates the increase in LDL-C |**29**|. This too is predictable as the impaired conversion of remnant lipoproteins to LDL in type III diminishes the load presented to the LDLRs.

Familial defective apolipoprotein B-100 in a group of hypercholesterolaemic patients in Poland. Identification of a new mutation Thr3492Ile in the apolipoprotein B gene.

Bednarska-Makaruk M, Bisko M, Pulawska MF, *et al. Eur J Hum Genet* 2001; **9**: 836–42.

BACKGROUND. The prevalence of the FDB Arg3500Gln mutation in 525 unrelated hypercholesterolaemic Polish subjects was evaluated. DNA samples were screened for FDB mutation using the SSCP method. The presence of mutation was confirmed using a mismatch MspI PCR strategy. Plasma lipid levels and clinical characteristics of 13 patients identified as carriers of the mutation and of their 23 affected relatives were analyzed and compared with non-affected ones. In the affected individuals a variable expression of lipid concentrations and of atherosclerosis symptoms were observed. The prevalence of FDB Arg3500Gln mutation in hypercholesterolaemic Polish subjects (3.7%) seems to be similar to the frequency reported in other Caucasian hypercholesterolaemic populations. The estimated prevalence of the mutation in the general Polish population was relatively high being 1/250. The same haplotype at the apoB locus in the carriers of this mutation in Poland as in other populations from Western Europe suggests its common origin.

INTERPRETATION. In one hypercholesterolaemic subject a non-hitherto described mutation was identified. It consisted in C→T transition in apoB codon 3492 leading to threonine to isoleucine substitution in 3492 position of apoB gene (Thr3492Ile).

Comment

The number of mutations leading to FDB are few but an additional case with a single amino acid change at a site close to that in the original cases is described here. In this Polish population, FDB was found in 10.8% of clinically diagnosed FH cases, compared with 2–3% in other communities. Similar high prevalences have been reported from Switzerland and in another Polish study.

Autosomal recessive hypercholesterolaemia (ARH)

In a description of 52 patients with phenotypic homozygous FH, Khachadurian and Uthman [30] described four children with spectacular xanthomata but less severe hypercholesterolaemia and apparent recessive inheritance. The parents were first cousins and had relatively normal plasma cholesterol concentrations (<6.2 mmol/l).

Several groups have since described recessive inheritance of clinical homozygous FH [31,32]. Homozygotes for ARH tend to have higher cholesterol levels than patients with homozygous FDB but lower than those with homozygous FH.

Although no defect has been identified in either the LDL or apoB genes, LDL clearance is abnormal in the liver |**33**| and in lymphoblasts but is relatively normal in fibroblasts. LDL binds normally but is not internalized or degraded. The defect may be specific to the LDLR as the transferrin receptor is internalized normally |**32**|. It has been proposed that the defect lies on chromosome 15 |**34**|, but two recent studies in a disparate group of cases, described below, have localized the mutation to chromosome 1.

Use of homozygosity mapping to identify a region on chromosome 1 bearing a defective gene that causes autosomal recessive homozygous hypercholesterolemia in two unrelated families.

Eden ER, Naoumova RP, Burden JJ, McCarthy MI, Soutar AK. *Am J Hum Genet* 2001; **68**: 653–60.

BACKGROUND. FH is a common inherited disorder of metabolism characterized clinically by high levels of LDL in plasma owing to reduced catabolism. This leads to accelerated atherosclerosis and thus to an increased risk of CHD. FH is usually caused by defects in the gene for either the LDLR or apoB, the ligand for the LDLR. Elsewhere, we have described two unrelated patients with phenotypic homozygous FH. Both patients were offspring of consanguineous unions, and linkage to either the gene for the LDLR or the gene for apoB was excluded in both. Their cells in culture do not degrade LDL, despite the presence of normal surface binding of LDL to the LDLR. This observation suggests that the patients may be homozygous for a defective gene that encodes a component of the internalization pathway. This investigation first excluded linkage of the defect to known genes for proteins reported to be involved in internalization of receptors in clathrin-coated pits. Genomewide homozygosity mapping was then performed.

INTERPRETATION. Genotyping of 500 polymorphic markers in three affected and seven unaffected members of the first pedigree showed that recessive hypercholesterolaemia in this family was localized to a single chromosomal region on 1p36–p35. Genotyping of two affected and five unaffected members of the second pedigree provided further evidence of linkage to this locus, thereby mapping the disease-causing gene to a 12-cM region on chromosome 1p36–p35, with a combined LOD (logarithm of the odds) score of 5.3 in these unrelated families. Identification of the gene in this region may lead to new insights into the mechanisms of LDLR-mediated uptake of LDL by cells and may help to identify further genetic risk factors for premature atherosclerosis.

Comment

This study in patients of Turkish and Asian Indian origin, previously described by the same group |**32**|, localized the defect in ARH to a small region of chromosome 1 in both families.

Autosomal recessive hypercholesterolemia caused by mutations in a putative LDL receptor adaptor protein.

Garcia CK, Wilund K, Arca M, *et al.* *Science* 2001; **292**(5520): 1394–8.

BACKGROUND. Atherogenic LDLs are cleared from the circulation by hepatic LDLRs. Two inherited forms of hypercholesterolaemia result from loss of LDLR activity: autosomal dominant FH, caused by mutations in the LDLR gene, and ARH, of unknown aetiology. The ARH locus is mapped to approximately a 1-centimorgan interval on chromosome 1p35 and identifies six mutations in a gene encoding a putative adaptor protein (ARH). ARH contains a phosphotyrosine binding (PTB) domain, which in other proteins bind NPXY motifs in the cytoplasmic tails of cell-surface receptors, including the LDLR. ARH appears to have a tissue-specific role in LDLR function, as it is required in the liver but not in fibroblasts.

INTERPRETATION. The molecular defect in ARH is identified and lies in a cytosolic protein (also termed ARH) that probably interacts with the cytoplasmic tails of LDL-receptors and may be involved in regulating their migration on the cell surface.

Comment

Of the four families studied in the current work, two were from Sardinia and two from Lebanon, including the family in which the disorder was originally described |**30**|. The mutation mapped to an area on chromosome 1, which overlaps that identified by Eden *et al.* in the paper discussed immediately above. No linkage to chromosome 15 was identified even though one of the families studied was also examined by Ciccarese *et al.* |**34**|.

The precise role of the ARH adaptor protein in LDL is unclear and it is not known why it does not appear to be necessary for normal function of the LDLR in fibroblasts. The authors suggest that ARH may play a role in polarized cells such as hepatocytes in directing LDLRs to the basolateral surfaces of the cells.

Phytosterolaemia (sitosterolaemia)

Cholesterol is an essential component of animal membranes and is the precursor of bile acids, vitamin D and steroid hormones. It can be derived from local synthesis within tissues or from the diet. In man, cholesterol absorption is 40–60% efficient and about 0.5 g of cholesterol is taken in the diet daily. Plant cell membranes are made of very similar sterols and are taken in similar amounts in the diet to cholesterol. Remarkably, these are scarcely absorbed at all in normal subjects and much of those that are, are re-excreted in the bile. Sitosterol, the most abundant of the phytosterols, differs from cholesterol only by the addition of an ethyl group. Use is now being made of plant sterols and their derivatives to interfere with cholesterol absorption in man as a means of lowering plasma cholesterol and this is discussed in Chapter 3 – Diet, exercise and lipoprotein metabolism.

Very rarely the barrier to absorption of plant sterols fails. Beta-sitosterolaemia (phytosterolaemia) was first described in two sisters by Bhattacharrya and Connor almost 30 years ago [35] and the defect was localized to chromosome 2 in 1998 [36]. It is now apparent that the disorder results from mutations in the family of proteins known as ATP-binding cassette (ABC) transporters [37,38]. Plant sterols together with cholesterol accumulate in plasma and tissues, leading to tendon xanthoma formation and the development of premature coronary disease. Affected individuals may also suffer from arthritis and haemolysis. Children with the disorder may have severe hypercholesterolaemia and be thought to have homozygous FH. The disease exhibits autosomal recessive inheritance with obligate heterozygotes (parents and unaffected siblings of patients) having normal levels of circulating cholesterol and phytosterols.

ABC transporters are discussed further in Chapter 5 – Disorders which cause hypolipidaemia. Cholesterol efflux from cells has also recently been shown to be dependent upon an ABC transporter (ABCA1), which is defective in Tangier disease.

Accumulation of dietary cholesterol in sitosterolemia caused by mutations in adjacent ABC transporters.

Berge KE, Tian H, Graf GA, *et al. Science* 2000; **290**: 1771–5.

BACKGROUND. In healthy individuals, acute changes in cholesterol intake produce modest changes in plasma cholesterol levels. A striking exception occurs in sitosterolaemia, an autosomal recessive disorder characterized by increased intestinal absorption and decreased biliary excretion of dietary sterols, hypercholesterolaemia, and premature coronary atherosclerosis. This study identified seven different mutations in two adjacent, oppositely oriented genes that encode new members of the ABC transporter family (six mutations in ABCG8 and one in ABCG5) in nine patients with sitosterolaemia. The two genes are expressed at highest levels in liver and intestine and, in mice, cholesterol feeding upregulates expressions of both genes.

INTERPRETATION. These data suggest that ABCG5 and ABCG8 normally cooperate to limit intestinal absorption and to promote biliary excretion of sterols, and that mutated forms of these transporters predispose to sterol accumulation and atherosclerosis.

Comment

See next paper.

Identification of a gene, ABCG5, important in the regulation of dietary cholesterol absorption.

Lee MH, Lu K, Hazard S, Hidaka H, *et al. Nat Genet* 2001; **27**: 79–83.

BACKGROUND. The molecular mechanisms regulating the amount of dietary cholesterol retained in the body, as well as the body's ability to exclude selectively other dietary sterols, are poorly understood. An average western diet will contain about 250–500 mg of dietary cholesterol and about 200–400 mg of non-cholesterol sterols. About 50–60% of the dietary cholesterol is absorbed and retained by the normal human body, but less than 1% of the non-cholesterol sterols is retained. Thus, there exists a subtle mechanism that allows the body to distinguish between cholesterol and non-cholesterol sterols. In sitosterolaemia, a rare autosomal recessive disorder, affected individuals hyperabsorb not only cholesterol but also all other sterols, including plant and shellfish sterols from the intestine. The major plant sterol species is sitosterol; hence the name of the disorder. Consequently, patients with this disease have very high levels of plant sterols in the plasma and develop tendon and tuberous xanthomas, accelerated atherosclerosis, and premature coronary artery disease. The authors have previously mapped the STSL locus to human chromosome 2p21 and further localized it to a region of less than 2 cM bounded by markers D2S2294 and D2S2291.

INTERPRETATION. A new member of the ABC transporter family, ABCG5, is mutant in nine unrelated sitosterolaemia patients.

Comment

The eighteen patients with phytosterolaemia in these two studies had mutations in one of two ABC transporters (ABCG5 or ABCG8). The fact that cholesterol feeding in mice upregulates expression of the genes for both transporters supports the notion that they act in a coordinated fashion so that a defect in either can lead to the disease.

Cholesterol is passively taken up into enterocytes and this work is likely to lead to greatly improved understanding of the regulation of cholesterol absorption and excretion in man. The likelihood is that both cholesterol and plant sterols enter intestinal cells in an uncontrolled way but that some cholesterol and most of the phytosterols are returned actively to the lumen under the influence of the ABC transporters. A similar process probably occurs in hepatocytes, which excrete absorbed plant sterols and cholesterol, into bile. Failure in the transporter mechanism leads to accumulation of both phytosterols and cholesterol.

Treatment of hypercholesterolaemia

Dietary treatment

Dietary treatment of primary hyperlipidaemia, including FH, is discussed in Chapter 3 – Diet, exercise and lipoprotein metabolism.

Pharmacological treatment

Responsiveness to statin therapy

LDL-receptor gene mutations and the hypocholesterolemic response to statin therapy.

Brorholt-Petersen JU, Jensen HK, Raungaard B, Gregersen N, Faergeman O. *Clin Genet* 2001; **59**: 397–405.

BACKGROUND. Studies of the cholesterol-lowering effect of statin therapy as a function of LDL-receptor mutation type have not produced a clear picture, possibly because they included patients with several different kinds of LDL-receptor mutations. The authors studied the response to treatment with fluvastatin in 28 patients with heterozygous FH as a result of a receptor-negative mutation (Trp23-stop) and in 30 patients with a receptor-binding defective mutation (Trp66-Gly) to test the hypothesis that response to treatment depends on the type of mutation. Patients were randomized to 12 weeks of treatment with fluvastatin 40 mg daily and 12 weeks of placebo treatment, preceded by a placebo run-in period of 8 weeks in a double-blind, cross-over design. Untreated plasma concentrations of lipids and lipoproteins were similar in the two groups of patients. Plasma cholesterol and LDL cholesterol response to therapy tended to be less marked in receptor-binding defective patients, but the differences were not statistically significant.

INTERPRETATION. A tabulation of the results of the present and earlier studies suggests that differences in treatment response as an apparent function of LDLR gene mutational type occur mainly in populations with recent genetic admixture (<400 years). In such populations, persons with the same mutation in the LDLR gene are more likely to share other but undetermined genetic variations affecting the pharmacology of statins.

Comment

In nine previous studies five have shown some effect of the nature of the LDLR mutation on statin-responsiveness, but four have failed to do so. In this study also no significant difference emerged between patients with receptor-negative and receptor-defective mutations, though those who were receptor-negative tended, if anything, to have a better response to fluvastatin. In both groups those who had the highest untreated LDL-C had the greatest absolute reduction in cholesterol. The authors point out that the inclusion of several members of the same kindred in studies of this type is a potential source for confounding. They may share other genetic factors that influence statin-responsiveness and which override effects due to the nature of the LDLR. The degree to which the activity of the normal LDLR is modulated by drug treatment is presumably likely to be a predominant effect.

Determinants of variable response to statin treatment in patients with refractory familial hypercholesterolemia.

O'Neill FH, Patel DD, Knight BL, *et al. Arterioscler Thromb Vasc Biol* 2001; **21**: 832–7.

BACKGROUND. Inter-individual variability in LDL-C response during treatment with statins is well documented but poorly understood. To investigate potential metabolic and genetic determinants of statin-responsiveness, 19 patients with refractory heterozygous FH were sequentially treated with placebo, atorvastatin (10 mg/d), bile acid sequestrant, and the two combined, each for 4 weeks. Levels of LDL-C, mevalonic acid (MVA), 7-alpha-OH-4-cholesten-3-one, and leukocyte LDLR and hydroxymethylglutaryl coenzyme A reductase mRNA were determined after each treatment period. Atorvastatin (10 mg/d) reduced LDL-C by an overall mean of 32.5%. Above-average responders (LDL-C −39.5%) had higher basal MVA levels (34.4 ± 6.1 micromol/l) than did below-average responders (LDL-C −23.6%, *P* <0.02; basal MVA 26.3 ± 6.1 micromol/l, *P* <0.01). Fewer good responders compared with the poor responders had an apolipoprotein E4 allele (3 of 11 *versus* 6 of 8, respectively; P <0.05). There were no baseline differences between them in 7-alpha-OH-4-cholesten-3-one, hydroxymethylglutaryl coenzyme A reductase mRNA, or LDLR mRNA, but the latter increased in the good responders on combination therapy (*P* <0.05). Severe mutations were not more common in poor than in good responders.

INTERPRETATION. The authors conclude that poor responders to statins have a low basal rate of cholesterol synthesis that may be secondary to a genetically determined increase in cholesterol absorption, possibly mediated by apolipoprotein E4. If so, statin-responsiveness could be enhanced by reducing dietary cholesterol intake or inhibiting absorption.

Comment

It is only a little over a decade since physicians treating hyperlipidaemia were struggling to obtain adequate reductions of LDL-C, by cajoling reluctant patients to take increasing doses of resins sometimes in combination with fibrates or nicotininc acid. What would now be regarded as target levels for total and LDL-C were rarely achieved in patients with high-risk disorders such as FH. The average reduction of cholesterol in clinical trials was only about 10%. This situation has changed since the advent of the statins and FH patients with total cholesterol concentrations of 10 mmol/l or more often reduce this by 50%. There is still however a significant number of patients, some severely hypercholesterolaemic, who respond poorly, or scarcely at all to maximum doses of the most potent statins. They constitute a particularly difficult group amongst problem patients referred to lipid clinics. They provide a chastening reminder to the specialist lipidologist that his or her prescription for a statin is no more effective than that of the general physician or general practitioner with only a passing interest in the subject.

The nature of the LDL mutation in heterozygous FH is an obvious candidate for this variablility, but as we have seen in the discussion of the paper by Brorholt-Petersen *et al.* above, the data on this are controversial. Response to statins is very variable even in subjects with the same mutation (Karayan *et al.* 1994 |**39**|). ApoE genotype has also been reported to be a determinant of statin-responsiveness, those with an ε4 allele demonstrating poorer response. This too is controversial.

The present work suggests that poor responders, who were more likely to carry an ε4 allele, have a low basal rate of cholesterol synthesis as reflected by low plasma mevalonate levels. The low synthetic rate may reflect above average efficiency of cholesterol absorption leading to a liver relatively replete in cholesterol with down-regulated LDLR activity. Such subjects presumably have little scope to reduce hepatic cholesterol and upregulate their LDLRs by inhibiting HMGCoA-reductase if synthetic rates are already low. The low basal cholesterol synthetic rate in the poor responders does not seem to be the result of impaired conversion of hepatic cholesterol to bile acids. Basal concentrations of 7-α-hydroxy-4-chlolesten-3-one, a marker of the activity of the rate limiting enzyme in bile acid synthesis (7-α-hydroxylase), were similar in good and poor responders.

It remains to be seen if inhibitors of cholesterol absorption, such as plant stanol and sterol esters or ezetimibe, will prove to be particularly effective when combined with statins in poor responders by interfering with cholesterol absorption and upregulating HMG-CoA reductase so that the statin has more scope for action. Existing data suggest that on average inhibitors of cholesterol absorption and statins exert additive effects on LDL-C. Might they be more than additive in poor statin responders if these are looked at specifically? The whole question of variability of response to statins is examined in a recent review by Thompson *et al.* |**40**|.

Are statins safe?

Withdrawal of cerivastatin from the world market.

Furberg CD, Pitt B. *Curr Control Trials Cardiovasc Med* 2001; **2**: 205–7.

BACKGROUND. Cerivastatin was recently withdrawn from the market because of 52 deaths attributed to drug-related rhabdomyolysis that leads to kidney failure. The risk was found to be higher among patients who received the full dose (0.8 mg/day) and those who received gemfibrozil concomitantly. Rhabdomyolysis was ten times more common with cerivastatin than the other five approved statins. This study addresses three important questions raised by this withdrawal. Should drugs continue to be

approved on surrogate efficacy? Are all statins interchangeable? Do the benefits outweigh the risks of statins?

Interpretation. The study concludes that decisions regarding the use of drugs should be based on direct evidence from long-term clinical outcome trials.

Comment

The authors state that many of the 52 fatal cases who had been taking cerivastatin were taking the highest licensed dose of 0.8 mg/day or were taking cerivastatin in combination with gemfibrozil. This drug interaction is said to have been implicated in 12 of 31 fatalities in the USA. Gemfibrozil is widely used in the USA but has been little used in the UK. It is however difficult to be certain that a similar problem might not have emerged with cerivastatin and other fibrates had enough patients been treated for long enough. In addition to the fatalities the authors state that there were 358 non-fatal cases of rhabdomyolysis out of about 700 000 users in the USA, most of whom required hospital admission. US law firms apparently are advertising in the press for patients who have developed rhabdomyolysis while on a statin.

The recent withdrawal of cerivastatin certainly caused anxiety amongst clinicians and patients in relation to the use of other statins, particularly in conjunction with fibrates. In the case of the natural fermentation-derived statins, lovastatin, pravastatin and simvastatin, it seems clear from the end-point studies that benefit outweighs risk. The evidence from 4S (Simvastatin Scandinavian Survival Study |**41**|), WOSCOPS (West of Scotland Coronary Prevention Study |**42**|), CARE (Cholesterol and Recurrent Events study |**43**|), LIPID (Long-term Intervention with Pravastatin in Ischaemic Disease |**44**|) and AFCAPS/TexCAPS (Air Force/Texas Coronary Atherosclerosis Prevention Study |**45**|) has now been greatly strengthened by the data from HPS (Heart Protection Study) |**46**|.

The synthetic statins, fluvastatin and atorvastatin, have some end-point evidence based predominantly on angiographic studies and trials of early treatment in patients with ischaemic events. These studies are however of relatively short duration and involve small numbers when compared with the large placebo-controlled end-point studies listed above. It would not be ethical to repeat these with newer statins as they come to the market. There are other studies in progress with fluvastatin and atorvastatin. Moreover atorvastatin has been much more widely used than cerivastatin, and fluvastatin has been available for longer.

What should be done? Furberg and Pitt caution against licensing based upon evidence of surrogate efficacy – lipid-lowering in the case of the statins. The extent to which this is feasible will depend upon the availability and efficacy of other drugs to treat or prevent the relevant condition. When there are already several effective licensed drugs, as there are with the statins, the criteria can be more stringent for new products with the same indication. Even then there is a difficulty. Setting the standard too high will stifle drug development and may deprive the community of significant improvements within a given class of drug. Serious side-effects and fatalities however not only damage the individuals concerned but also erode the confidence of other patients under treatment with similar agents and they in turn may suffer as a result of lack of compliance. The demonstration of overall benefit may not be enough. How much harm is acceptable from serious adverse events caused by a drug which reduces morbidity and mortality overall? *Primum non nocere.* The regulatory authorities have a difficult and unenviable task.

Does vitamin E beneficially affect muscle pains during HMG-Co-enzyme-A-reductase inhibitors without CK-elevation?

Sinzinger H. *Atherosclerosis* 2000; **149**: 225.

BACKGROUND. A small number of patients on statins develop myalgia, which seems to be drug related, but without significant increase in serum creatine kinase (CK).

INTERPRETATION. Ingestion of 1000 IU vitamin E led to the resolution of statin-related myalgia in four patients. It may take 2–3 weeks for the benefit to be apparent.

Comment

This is a letter to the editor and not a placebo-controlled trial. The observation is of considerable interest and merits further study. Severe myositis and rhabdomyolysis induced by statins are very rare. Mild to moderate elevations of serum CK are nevertheless common in patients on statins, and indeed in many hyperlipidaemic patients even before they go onto lipid-lowering medication [47]. It is recommended in the British National Formulary that statins should be withdrawn in asymptomatic patients if the CK activity reaches ten times the upper limit of normal.

Patients on statins however not infrequently complain of myalgia in the presence of normal or only moderately raised serum CK. It can be helpful to ask such patients to discontinue the drug for a few weeks and then restart it to see if the muscle symptoms relate to the medication. If there is still doubt the cycle can be repeated. In many patients the myalgia persists regardless of statin withdrawal and is probably unrelated. In others where there seems to be a clear relationship an alternative statin can be tried and this is often successful as in at least one of the patients described here by Sinzinger. In the few where the problem recurs with all the statins there is a major therapeutic problem particularly if they are high-risk subjects with vascular disease. In these patients vitamin E may be worth a trial. It is unlikely to be harmful, and unfortunately equally unlikely to provide additional cardiovascular benefit, given the failure of antioxidant vitamins to show a significant clinical benefit in end-point trials hitherto. Might vitamin E reduce the risk of statin or fibrate-associated rhabdomyolysis?

Mechanisms for clinical benefit from statin use

The rapidity with which statin treatment reduces vascular events, within weeks or months, is impressive, particularly when one considers that angiographic studies show only modest benefits on lesion size over a few years. One major factor widely believed to be involved is plaque remodelling leading to a lower lipid content and a more robust fibrous cap, thus reducing the chances of plaque rupture and thrombosis.

Another mechanism relates to changes in endothelial function. Impaired endothelium-dependent vasodilatation is well documented in hypercholesterolaemia |**48,49**| and several groups have demonstrated an improvement in endothelial function and vascular reactivity with cholesterol reduction. The effect is rapid and has been demonstrated even after a single plasma exchange. Much of this work however has been short-term over a period of a few weeks. The following study shows that the effect is sustained for at least a year, both in terms of vascular reactivity (flow-mediated dilatation) and soluble E-selectin levels.

Sustained long-term improvement of arterial endothelial function in heterozygous familial hypercholesterolemia patients treated with simvastatin.

Alonso R, Mata P, De Andres R, Villacastin BP, Martinez-Gonzalez J, Badimon L. *Atherosclerosis* 2001; **157**: 423–9.

Background. Patients with heterozygous familial hypercholesterolaemia (hFH) are at very high risk for premature CHD. In the last decade, treatment with statins has reduced cardiovascular mortality in these patients. The aim of this study was to analyze arterial endothelial function assessed as flow-mediated dilatation (FMD) and soluble E-selectin (sE-selectin) levels in patients with hFH under a long-term lipid-lowering treatment. Twenty-five patients who completed the study received a dose of simvastatin to achieve a treatment goal of at least 30% reduction in serum LDL-C for 52 weeks. Functional and biochemical measurements were taken at entry, and at week 12 and 52 of treatment. FMD was measured by vascular ultrasound of the brachial artery. sE-selectin, tumour necrosis factor-→ (TNF-→) and interleukin-6 were determined by enzyme linked immunosorbent assay (ELISA). LDL-C levels were significantly reduced by treatment at week 12 and maintained at week 52 (reduction *vs* baseline, 44 ± 12 and 43 ± 11%, respectively, *P* <0.0001). A significant improvement in endothelial function, measured as FMD (baseline, 4.7 ± 6.2%; 12 weeks, 12.3 ± 5.9%; 52 weeks, 9.7 ± 4.7%; *P* <0.005) and a reduction in sE-selectin levels (baseline, 16.2 ± 3.4 ng/ml; 12 weeks, 11.0 ± 3.2 ng/ml; 52 weeks, 12.3 ± 4.2 ng/ml; *P* <0.01) were observed. Endothelial-independent relaxation induced by nitroglycerin was not modified during the study.

Interpretation. The results indicate that a long-term treatment with simvastatin produced a sustained beneficial effect in endothelial function in hFH patients.

Comment

There was no parallel placebo-treated control group in this study as it was not thought ethical to include one. Comparisons are therefore made with baseline data in the study group.

The patients in this study had severe hypercholesterolaemia with an average concentration of 11.2 mmol/l. About half the patients were taking 40 mg simvastatin daily with the remainder on 80 mg/day. Others have shown an increase in distensibility and compliance of the femoral artery with treatment of severe hypercholesterolaemia with statins over a period of a year |**50**|. Profound (>45%) reductions in LDL-C) were needed to see these effects, which is similar to the average reduction in the present study.

Alternatives to statin treatment in FH

Despite the major success of the statins there are many patients with FH, particularly homozygotes who cannot attain ideal LDL-C concentrations even with maximum dosage. A more powerful statin, rosuvastatin |**51**|, is due to be launched in the near future in the UK. Combination of statins with inhibitors of cholesterol absorption, such as ezetimibe (not yet available in the UK), is also likely to produce additional LDL-lowering of about 15% |**52,53**|.

Other experimental approaches include the use of acyl-CoA cholesterol:acyltransferase (ACAT) inhibitors to prevent storage of cholesteryl esters and the development of foam cells in the arterial wall |**54**| and inhibitors of microsomal triglyceride transfer protein (MTP) |**55**|. The following study demonstrates the profound effects of an MTP inhibitor in the Watanabe rabbit, a model for homozygous FH. This strategy is in effect the induction of iatrogenic abetalipoproteinaemia, a hypolipidaemic disorder in which MTP is defective (see Chapter 4 – Disorders which cause hypolipidaemia).

MTP inhibitor decreases plasma cholesterol levels in LDL receptor-deficient WHHL rabbits by lowering the VLDL secretion.

Shiomi M, Ito T. *Eur J Pharmacol* 2001; **431**: 127–31.

Background. To examine whether a microsomal triglyceride transfer protein (MTP) inhibitor is effective in patients with homozygous FH, the authors administered (2S)-2-cyclopentyl-2-{4-[(2,4-dimethyl-9H-pyrido[2,3-b]indol-9-yl)methyl]p henyl}-*N*-[(1S)-2-hydroxy-1-phenylethyl]ethanamide (implitapide), a new MTP inhibitor, to LDLR-deficient Watanabe heritable hyperlipidaemic (WHHL) rabbits at doses of 3, 6, and 12 mg/kg for 4 weeks. In the 12 mg/kg group, the plasma cholesterol and TG levels were decreased by 70% and 45%, respectively, and the VLDL secretion rate was decreased by 80%. The composition of newly secreted VLDL was similar in each group.

INTERPRETATION. This suggests that implitapide diminished the number of VLDL particles secreted from the liver. Although the ratio of vitamin E/LDL was not altered by implitapide, TG accumulation and a decrease in vitamin E were observed in the liver. In conclusion, an inhibition of VLDL secretion led to a decrease of plasma LDL in WHHL rabbits, and MTP inhibitors should have hypolipidaemic effects against homozygous FH.

Comment

The WHHL rabbit lacks LDLRs and is a useful model for homozygous FH. The degree of LDL-lowering in this study is impressive but there is a question about the likely safety of MTP inhibitors in man given that they tend to cause fat deposition in the liver. Fatty liver in non-alcoholic steatohepatitis (NASH) is thought to progress to cryptogenic cirrhosis in some patients |**56**|. The frequency and severity of liver disease in abetalipoproteinaemia, which is caused by defective MTP function, is uncertain. Fatty deposition is common, but abnormalities of liver function tests are not usually severe. Cirrhosis has been described though the consumption of large amounts of medium chain TGs may have been implicated in this |**57**|. A small risk of chronic liver disease in later life may be acceptable in a disorder of the severity of homozygous FH, but would not be acceptable if MTP inhibitors are to find a place in the treatment of large numbers of people with more moderate hypercholesterolaemia.

Non-pharmacological treatments

Surgical management of dyslipidemia: clinical and experimental evidence.

Moghadasian MH, Frohlich JJ, Saleem M, Hong JM, Qayumi K, Scudamore CH. *J Invest Surg* 2001; **14**: 71–8.

BACKGROUND. Coronary artery disease (CAD) is still a major cause of mortality in developed countries, and dyslipidaemia is one of its major causes. In an attempt to reduce both mortality and morbidity from CAD, several dietary, pharmacological, and surgical approaches have been used to reduce plasma cholesterol levels. In this brief review, the evidence for cholesterol-lowering effects and safety of partial ileal bypass (PIB) procedure in both human and animal studies is summarized. The results of the Program on the Surgical Control of the Hyperlipidaemias (POSCH), which involved a total of 838 subjects with myocardial infarction (MI), are promising. A 5-year follow-up of this study revealed significant reductions of up to 27% in total cholesterol (TC) and up to 42% in LDL cholesterol levels along with an increase of up to 8% in HDL cholesterol levels as compared to controls. These changes were associated with other benefits such as increased HDL/TC and HDL/LDL ratios, and a significant decrease in apoB-100 and increase in apoA-I levels. Similar results were also demonstrated by other studies. PIB surgery is one of the most effective methods for reduction of plasma cholesterol levels, particularly in patients with heterozygous FH. This procedure is also

applicable to treatment of sitosterolaemia, a rare genetic disorder in which the absorption of plant sterols is abnormally high.

Interpretation. Although no major complications of PIB have been reported, more extensive studies are required to evaluate its long-term effects on renal and hepatic function. Similarly, long-term impact of this procedure on progression/regression of atherosclerotic lesions must be documented. Finally, indications for this procedure should be carefully considered, particularly in view of availability of other treatments of dyslipidaemia.

Comment

The need for PIB surgery has greatly diminished since the advent of statin therapy. It was never very effective for homozygous FH but there may still be a small proportion of FH heterozygotes who respond poorly to statin treatment and for whom its use may still be justified.

Low-density lipoprotein apheresis for the treatment of refractory hyperlipidemia.

Vella A, Pineda AA, O'Brien T. *Mayo Clin Proc* 2001; **76**: 1039–46.

Background. The advent of treatment with 3-hydroxy-3-methylglutaryl coenzyme A (HMG-CoA) inhibitors has meant that, with a combination of diet and drug therapy, adequate control of serum cholesterol concentrations can be achieved in most patients with hypercholesterolaemia. However, some patients, primarily those with FH, may require additional therapy to lower their cholesterol levels. In recent years, LDL-apheresis has emerged as an effective method of treatment in these patients. The criteria for commencement of LDL-apheresis are LDL cholesterol levels of 500 mg/dl or higher for homozygous FH patients, 300 mg/dl or higher for heterozygous FH patients in whom medical therapy has failed, and 200 mg/dl or higher for heterozygous FH patients with documented coronary disease and in whom medical therapy has failed. In addition to cholesterol-lowering in patients with FH, other indications for LDL-apheresis are emerging. These include its use in the treatment of graft vascular disease in patients receiving cardiac transplants as well as in the treatment of certain glomerulonephritides.

Interpretation. This review examines the role of LDL-apheresis in the management of lipid disorders and the evidence available to support its use in clinical practice.

Comment

Plasma exchange has been used for about 30 years for the physical removal of lipoproteins from the circulation in patients with primary biliary cirrhosis and painful xanthomatous neuropathy |**58**| and for the treatment of FH |**59,60**|. It is still used but has the disadvantage of removing *inter alia* HDL, albumin and immunoglobulins as well. Selective extracorporeal LDL-apheresis is therefore preferred when available. Several techniques have been used including binding to monoclonal or polyclonal antibodies to apoB-100, or to dextran sulphate. Heparin-induced extracorporeal LDL precipitation (HELP) has the putative advantage of lowering fibrinogen and Lp(a) |**61**|. This is a useful update on extracorporeal LDL removal.

Apheresis technology for prevention and regression of atherosclerosis.

Yamamoto A, Harada-Shiba M, Kawaguchi A, Tsushima M. *Ther Apher* 2001; **5**: 221–5.

BACKGROUND. FH is a congenital disorder of cholesterol metabolism, which is due to a deficiency in LDLRs. The homozygous form of FH is especially liable to CAD in youth because of the very high LDL-C levels. It is resistant to drug therapy, and LDL-apheresis is the only practical way of treatment for these patients. Some patients with heterozygous FH also have high LDL-C levels that cannot be brought down into the optimum range by any combination drug therapy. The authors treated or are treating 10 homozygous and 28 heterozygous FH patients in their hospital or in affiliated hospitals expert in blood purification. Among the ten homozygous patients, two died of MI. Only one young female patient is still free of symptoms, and the other patients have been suffering from regurgitation through the aortic valve although they have not experienced MI. Rapid rebound of LDL-C after each apheresis treatment limits the period during which LDL-C is in the optimum range. The use of atorvastatin at a high dose (40 mg/day) was attempted to suppress this rebound. In contrast with good results in receptor-defective-type patients, receptor-negative-type patients did not show a response in LDL-cholesterol levels to the statin therapy although there was a slight increase in HDL-C with a decrease in VLDL-triglyceride and-cholesterol. Follow-up study of the patients with heterozygous FH revealed that LDL-apheresis was effective in lengthening the life expectancy of the patients with pre-existing CAD, especially those who had received intervention coronary artery bypass grafting (CABG) or percutaneous transluminal coronary angioplasty (PTCA). It was also shown that the use of probucol in combination with LDL-apheresis was effective in reducing coronary events as shown by the necessity of CABG or PTCA. Clinical data on the effect of LDL-apheresis, recently reported from some other institutions in Japan, is also reviewed.

Comment

This study gives a brief review of the authors' experience and of LDL-apheresis generally for the treatment of homozygous and heterozygous FH.

Acute and long-term effects of low-density lipoprotein (LDL)-apheresis on oxidative damage to LDL and reducing capacity of erythrocytes in patients with severe familial hypercholesterolaemia.

Stefanutti C, Di Giacomo S, Vivenzio A, *et al. Clin Sci* 2001; **100**: 191–8.

BACKGROUND. Several studies have suggested that the oxidative modification of LDL could play a key role in the early stages of atherosclerosis. The susceptibility of LDL to oxidation has been found to be greater in patients with CHD. FH is a powerful clinical model in which to study the predictive role of LDL in atherogenesis. LDL-apheresis is a treatment that is able to decrease lipid levels in plasma. This study was aimed at investigating the reducing capacity of erythrocytes and the *in vitro* susceptibility to oxidation of LDL isolated from patients with homozygous, heterozygous and double-heterozygous FH, who were treated fortnightly with LDL-apheresis or left untreated. In 14 FH patients, at baseline and after a cycle of treatment, the susceptibility of LDL to oxidative modification was analyzed by studying the kinetics of conjugate diene formation. Plasma hydroperoxides, polyunsaturated fatty acid content, LDL electrophoretic mobility on agarose, the titre of auto-antibodies against oxidized LDL and serum paraoxonase activity were also measured. Furthermore, in order to evaluate a potential relationship between LDL oxidation and redox status, erythrocyte GSH and ATP levels were determined in FH patients treated regularly or never treated previously by LDL-apheresis. Unlike in the control group, the oxidative status of LDL in all FH patients was modified by LDL-apheresis, as revealed by the higher negative charge and the increase in levels of hydroperoxides and antibodies against oxidized LDL in the plasma.

INTERPRETATION. These findings suggest both an acute effect and a long-term effect of LDL-apheresis in FH patients treated with dextran sulphate cellulose apheresis. The acute effect of LDL-apheresis on the susceptibility to oxidation of plasma and LDL was demonstrated by significant decreases in plasma hydroperoxide content, total LDL concentration and polyunsaturated fatty acid content. The increased resistance of LDL to oxidation was shown by prolongation of the lag time ($P < 0.05$) in samples after a single cycle of treatment. The long-term effect of LDL-apheresis was demonstrated by the comparable values for lag phases (obtained from the kinetics of conjugate diene formation) in patients under active treatment and controls. Compared with healthy controls and untreated patients, the erythrocyte GSH content was significantly higher ($P < 0.001$) in the treated group, suggesting the activation of reducing mechanisms.

Comment

Previous work has suggested that LDL-apheresis can reduce the susceptibility of LDL to oxidative modification by removing small dense LDL which are particularly susceptible to oxidation and by increasing the activity of free radical scavenging enzyme systems. These observations are confirmed and extended in the current study. It has to be remembered however that, despite the attractions of the hypothesis that oxidative stress increases the atherogenicity of LDL and the wealth of supporting experimental work, intervention studies with antioxidant vitamins have so far proved disappointing |**62,63**|.

References

1. Goldstein JL, Brown MS. Molecular medicine. The cholesterol quartet. *Science* 2001; **292**: 1310–12.

2. Goldstein JL, Hobbs HH, Brown MS. Familial hypercholesterolemia. In: Scriver CR, Beaudet AL, Sly WS, Valle D (eds). *The Metabolic and Molecular Bases of Inherited Disease*, 8th edn. New York: McGraw-Hill, 2001: 2863–913.

3. Muller C. Angina pectoris in hereditary xanthomatosis. *Arch Int Med* 1939; **64**: 675–700.

4. Vega GL, Grundy SM. In vivo evidence for reduced binding of low density lipoproteins to receptors as a cause of primary moderate hypercholesterolemia. *J Clin Invest* 1986; **78**: 1410–14.

5. Salen G, Shefer S, Nguyen LB, Ness GC, Tint GS. Sitosterolaemia. In: Betteridge DJ, Illingworth DR, Shepherd J (eds). *Lipoproteins in Health and Disease*, 1st edn. London: Arnold, 1999: 815–27.

6. Steering Committee on behalf of the Simon Broome Register Group. Mortality in treated heterozygous familial hypercholesterolaemia: implications for clinical management. *Scientific Atherosclerosis* 1999; **142**: 105–12.

7. Heath KE, Gahan M, Whittall RA, Humphries SE. Low-density lipoprotein receptor gene (LDLR) world-wide website in familial hypercholesterolaemia: update, new features and mutation analysis. *Atherosclerosis* 2001; **154**: 243–6.

8. Wood D, Durrington P, Poulter N, McInnes G, Rees A, Wray R. Joint British recommendations on prevention of coronary heart disease in clinical practice. *Heart* 1998; **80**(Suppl 2): S1–29.

9. Haddad L, Day IN, Hunt S, Williams RR, Humphries SE, Hopkins PN. Evidence for a third genetic locus causing familial hypercholesterolemia. A non-LDLR, non-APOB kindred. *J Lipid Res* 1999; **40**: 1113–22.

10. Varret M, Rabes JP, Saint-Jore B, Cenarro A, Marinoni JC, Civeira F, Devillers M, Krempf M, Coulon M, Thiart R, Kotze MJ, Schmidt H, Buzzi JC, Kostner GM, Bertolini S, Pocovi M, Rosa A, Farnier M, Martinez M, Junien C, Boileau C. A third major locus for autosomal dominant hypercholesterolemia maps to 1p34.1-p32. *Am J Hum Genet* 1999; **64**: 1378–87.

11. Hunt SC, Hopkins PN, Bulka K, McDermott MT, Thorne TL, Wardell BB, Bowen BR, Ballinger DG, Skolnick MH, Samuels ME. Genetic localization to chromosome 1p32 of the third locus for familial hypercholesterolemia in a Utah kindred. *Arterioscler Thromb Vasc Biol* 2000; **20**: 1089–93.

12. Neil HA, Hammond T, Huxley R, Matthews DR, Humphries SE. Extent of underdiagnosis of familial hypercholesterolaemia in routine practice: prospective registry study. *Br Med J* 2000; **321**: 148.

13. Bhatnagar D, Morgan J, Siddiq S, Mackness MI, Miller JP, Durrington PN. Outcome of case finding among relatives of patients with known heterozygous familial hypercholesterolaemia. *Br Med J* 2000; **321**: 1497–500.

14. Durrington PN, Adams JE, Beastall MD. The assessment of Achilles tendon size in primary hypercholesterolaemia by computed tomography. *Atherosclerosis* 1982; **45**: 345–58.

15. Heiberg A, Slack J. Family similarities in the age at coronary death in familial hypercholesterlaemia. *Br Med J* 1977; **ii**: 493–5.

16. Bertolini S, Cassanelli R, Garuti M, Ghisellini M, Simone ML, Rolleri M, Mastruzo P, Calandra S. Analysis of LDL receptor gene mutations in Italian patients with homozygous familial hypercholesterolemia. *Arterioscl Thromb Vasc Biol* 1999; **19**: 408–18.

17. Kraft HG, Lingenhel A, Raal FJ, Hohenegger M, Utermann G. Lipoprotein(a) in homozygous familial hypercholesterolemia. *Arterioscler Thromb Vasc Biol* 2000; **20**: 522–8.

18. Vuorio AF, Turtola H, Piilahti KM, Repo P, Kanninen T, Kontula K. Familial hypercholesterolemia in the Finnish north Karelia. A molecular, clinical, and genealogical study. *Arterioscler Thromb Vasc Biol* 1997; **17**: 3127–38.

19. Williams RR, Hasstedt SJ, Wilson DE, Ash KO, Yanowitz FF, Reiber GE, Kuida H. Evidence that men with familial hypercholesterolemia can avoid early coronary death. An analysis of 77 gene carriers in four Utah pedigrees. *JAMA* 1986; **255**: 219–24.

20. Ordovas JM, Cupples LA, Corella D, Otvos JD, Osgood D, Martinez A, Lahoz C, Coltell O, Wilson PW, Schaefer EJ. Association of cholesteryl ester transfer protein-TaqIB polymorphism with variations in lipoprotein subclasses and coronary heart disease risk: the Framingham study. *Arterioscler Thromb Vasc Biol* 2000; **20**: 1323–9.

21. Havel RJ. Remnant lipoproteins as therapeutic targets. *Curr Opin Lipidol* 2000; **11**: 615–20.

22. Karpe F. Postprandial lipoprotein metabolism and atherosclerosis. *J Intern Med* 1999; **246**: 341–55.

23. Watts GF. Postprandial lipaemia in familial hypercholesterolaemia: clinical and metabolic significance. *Atherosclerosis* 2000; **148**: 426–8.

24. Rubinsztein DC, Cohen JC, Berger GM, van der Westhuyzen DR, Coetzee GA, Gevers W. Chylomicron remnant clearance from the plasma is normal in familial hypercholesterolemic homozygotes with defined receptor defects. *J Clin Invest* 1990; **86**: 1306–12.

25. Mamo JC, Smith D, Yu KC, Kawaguchi A, Harada–Shiba M, Yamamura T, Yamamoto A. Accumulation of chylomicron remnants in homozygous subjects with familial hypercholesterolaemia. *Eur J Clin Invest* 1998; **28**: 379–84.

26. Cabezas MC, de Bruin TW, Westerveld HE, Meijer E, Erkelens DW. Delayed chylomicron remnant clearance in subjects with heterozygous familial hypercholesterolaemia. *J Intern Med* 1998; **244**: 299–307.

27. Redgrave TG, Watts GF, Martins IJ, Barrett PH, Mamo JC, Dimmitt SB, Marais AD. Chylomicron remnant metabolism in familial dyslipidemias studied with a remnant-like emulsion breath test. *J Lipid Res* 2001; **42**: 710–15.

28. Eriksson M, Angelin B, Henriksson P, Ericsson S, Vitols S, Berglund L. Metabolism of lipoprotein remnants in humans. Studies during intestinal infusion of fat and cholesterol in subjects with varying expression of the low density lipoprotein receptor. *Arterioscler Thromb* 1991; **11**: 827–37.

29. Carmena R, Roy M, Roederer G, Minnich A, Davignon J. Coexisting dysbetalipoproteinemia and familial hypercholesterolemia. Clinical and laboratory observations. *Atherosclerosis* 2000; **148**: 113–24.

30. Khachadurian AK, Uthman SM. Experiences with the homozygous cases of familial hypercholesterolemia. A report of 52 patients. *Nutr Metab* 1973; **15**: 132–40.

31. Zuliani G, Vigna GB, Corsini A, Maioli M, Romagnoni F, Fellin R. Severe hypercholesterolaemia: unusual inheritance in an Italian pedigree. *Eur J Clin Invest* 1995; **25**: 322–31.

32. Norman D, Sun XM, Bourbon M, Knight BL, Naoumova RP, Soutar AK. Characterization of a novel cellular defect in patients with phenotypic homozygous familial hypercholesterolemia. *J Clin Invest* 1999; **104**: 619–28.

33. Zuliani G, Arca M, Signore A, Bader G, Fazio S, Chianelli M, Bellosta S, Campagna F, Montali A, Maioli M, Pacifico A, Ricci G, Fellin R. Characterization of a new form of inherited hypercholesterolemia: familial recessive hypercholesterolemia. *Arterioscler Thromb Vasc Biol* 1999; **19**: 802–9.

34. Ciccarese M, Pacifico A, Tonolo G, Pintus P, Nikoshkov A, Zuliani G, Fellin R, Luthman H, Maioli M. A new locus for autosomal recessive hypercholesterolemia maps to human chromosome 15q25–q26. *Am J Hum Genet* 2000; **66**: 453–60.

35. Bhattacharyya AK, Connor WE. Beta–sitosterolemia and xanthomatosis. A newly described lipid storage disease in two sisters. *J Clin Invest* 1974; **53**: 1033–43.

36. Patel SB, Salen G, Hidaka H, Kwiterovich PO, Stalenhoef AF, Miettinen TA, Grundy SM, Lee MH, Rubenstein JS, Polymeropoulos MH, Brownstein MJ. Mapping a gene involved in regulating dietary cholesterol absorption. The sitosterolemia locus is found at chromosome 2p21. *J Clin Invest* 1998; **102**: 1041–4.

37. Allayee H, Laffitte BA, Lusis AJ. Biochemistry. An absorbing study of cholesterol. *Science* 2000; **290**: 1709–11.

38. Lee MH, Lu K, Patel SB. Genetic basis of sitosterolemia. *Curr Opin Lipidol* 2001; **12**: 141–9.

39. Karayan L, Qiu S, Betard C, Dufour R, Roederer G, Minnich A, Davignon J, Genest J Jr. Response to HMG CoA reductase inhibitors in heterozygous familial hypercholesterolemia due to the 10-kb deletion (French Canadian mutation) of the LDL receptor gene. *Arterioscler Thromb* 1994; **14**(8): 1258–63.

40. Thompson GR, O'Neill F, Seed M. Why some patients respond poorly to statins and how this might be remedied. *Eur Heart J* 2002; **23**: 200–6.

41. Scandinavian Simvastatin Survival Study (4S). Randomized trial of cholesterol lowering in 4444 patients with coronary heart disease. *Lancet* 1994; **344**: 1383–9.

42. Shepherd J, Cobbe SM, Ford I, Isles CG, Lorimer AR, Macfarlane PW, McKillop JH, Packard CJ, Group ftWoSCPS. Prevention of coronary heart disease with pravastatin in men with hypercholesterolemia. *N Engl J Med* 1995; **333**: 1301–7.

43. Sacks FM, Pfeffer MA, Moye LA, Rouleau JL, Rutherford JD, Cole TG, Brown L, Warnica JW, Arnold JMO, Wun C–C, Davis BR, Braunwald E. The effect of pravastatin

on coronary events after myocardial infarction in patients with average cholesterol levels. *N Engl J Med* 1996; **335**: 1001–9.

44. The Long–Term Intervention with Pravastatin in Ischaemic Disease (LIPID) Study Group. Prevention of cardiovascular events and death with pravastatin in patients with coronary heart disease and a broad range of initial cholesterol levels. *N Engl J Med* 1998; **339**: 1349–57.

45. Downs JR, Clearfield M, Weis S, Whitney E, Shapiro DR, Beere PA, Langendorfer A, Stein EA, Kruyer W, Gotto AM Jr. Primary prevention of acute coronary events with lovastatin in men and women with average cholesterol levels: results of AFCAPS/TexCAPS. Air Force/Texas Coronary Atherosclerosis Prevention Study. *JAMA* 1998; **279**: 1615–22.

46. Heart Protection Study Collaborative Group. MRC/BHF Heart Protection Study of cholesterol lowering with simvastatin in 20 536 high-risk individuals: a randomized placebo-controlled trial. *Lancet* 2002; **360**: 7–22.

47. Bhatnagar D, Durrington PN, Neary R, Miller JP. Elevation of skeletal muscle isoform of serum creatine kinase in heterozygous familial hypercholesterolaemia. *J Intern Med* 1990; **228**: 493–5.

48. Celemajer DS, Sorensen KE, Gooch VM. Noninvasive detection of endothelial dysfunction in children and adults at risk of atherosclerosis. *Lancet* 1992; **340**: 1111–5.

49. Chowienczyk PJ, Watts GF, Cockroft JR. Impaired endothelium–dependent vasodilation of forearm resistance vessels in hypercholesterolaemia. *Lancet* 1992; **340**: 1430–2.

50. Smilde TJ, van den Berkmortel FW, Wollersheim H, van Langen H, Kastelein JJ, Stalenhoef AF. The effect of cholesterol lowering on carotid and femoral artery wall stiffness and thickness in patients with familial hypercholesterolaemia. *Eur J Clin Invest* 2000; **30**: 473–80.

51. Olsson AG. A new statin: a new standard. *Clin Cardiol* 2001; **24**(8 Suppl III): 18–23.

52. Miettinen TA. Cholesterol absorption inhibition: a strategy for cholesterol-lowering therapy. *Int J Clin Pract* 2001; **55**: 710–16.

53. Meng CQ. Ezetimibe (Schering–Plough). *Curr Opin Investig Drugs* 2001; **2**: 389–92.

54. Brown WV. Therapies on the horizon for cholesterol reduction. *Clin Cardiol* 2001; **24**(Suppl III): 24–7.

55. Best JD, Jenkins AJ. Novel agents for managing dyslipidaemia. *Expert Opin Investig Drugs* 2001; **10**: 1901–11.

56. Caldwell SH, Oelsner DH, Iezzoni JC, Hespenheide EE, Battle EH, Driscoll CJ. Cryptogenic cirrhosis: clinical characterization and risk factors for underlying disease. *Hepatol* 1999; **29**: 664–9.

57. Malloy MJ, Kane JP. Disorders involving deficiencies of lipoproteins that contain B apolipoproteins. In: Betteridge DJ, Illingworth DR, Shepherd J (eds). *Lipoproteins in Health and Disease*, 1st edn. London: Arnold, 1999: 863–77.

58. Turnberg LA, Mahoney MP, Gleeson MH, Freeman CB, Gowenlock AH. Plasmaphoresis and plasma exchange in the treatment of hyperlipaemia and xanthomatous neuropathy in patients with primary biliary cirrhosis. *Gut* 1972; **13**: 976–81.

59. de Gennes JL, Touraine R, Maunand B, Truffert J, Laudat P. Formes homozygotes cutaneo–tendineuses de xanthomatose hypercholestérolémique dans une observation familiale exemplaire. Essai de plasmapherese à titre de traitement héroique. *Bull Mem Soc Med Hop Paris* 1967; **118**: 1377–402.

60. Thompson GR, Lowenthal R, Myant NB. Plasma exchange in the management of homozygous familial hypercholesterolaemia. *Lancet* 1975; **1**: 1208–11.

61. Seidel D. Non–pharmacological procedures: treatment of severe hypercholesterolaemia in patients with coronary heart disease by means of heparin–induced extracorporeal low–density lipoprotein plasmapheresis. In: Betteridge DJ, Illingworth DR, Shepherd J (eds). *Lipoproteins in Health and Disease*, 1st edn. London: Arnold, 1999: 1255–66.

62. Gruppo Italiano per lo Studio della Sopravvivenza nell'Infarto miocardico. Dietary supplementation with n–3 polyunsaturated fatty acids and vitamin E after myocardial infarction: results of the GISSI–Prevenzione trial. *Lancet* 1999; **354**: 447–55.

63. Yusuf S, Dagenais G, Pogue J, Bosch J, Sleight P. Vitamin E supplementation and cardiovascular events in high-risk patients. The Heart Outcomes Prevention Evaluation Study Investigators. *N Engl J Med* 2000; **342**: 154–60.

2

Hypertriglyceridaemia and mixed hyperlipidaemia

Plasma triglycerides as a risk factor

The status of plasma triglycerides (TG), and hence the lipoproteins in which they are carried, has been controversial. Most studies have found TG to be a risk factor for CHD events on univariate analysis, but the significance of this relationship was generally lost on multivariate analysis when other factors, particularly high-density lipoprotein cholesterol (HDL-C), were taken into account [**1**]. Some more recent studies however find TG to remain an independent risk factor even after multivariate analysis [**2**].

Triglyceride-rich lipoproteins (TGRL) vary enormously in size and composition and it is likely that they also vary in their atherogenic potential. Chylomicrons and very low-density lipoproteins (VLDL), secreted by the intestine and the liver respectively, are acted upon by lipoprotein lipase (LPL). This partially removes TG to form remnant lipoproteins, which still carry substantial TG but are also relatively enriched in cholesterol. These remnants are widely believed to be atherogenic when compared with their newly secreted parent TGRL [**3–6**]. This is supported by the propensity of patients with type III hyperlipoproteinaemia (familial dysbetalipoproteinaemia or remnant removal disease) to develop vascular disease [**7**] despite low low-density lipoprotein (LDL) concentrations.

The following case report supports the notion that, even when present in very high concentrations, nascent chylomicrons and VLDL may not be atherogenic, though this is contentious [**8**].

No evidence of accelerated atherosclerosis in a 66-year-old chylomicronemia patient homozygous for the nonsense mutation (Tyr61→stop) in the lipoprotein lipase gene.

Ebara T, Okubo M, Horinishi A, Adachi M, Murase T, Hirano T.
Atherosclerosis 2001; **159**: 375–9.

Background. Whether chylomicronaemia is atherogenic or not has yet to be determined in humans. This study investigated a 66-year-old female with severe chylomicronaemia resulting from an LPL deficiency. The patient's plasma TG level was

approximately 2000 mg/dl. Both LPL activity and the mass of post-heparin plasma in this patient were virtually absent. A nonsense mutation in exon 3 (Tyr61 Stop) was identified in the patient's LPL gene, and a restriction fragment length polymorphism analysis established that the patient was homozygous for this mutation. The patient was neither a diabetic nor a smoker. Clinically, the patient had never experienced pancreatitis or angina pectoris. An examination of her carotid, femoral and coronary arteries by ultrasonogram and electrocardiogram after exercise-tolerance testing showed no accelerated atherosclerosis.

INTERPRETATION. This case suggests that atherosclerosis may not occur despite massive hyperlipidaemia, when LPL bridging was not present due to the absence of LPL secretion and circulating mass.

Comment

Lipoprotein lipase deficiency may cause pancreatitis but it is often stated that it does not carry a high risk of atherosclerotic vascular disease. The reasons cited are that chylomicrons are too large to gain access to the vascular intima and LDL levels are extremely low. HDL levels however can also be very low. In addition, the patients often take extremely low-fat diets in order to avoid attacks of abdominal pain. The current study supports the general view, but conflicts with the work of Benlian *et al.* |**8**| who described four patients with functional deficiency of LPL all of whom had evidence of vascular disease before the age of 55 years. The authors of the present study note that their patient had no other risk factors for vascular disease whereas three of Benlian's cases had at least one other risk factor. In addition, the case in the current study had a nonsense mutation leading to virtually undetectable LPL in terms of both mass and activity. Benlian's cases had a missense mutation leading to defective activity and a low but significant mass of inactive enzyme. The role of LPL in lipoprotein metabolism is complex |**9,10**|. It not only hydrolyzes TG but also serves to retain apoprotein B (apoB)-containing particles on vessel walls by providing a link with proteoglycans, a property which is likely to be pro-atherogenic and which may have been retained by the enzymatically inactive lipase in the cases of Benlian *et al.* |**8**|.

One needs to be cautious in advising patients with severe hypertriglyceridaemia about the risk of vascular disease given the conflicting findings in the literature. Moreover, gross hypertriglyceridaemia due to an inborn error such as LPL or apoC-II deficiency, in which LDL concentrations are very low, is a fundamentally different situation from the much more common patient in the lipid clinic. The latter is the patient with moderate primary hypertriglyceridaemia who develops a severe exacerbation as a result of a superimposed secondary factor, such as poorly controlled diabetes, alcohol or medications. Agents which may exacerbate hypertriglyceridaemia in this way include oral oestrogens, β-blockers and protease inhibitors in patients with HIV. It would be unwise to assume that such patients are at low risk of vascular disease.

Remnant lipoprotein metabolism

The partially delipidated remnants of TGRL have long been thought to be atherogenic |**3**|, but the difficulty of distinguishing remnant particles from newly secreted TGRL in the VLDL–IDL (intermediate density lipoprotein) density range has greatly hampered research in this area. A variety of techniques has been used but none is completely satisfactory. Early work mainly used vitamin A labelling, since oral vitamin A is secreted from the intestine in chylomicrons and largely remains with the particles during their metabolism, but there is some exchange with other lipoprotein classes. The ability to distinguish apoB-48 (in chylomicrons) from apoB-100 (in hepatic VLDL) has greatly facilitated work in this field. Another method using antibodies to apoA-I and apoB-100 to remove non-remnant particles was used in the work by Ooi *et al.* and Stein *et al.* discussed below. A new approach using a stable isotope breath test to study chylomicron remnant metabolism is described in the following paper.

Chylomicron remnant metabolism in familial dyslipidemias studied with a remnant-like emulsion breath test.

Redgrave TG, Watts GF, Martins IJ, *et al. J Lipid Res* 2001; **42**: 710–15.

BACKGROUND. A stable isotope breath test was developed for the assessment of chylomicron remnant metabolism and the results are reported from the breath test in human subjects selected for disorders of chylomicron or remnant metabolism. In type I hyperlipaemia, the phenotype is extreme hypertriglyceridaemia due to a lack of LPL activity, which causes the failure of remnant formation. The type III dyslipidaemia phenotype is caused by the inefficient removal of chylomicron remnants from plasma, generally because of homozygosity for apolipoprotein E2 alleles. The breath test was predicted to be abnormal in type III hyperlipaemia, whereas *a priori* in type I hyperlipaemia, defective remnant clearance was not anticipated. Subjects were injected with lipid emulsions prepared with a composition similar to normal chylomicron remnants. The emulsions contained cholesteryl ester incorporating the stable non-radioactive isotope ^{13}C in the fatty acid moiety. End exhalation breath was collected at intervals after intravenous injection of the remnant-like emulsions and analyzed for ^{13}C enrichment by isotope–ratio mass spectrometry.

INTERPRETATION. Compared with the group of normolipaemic men, the fractional catabolic rate of remnants measured by the breath test was significantly decreased ($P = 0.006$) in subjects with type III dyslipidaemia. In the group with type I hyperlipaemia, the fractional catabolic rate was not different ($P = 0.233$) from the control group. Therefore, the underlying capacity for remnant catabolism was normal in this group of markedly hypertriglyceridaemic subjects. By short-circuiting the step of lipolysis, the remnant-like emulsion breath test provides direct information about remnant clearance and metabolism, which should assist in investigations of post-prandial lipid metabolism.

Comment

This method examines remnant removal and clearance using a synthetic lipid emulsion with a composition similar to that of chylomicron remnants labelled with ^{13}C-oleic acid in the cholesteryl esters. The artificial remnants acquire apoE and are cleared by the liver where the cholesteryl esters are hydrolyzed. The $^{13}CO_2$ released by fatty acid oxidation can be recovered in the expired air. The possible limitations of the method are discussed, but the finding of normal remnant clearance in lipoprotein lipase deficiency and defective clearance in type III support the validity of the method. The test enables distinction to be made between defects in lipolysis and defects in remnant clearance.

Post-prandial remnant-like lipoproteins in hypertriglyceridemia.

Ooi TC, Cousins M, Ooi DS, *et al. J Clin Endocrinol Metab* 2001; **86**: 3134–42.

BACKGROUND. It has been proposed that remnants of chylomicrons and VLDL are atherogenic. This investigation used an immunochemical method to isolate remnant-like particles (RLP) and measured them in terms of their cholesterol and TG. RLP consist of apoB-48-containing TG-rich lipoproteins and remnant-like VLDL containing apoB-100. The study aim was to look for information from post-prandial RLP data that could not be known from other markers of TG-rich lipoproteins and fasting TG and RLP data alone. A total of 41 subjects were studied. Eight subjects had hypertriglyceridaemia (HTG) and low HDL, 14 had combined hyperlipidaemia (CH), 5 had the apoE-2/2 genotype receiving gemfibrozil, 10 were normolipidaemic (NL) controls, and 4 had hypercholesterolaemia. As a whole group, there was correlation among (1) fasting TG, RLP cholesterol (RLP-C), and RLP-TG but not VLDL apoB-100, VLDL apoB-48 and their respective post-prandial responses measured as incremental area under the curve (IAUC); (2) fasting TG and post-prandial IAUC of RLP-C and RLP-TG, (3) RLP-C IAUC, RLP-TG IAUC, and TG IAUC, retinyl palmitate (RP) IAUC, and VLDL apoB-48 IAUC but not VLDL apoB-100 IAUC. The HTG/low HDL-C and CH groups had higher IAUC for RLP-C, RLP-TG, TG, and RP than the NL group. Fasting and post-prandial RLP were TG-enriched in the HTG/low HDL-C group and to a lesser extent in the CH group. The HTG/low HDL-C and CH groups had a delay in their RLP-C but not RLP-TG peaks suggesting a delay in hepatic clearance of RLP and/or a protracted period of lipolysis and/or processing of RLP. The fasting and post-prandial RLP-C/RLP-TG and RLP-C/TG ratios were elevated in the apoE-2/2 group in spite of gemfibrozil therapy. The increment in post-prandial RLP was, however, not exaggerated.

INTERPRETATION. These data indicate that (1) post-prandial RLP lipaemia is enhanced in HTG subjects when compared with NL subjects, (2) post-prandial RLP lipaemia is proportional to fasting RLP and TG levels and mirrors, to a large extent, increases in post-prandial TG, RP, and VLDL apoB-48 but not VLDL apoB-100, (3) there are compositional differences in fasting and post-prandial RLP in the three forms of HTG studied, RLP being TG-enriched in the HTG/low HDL-C group and to a lesser extent in the CH group, and cholesterol-enriched in the apoE-2/2 group, and (4) apoE-2/2 subjects had high fasting and post-prandial RLP-C concentrations in spite of being on treatment with gemfibrozil and having normal fasting and post-prandial TG concentrations.

Comment

Remnant particles are difficult to separate from newly secreted TGRL. The technique used here, and in the following paper by Stein *et al.*, is to pass plasma through a gel containing antibodies to apoA-I and a specific monoclonal antibody to apoB-100. The former removes all HDL and newly secreted chylomicrons containing apoA-I. Remnants which lose apoA-I during lipolysis pass through the gel. The latter removes newly secreted VLDL, and LDL, but does not recognize partially metabolized remnants of VLDL which have acquired apoE. Neither does it recognize chylomicron remnants, which contain apoB-48 rather than apoB-100.

This technique has previously been used to show increased amounts of post-prandial RLP-C in normolipidaemic subjects with coronary heart disease (CHD). This study examines post-prandial RLP in subjects with different types of HTG. In the whole group of patients there is a striking correlation between fasting TG levels and the post-prandial IAUC for remnant cholesterol and TG. It is notable that in all groups the major proportion of the post-prandial increase in plasma TG levels was in remnant particles confirming other recent work |**11**|.

The fact that type III patients have cholesterol-enriched TGRL in the fasting state even when on treatment with fibrates has long been known but their post-prandial remnant responses while on drug therapy have not previously been reported. The apoE-2 homozygotes in this study have remnants which are greatly cholesterol-enriched both in the fasting and post-prandial states even when normotriglyceridaemic on fibrate therapy. This study emphasizes major differences in the composition of the remnants which accumulate post-prandially in different HTG states and which may presumably have different atherogenic potential.

Effect of statin therapy on remnant lipoprotein cholesterol levels in patients with combined hyperlipidemia.

Stein DT, Devaraj S, Balis D, Adams-Huet B, Jialal I. *Arterioscler Thromb Vasc Biol* 2001; **21**: 2026–31.

Background. Clinical trials with statins have demonstrated significant reductions in cardiovascular events. Remnant lipoproteins are independent predictors of cardiovascular events. Because of the paucity of data on the effect of statins on

remnant lipoproteins, this investigation tested the effect of pravastatin, simvastatin, and atorvastatin on remnant lipoprotein cholesterol (RLP-C) levels in a randomized cross-over study in patients with CH. After a 6-week diet phase, patients (*n* = 22) were randomized to pravastatin (40 mg/d), simvastatin (20 mg/d), or atorvastatin (10 mg/d) for 6 weeks, with a 3-week washout between each drug. All three drugs significantly decreased total and LDL cholesterol (*P* <0.001). Mean reduction in LDL cholesterol with pravastatin, simvastatin, and atorvastatin was 21%, 29%, and 32%, respectively. None of the drugs affected HDL cholesterol levels. Median levels of TGs were significantly reduced with simvastatin (26%, *P* = 0.001) and atorvastatin (24%, *P* = 0.0001) but not with pravastatin (9%, *P* = 0.18). Non-HDL cholesterol decreased significantly with all three statins (20%, 29%, and 32% with pravastatin, simvastatin, and atorvastatin, respectively; *P* <0.001). Median RLP-C levels were significantly reduced with simvastatin (6%, *P* <0.05) and atorvastatin (25.9%, *P* <0.001) but not with pravastatin (2.9%, *P* = 0.58).

INTERPRETATION. Atorvastatin and simvastatin, in addition to reducing LDL cholesterol and TG levels, significantly reduced RLP-C levels. This could be another potential mechanism to explain their cardiovascular benefits.

Comment

The three statins investigated in this study were used in different doses with a view to producing similar reductions in LDL-C, though this was only partially successful. The dose of pravastatin used was the maximum licensed in the UK. Nevertheless it did not reduce TG and the reduction in LDL-C was less than with the other two drugs. This may be relevant to the failure to see a reduction in RLP-C with pravastatin. By contrast a modest reduction in plasma TG (11–14%) was seen in large clinical end-point trials with pravastatin. A re-analysis of the West of Scotland Coronary Prevention Study (WOSCOPS) has suggested that the benefit seen with pravastatin exceeds that which would be predicted from lowering LDL-C alone |**12**|. If this is correct then the current study implies that any extra benefit does not relate to reduction of the concentration of remnant lipoproteins.

It is uncertain whether the reduction in RLP-C seen with simvastatin and atorvastatin in this study is the result of a reduction in TGRL secretion, an increase in remnant clearance or a combination of the two.

The effects of exercise on post-prandial lipaemia and remnant lipoprotein metabolism are discussed in Chapter 3 – Diet, exercise and lipoprotein metabolism.

Familial combined hyperlipidaemia and familial hypertriglyceridaemia

Familial hypertriglyceridaemia (FHTG) and familial combined hyperlipidaemia (FCH), were described about 30 years ago in studies of the families of survivors of myocardial infarction (MI) |**13,14**|. The degree of risk which FHTG confers with respect to CHD is uncertain, though it appears to be less than that associated with

FCH [15]. The metabolic defect is thought to be overproduction of TG leading to secretion of large TG-rich VLDL. The genetic defect however is unknown. FCH appeared to be an autosomal dominant condition characterized by the appearance of multiple lipoprotein phenotypes within an affected family (usually types IIa, IIb or IV). It is found in about 1% of the general population in some Western societies and up to 20% of survivors of premature MI. A single metabolic defect or underlying genetic locus has however not been identified and increasingly it appears that the disorder is heterogeneous [16].

Lipoprotein and apolipoprotein abnormalities in familial combined hyperlipidemia: a 20-year prospective study.

McNeely MJ, Edwards KL, Marcovina SM, Brunzell JD, Motulsky AG, Austin MA. *Atherosclerosis* 2001; **159**: 471–81.

BACKGROUND. In order to characterize the lipoprotein abnormalities in FCH and to describe factors associated with the stability of the FCH phenotype during 20-year follow-up, 287 individuals from 48 families with FCH originally identified in the early 1970s (baseline) were studied. Hyperlipidaemia was defined as lipid-lowering medication use, or >age- and sex-specific 90th percentile for TGs or cholesterol. TG, cholesterol and medical history data were obtained at baseline and 20-year follow-up. Additional follow-up measures included HDL-C, LDL-C, LDL particle size, lipoprotein(a), apolipoprotein (apo)A-I, apoB, and apoE polymorphism. Longitudinally, two-thirds of relatives were consistently normolipidaemic or hyperlipidaemic, and one-third were discordant for hyperlipidaemic status at baseline and 20-year follow-up. Individuals with hyperlipidaemia at baseline and/or follow-up had higher apoB levels than those with consistently normal lipids (*P* <0.05), whereas small LDL size was associated with concurrent hyperlipidaemia. Among individuals who were normolipidaemic at baseline, the following variables were independently associated with development of hyperlipidaemia over 20 years: older age at baseline, male gender, greater increase in body mass index (BMI) during follow-up, and apoE alleles epsilon 2 or epsilon 4.

INTERPRETATION. In conclusion, apoB is associated with hyperlipidaemia and apoE polymorphism is associated with later onset of hyperlipidaemia in FCH.

Comment

This observational study followed patients defined as having FCH in two studies reported more than 25 years ago. In the ensuing 20 years about 17% changed from normolipidaemia to hyperlipidaemia and an equal proportion changed in the opposite direction. This lack of concordance on follow-up is not an uncommon phenomenon in follow-up studies and is discussed by the authors. It is not a function of ageing as equal numbers move in each direction and the definition of hyperlipidaemia was age-related. It probably relates to a number of factors including regression to the mean, and biological and analytical variability in repeated lipid measurements. True changes in mean lipid levels (secular trends) in the populations from which the subjects were drawn probably also play a part.

Small dense LDL appeared to be related to the presence of hyperlipidaemia rather than to a diagnosis of FCH *per se*, whereas raised levels of apoB seemed to more closely related to the FCH phenotype. Even though about half of those with a first-degree affected relative in families with FCH have hypercholesterolaemia and/or hypertriglyceridaemia consonant with monogenic dominant inheritance, the authors conclude that FCH may well be an oligogenic disorder. ApoB levels and LDL subclass are two possibly implicated gene effects, though the genes concerned remain unidentified.

A major gene effect on fasting insulin and insulin sensitivity in familial combined hyperlipidemia.

Pihlajamaki J, Austin M, Edwards K, Laakso M. *Diabetes* 2001; **50**: 2396–401.

Background. The most common inherited dyslipidaemia, FCH, is associated with insulin resistance. Whether insulin sensitivity in these families is inherited is not known. Therefore, the inheritance of insulin sensitivity was investigated in 352 non-diabetic family members from 37 families with FCH, 105 of whom had undergone testing using the hyperinsulinaemic–euglycaemic clamp technique for the measurement of insulin sensitivity. First, complex segregation analysis of fasting insulin levels (both unadjusted and age-, age(2)-, and BMI-adjusted) was used for modelling of the variance in fasting insulin levels. In these analyses, Mendelian codominant inheritance ($P = 0.320$ for unadjusted and $P = 0.295$ for adjusted insulin values) was not rejected over the most general model and fitted the data significantly better than the sporadic model ($P < 0.001$). Polygenic and environmental models were rejected ($P < 0.001$). The Mendelian codominant model explained 44 and 45% of the variance in unadjusted and adjusted fasting insulin levels, respectively. The proposed genotypes of this locus, based on segregation analysis, were associated with directly measured insulin sensitivity in 105 FCH family members who underwent the hyperinsulinaemic–euglycaemic clamp ($P < 0.001$).

Interpretation. These results provide evidence for a major gene regulating insulin sensitivity in FCH families. Possible pleiotropic effects of this insulin sensitivity locus on dyslipidaemias in FCH remain to be elucidated.

Comment

It is now regarded as unlikely that a single genetic locus will be identified which underlies the manifestations of FCH. It is thought most probably to be an oligogenic disorder. Major genetic effects have been reported with regard to several of its components including circulating levels of TG, apoB and small, dense LDL, though the genes concerned have not been identified. The current work, in FCH patients in whom diabetes was excluded, suggests a major genetic contribution to fasting insulin levels, and hence to insulin resistance, in FCH patients. Still no gene has been identified, though previous work by the same investigators has excluded some possible candidates including the gene for peroxisome proliferator-activated receptor-γ (PPAR-γ).

Analysis of the ileal bile acid transporter gene, SLC10A2, in subjects with familial hypertriglyceridemia.

Love MW, Craddock AL, Angelin B, Brunzell JD, Duane WC, Dawson PA.
Arterioscler Thromb Vasc Biol 2001; **21**: 2039–45.

Background. FHTG, a disease characterized by elevated plasma VLDL TG levels, has been associated with impaired intestinal absorption of bile acids. The aim of this study was to test the hypothesis that defects in the active ileal absorption of bile acids are a primary cause of FHTG. Single-stranded conformation polymorphism analysis was used to screen the ileal Na^+/bile acid cotransporter gene (SLC10A2) for FHTG-associated mutations. Analysis of 20 hypertriglyceridaemic patients with abnormal bile acid metabolism revealed three missense mutations (V98I, V159I, and A171S), a frame-shift mutation (646insG) at codon 216, and 4 polymorphisms in the 5′ flanking sequence of SLC10A2. The SLC10A2 missense mutations and 5′ flanking sequence polymorphisms were not correlated with bile acid production or turnover in the hypertriglyceridaemic patients and were equally prevalent in the unaffected control subjects. In transfected COS cells, the V98I, V159I, and A171S isoforms all transported bile acids similar to the wild-type SLC10A2. The 646insG frame-shift mutation abolished bile acid transport activity in transfected COS cells but was found in only a single FHTG patient.

Interpretation. These findings indicate that the decreased intestinal bile acid absorption in FHTG patients is not commonly associated with inherited defects in SLC10A2.

Comment

The genetic basis for FHTG is unknown. Candidate genes include those directly involved in TGRL metabolism and those concerned with bile acid metabolism. Factors which increase endogenous bile acid synthesis such as cholestyramine and ileal exclusion increase TG concentrations, whereas feeding bile acid supplements decreases both bile acid synthesis and TG levels. Moreover many patients with type IV hyperlipoproteinaemia and FHTG have increased bile acid synthesis. It has been suggested therefore that the underlying defect in these patients relates to bile acid malabsorption [**17,18**]. Mutations in the sodium/bile acid cotransporter gene studied here however do not appear to be commonly responsible for the lipoprotein abnormalities in FHTG. The authors suggest as an alternative defects in the farnesoid X-activated receptor (FXR), which regulates bile acid synthesis and transport.

Pharmacological treatment of combined hyperlipidaemia

When patients with mixed hyperlipidaemia are treated with statins the resultant lipid profile often leaves something to be desired particularly in terms of the concentrations of TG and HDL-C. When the residual abnormality cannot be resolved by means of dietary change or weight loss (as is often the case) it is not uncommon to add a fibrate drug which usually gives additional biochemical benefit. Additional clinical benefit however has not been demonstrated in end-point trials, and it is likely to be a considerable time before such data are available. Combinations of statins and fibrates seem generally to be safe though there is understandable anxiety about combining two drug classes that can give rise to myositis and even rhabdomyolysis. This was heightened by the recent withdrawal of cerivastatin, after a number of the fatalities occurred with a combination of cerivastatin and gemfibrozil |**19**|.

Fibrates have been shown to reduce cardiovascular end-points |**20,21**| and to have favourable effects on sequential coronary angiograms in clinical trials |**22,23**|, but only statins have been shown to reduce all-cause mortality. Fibrates however are more potent in reducing plasma TG. It is not always easy in patients with HTG to decide when to use fibrates and when to use statins. It seems reasonable in those who have elevated LDL-C to use a statin in the first instance. This will usually produce a modest reduction in TG. A decision can then be made about the need for combination therapy with a fibrate |**24**|.

Combination lipid-altering therapy: an emerging treatment paradigm for the 21st century.

Jacobson TA. *Curr Atheroscler Rep* 2001; **3**: 373–82.

Background. For the care of an expanding segment of the US population with multiple coronary risk factors, combination lipid-altering therapy is emerging as a treatment imperative. The most recent National Cholesterol Education Program (NCEP)'s consensus guidelines emphasize long-term global CHD risk status, designate patients with CHD risk equivalents (e.g. diabetes, peripheral arterial disease, 20% or more 10-year absolute CHD risk) for aggressive lipid-altering therapy, and deem the metabolic syndrome (e.g. obesity, insulin resistance, hypertension, elevated TGs, low levels of HDL-C, small dense LDL particles) as a secondary target for intervention. With the advancing age of the US population and the high prevalence of diabetes, the metabolic syndrome, and CHD, increasing numbers of patients will require a more balanced metabolic attack attainable only through combination lipid-altering regimens.

Interpretation. Many of these patients, as well as persons at heightened risk for cardiovascular disease because of a range of heritable conditions (e.g. familial hypercholesterolaemia, familial combined hyperlipidaemia), will undoubtedly require binary or ternary regimens involving statins in concert with niacin, fibric–acid derivatives, or bile acid resins. Such approaches enable the clinician to exploit the complementary effects of these agents, allowing them to be administered at low, optimally tolerable doses that are consistent with superior efficacy and a lower risk of adverse events as compared with escalating doses of monotherapy.

Comment

This thoughtful review is predicated on the need to find more effective treatments for the 70% or so of patients whose cardiovascular events are not prevented by statins. It provides a useful summary of trials in which statins have been used in combination with nicotinic acid, bile acid sequestrants and fibrates. It was written at a time when there was already concern about a higher incidence of rhabdomyolysis with cerivastatin than with other statins, particularly when used in combination with gemfibrozil, though before cerivastatin was withdrawn. The recommendations of the Third Report of the NCEP Adult Treatment Panel (ATP-III) for use of statin–fibrate combinations can be found at www.nhlbi.nih.gov and can be summarized as:

1. normal renal function
2. no other medications that could increase the concentration of statin or fibrate
3. statin dose limited to starting or intermediate level when combined with fibrate; then increase statin dose cautiously if necessary
4. teach the patient to recognize and report muscle soreness, tenderness and pain
5. creatine kinase (CK) measured at baseline and on any complaint of muscle symptoms
6. statin–fibrate therapy discontinued if CK reaches ten times upper limit of normal in a patient with muscle symptoms; wait for symptoms to disappear and CK to return to normal before reinitiating therapy with either drug and use lower dose
7. if patient experiences muscle soreness, tenderness or pain with moderate increase in CK (3–10-fold) monitor symptoms and CK until they return to normal or until situation worsens mandating discontinuation of therapy.

Long-chain ω–3 fatty acids provide an alternative to fibrates in patients treated with statins who have unacceptable residual HTG. Reduction of plasma TGs is well established but there may be little or no change in HDL-C. Fish oil preparations have not been extensively used in the UK for this purpose, perhaps partly because ingestion of sufficient quantities as either capsules or as liquid oil is inconvenient or unpalatable compared to taking a single fibrate tablet. This has been compounded by doubts about the significance of plasma TGs as an independent risk factor which have led to their neglect, the emphasis being upon total and LDL-C, particularly since the publication of the statin trials beginning with the Simvastatin Scandinavian Survival Study (4S) in 1994.

Nevertheless there is evidence that long-chain marine fatty acids, predominantly eicosapentaenoic (EPA, 20:5) and docosahexaenoic (DHA; 22:6) acids, can prevent cardiovascular events. This was originally suggested in epidemiological studies |**25**|, and has been supported in subsequent clinical trials |**26,27**|. Several mechanisms that might explain the benefit have been proposed |**28,29**|. Animal work and the trials in man |**26,27,30**| suggest that at least part of the benefit relates to stabilization of the cardiac myocyte and the prevention of fatal arrhythmias.

The effects of marine oils on plasma lipoproteins are subtle and complex, as the following study, *inter alia*, shows.

Effects of atorvastatin and omega–3 fatty acids on LDL subfractions and post-prandial hyperlipemia in patients with combined hyperlipemia.

Nordoy A, Hansen JB, Brox J, Svensson B. *Nutr Metab Cardiovasc Dis* 2001; **11**: 7–16.

BACKGROUND. The aim of the present study was to see whether a moderate dose of omega–3 fatty acids (FA) potentiates the beneficial effects of statins on the high risk for CHD in patients with CH. In the present double-blind parallel study, 42 patients with CH with serum TGs 2–15 mmol/l and serum total cholesterol >5.3 mmol/l at the end of a three-month dietary run-in period were treated with 10 mg/d atorvastatin for ten or more weeks. During the last 5 weeks they were randomized into two groups that received either 1.68 g/d omega–3 FA as ethylesters of EPA (45%) and DHA (39%), or placebo (corn oil). As expected, atorvastatin significantly reduced serum total LDL-C, TGs and apolipoproteins B, E, C-II and C-III, whereas HDL-C was increased. Addition of omega–3 FA further increased HDL-C (*P* <0.03), and reduced systolic blood pressure (*P* <0.03), while the small dense LDL-particles (LDL III) (*P* <0.05) and post-prandial hypertriglyceridaemia (*P* <0.01) were reduced compared with the baseline, though there were no significant differences to the placebo group. This may be related to the large individual variation in these parameters and the small number of patients. No significant effects on basic or post-heparin activities of lipoprotein lipase or hepatic lipase were observed after atorvastatin with or without addition of omega–3 FA.

INTERPRETATION. This study indicates that addition of a low dose of omega–3 FA may further improve the risk profile for CHD in patients with combined hyperlipaemia treated with atorvastatin. The effect is related to reduction of post-prandial hyperlipaemia and redistribution of LDL subfractions.

Comment

This study confirms earlier work demonstrating a beneficial effect when ω–3 fatty acids are added to statins in patients with mixed hyperlipidaemia, but the results need to be interpreted with some caution. Although supplementation with ω–3 fatty acids reduced alimentary lipaemia and the concentration of small dense LDL when compared with baseline values, these variables were not significantly affected when compared with the parallel control group which received a corn oil supplement. Surprisingly ω–3 fatty acids did not in this study produce a significant reduction in fasting TG over and above that produced by the statin. The authors discuss these results which may reflect the fact that their study lacked sufficient power. If ω–3 fatty acids do indeed reduce post-prandial lipaemia and small dense LDL when added to statin treatment these may be important additional mechanisms underlying the benefits of fish oil. Small dense LDL are believed to be particularly atherogenic. They are poorly taken up by the LDL-receptor, have a prolonged residence time in the circulation and are susceptible to oxidation |**4**|.

Combinations of statins with ω–3 fatty acids deserve further study. There will inevitably be some difficulty in persuading clinicians to prescribe agents for putative benefits on variables that are not readily measurable, such as post-prandial triglyceride curves and LDL size, if these are not accompanied by significant changes in the routine fasting lipid profile.

An omega–3 polyunsaturated fatty acid concentrate administered for one year decreased triglycerides in simvastatin treated patients with coronary heart disease and persisting hypertriglyceridaemia.

Durrington PN, Bhatnagar D, Mackness MI, *et al. Heart* 2001; **85**: 544–8.

BACKGROUND. Omega–3 fatty acids, such as those present in fish oil, have been reported to prolong life in MI survivors. These fatty acids can decrease serum TG concentrations, but so far the doses used in trials examining their effects on coronary end-points have had only minimal TG lowering effects. The aim of the current study was to examine the TG-lowering effectiveness, safety, and tolerability of Omacor, a concentrate of omega–3, long-chain, polyunsaturated fatty acids from fish oil (84% of the total as opposed to an average of 35% in fish oil) over one year in patients with established CHD and persisting hypertriglyceridaemia, despite receiving simvastatin in doses similar to those employed in the 4S. Fifty-nine patients with CHD, receiving simvastatin 10–40 mg daily with serum TGs >2.3 mmol/l, were randomized to receive Omacor 2 g twice a day or placebo for 24 weeks in a double blind trial. Forty-six patients accepted the offer of active treatment for a further 24 weeks in an open phase of the trial. There was a sustained significant decrease in serum TGs by 20–30% (*P* <0.005) and in VLDL cholesterol by 30–40% (*P* <0.005) in patients receiving active Omacor at three, six, and twelve months compared either to baseline or placebo. Omacor did not have any deleterious effect on LDL or HDL cholesterol or on biochemical and haematological safety tests. There was no adverse effect on glycaemic control in patients with diabetes, who showed a decrease in serum TG, which was at least as great as in non-diabetic patients. One patient receiving placebo died of acute MI. Three patients withdrew from the trial (two on placebo and one on active treatment). Omacor was generally well tolerated.

Interpretation. Omacor was found to be a safe and effective means of lowering serum TGs over one year in patients with CHD and CH whose TGs remained elevated despite simvastatin treatment.

Comment

The dose of ω–3 fatty acids used in this study was twice that in the study of Nordoy *et al.* described above and may account for the significant additional reduction in serum TG produced by addition to a statin drug. Moreover, Nordoy *et al.* used atorvastatin which has been suggested to have a greater hypotriglyceridaemic effect than other statins, possibly making it more difficult to demonstrate an additional effect of ω–3 fatty acids. They used only 10 mg daily of atorvastatin however and in general it appears that in hypertriglyceridaemic subjects the degree of TG-lowering correlates with the reduction in LDL |**31**|. The current study was also of longer duration than that of Nordoy *et al.*, examining the effects of marine fatty acids over a period of 6–12 months, compared with 5 weeks.

Homocysteine and lipid-lowering agents; a comparison between atorvastatin and fenofibrate in patients with mixed hyperlipidemia.

Giral P, Bruckert E, Jacob N, Chapman MJ, Foglietti MJ, Turpin G.
Atherosclerosis 2001; **154**: 421–7.

Background. Hyperhomocysteinaemia is a risk factor for cardiovascular disease. Elevation in homocysteine levels has recently been demonstrated during lipid-lowering treatment with fibrates. This study compared the effect of a statin and a fibrate (atorvastatin and fenofibrate) on plasma levels of homocysteine and other thiol compounds in hyperlipidaemic patients. The study was of open randomized, parallel design with a preliminary screening phase, and a six-week placebo period. After the placebo period, patients were allocated randomly to atorvastatin or fenofibrate for a six-month period. Plasma thiols were assayed by high pressure liquid chromatography with fluorescence detection. There were 29 patients in the fenofibrate group and 24 in the atorvastatin group. Fenofibrate induced a significant increase in both homocysteine and cysteine plasma levels (+35.8 and +18%, respectively, $P<0.0001$); by contrast, cysteinylglycine remained stable. There were no significant changes in any thiol compounds in the atorvastatin group. Both treatments induced a significant decrease in uric acid, although fenofibrate was noticeably more effective than atorvastatin (–22.8 and –6.4%, respectively). Fenofibrate induced a non-significant increase in creatinine (12%) while atorvastatin reduced it (4.7%, not significant).

Interpretation. This study confirms that the induction of elevations in plasma homocysteine and cysteine levels is a distinct feature of the pleiotropic effects of fibrates. Further studies are needed not only to investigate the potential deleterious effects of this modification, but also to define the specific mechanism which underlies such fibrate-mediated action.

Comment

There is evidence that nicotinic acid, cholestyramine, clofibrate and fenofibrate may increase plasma homocysteine levels, whereas simvastatin was without effect |**32,33**|. Fish oil has been shown to reduce homocysteine levels |**34**|. This study confirms and extends the observations that fenofibrate but not atorvastatin produces an increase in plasma thiol compounds. Fenofibrate increased both homocysteine and cysteine but not cysteinylglycine levels. The present authors have previously demonstrated that cysteine may also be a risk factor in hyperlipidaemic patients |**35**|. The mechanisms by which fenofibrate might increase homocysteine levels are discussed but remain unclear at this stage. It is also unclear whether the increase in homocysteine levels with at least some fibrates and some other classes of lipid-lowering drug attenuates their benefits and whether supplementation with folic acid is advisable. Indeed it remains to be established if lowering homocysteine with vitamin supplementation has cardiovascular benefits. This is being addressed in the Study of Effectiveness of Additional Reductions in Cholesterol and Homocysteine (SEARCH) which is expected to report at the end of 2004.

References

1. Hulley SB, Rosenman RH, Bawol RD, Brand RJ. Epidemiology as a guide to clinical decisions. The association between triglyceride and coronary heart disease. *N Engl J Med* 1980; **302**: 1383–9.
2. Hokanson JE, Austin ME. Plasma triglyceride level is a risk factor for cardiovascular disease independent of high-density lipoprotein cholesterol level: a meta-analysis of population-based prospective studies. *J Cardiovasc Risk* 1996; **3**: 213–19.
3. Zilversmit DB. Atherogenesis: a post-prandial phenomenon. *Circulation* 1979; **60**: 473–85.
4. Krauss RM. Atherogenicity of triglyceride-rich lipoproteins. *Am J Cardiol* 1998; **81**: 13B–17B.
5. Hodis HN. Triglyceride-rich lipoprotein remnant particles and risk of atherosclerosis. *Circulation* 1999; **99**: 2852–4.
6. Havel RJ. Remnant lipoproteins as therapeutic targets. *Curr Opin Lipidol* 2000; **11**: 615–20.
7. Mahley RW, Rall SC. Type III hyperlipoproteinaemia (dysbetalipoproteinaemia; remnant particle disease). In: Betteridge DJ, Illingworth DR, Shepherd J (eds). *Lipoproteins in Health and Disease*, 1st edn. London: Arnold, 1999: 719–36.

8. Benlian P, De Gennes JL, Foubert L, Zhang H, Gagne SE, Hayden M. Premature atherosclerosis in patients with familial chylomicronemia caused by mutations in the lipoprotein lipase gene. *N Engl J Med* 1996; **335**: 848–54.

9. Goldberg IJ. Lipoprotein lipase and lipolysis: central roles in lipoprotein metabolism and atherogenesis. *J Lipid Res* 1996; **37**: 693–707.

10. Clee SM, Bissada N, Miao F, Miao L, Marais AD, Henderson HE, Steures P, McManus J, McManus B, LeBoeuf RC, Kastelein JJ, Hayden MR. Plasma and vessel wall lipoprotein lipase have different roles in atherosclerosis. *J Lipid Res* 2000; **41**: 521–31.

11. Marcoux C, Hopkins PN, Wang T, Leary ET, Nakajima K, Davignon J, Cohn JS. Remnant-like particle cholesterol and triglyceride levels of hypertriglyceridemic patients in the fed and fasted state. *J Lipid Res* 2000; **41**: 1428–36.

12. Packard CJ, Shepherd J, Cobbe SM, Ford I, Isles CG, McKillop JH, Macfarlane PW, Lorimer AR, Norrie J. Influence of pravastatin and plasma lipids on clinical events in the West of Scotland Coronary Prevention Study (WOSCOPS). *Circulation* 1998; **97**: 1440–5.

13. Nikkila EA, Aro A. Family study of serum lipids and lipoproteins in coronary heart disease. *Lancet* 1973; **1**: 954–9.

14. Goldstein JL, Schrott HG, Hazzard WR, Bierman EL, Motulsky AG. Hyperlipidemia in coronary heart disease. II. Genetic analysis of lipid levels in 176 families and delineation of a new inherited disorder, combined hyperlipidemia. *J Clin Invest* 1973; **52**: 1544–68.

15. Austin MA, McKnight B, Edwards KL, Bradley CM, McNeely MJ, Psaty BM, Brunzell JD, Motulsky AG. Cardiovascular disease mortality in familial forms of hypertriglyceridemia: A 20-year prospective study. *Circulation* 2000; **101**: 2777–82.

16. Jarvik GP, Austin MA, Brunzell JD. Familial combined hyperlipidaemia. In: Betteridge DJ, Illingworth DR, Shepherd J (eds). *Lipoproteins in Health and Disease,* 1st edn. London: Arnold, 1999: 693–9.

17. Angelin B, Hershon KS, Brunzell JD. Bile acid metabolism in hereditary forms of hypertriglyceridemia: evidence for an increased synthesis rate in monogenic familial hypertriglyceridemia. *Proc Nat Acad Sci USA* 1987; **84**: 5434–8.

18. Duane WC. Abnormal bile acid absorption in familial hypertriglyceridemia. *J Lipid Res* 1995; **36**: 96–107.

19. Furberg CD, Pitt B. Withdrawal of cerivastatin from the world market. *Curr Control Trials Cardiovasc Med* 2001; **2**: 205–7.

20. Rubins HB, Robins SJ, Collins D, Fye CL, Anderson JW, Elam MB, Faas FH, Linares E, Schaefer EJ, Schectman G, Wilt TJ, Wittes J. Gemfibrozil for the secondary prevention of coronary heart disease in men with low levels of high-density lipoprotein cholesterol. Veterans Affairs High-Density Lipoprotein Cholesterol Intervention Trial Study Group. *N Engl J Med* 1999; **341**: 410–18.

21. Frick MH, Elo O, Haapa K, Heinonen OP, Heinsalmi P, Helo P, Huttunen JK, Kaitaniemi P, Koskinen P, Manninen V, Maenpaa H, Malkonen M, Manttari M, Norola S, Pasternack A, Pikkarainen J, Romo M, Sjoblom T, Nikkila EA. Helsinki heart study: primary prevention trial with gemfibrozil in middle-aged men with hyperlipidemia. *N Engl J Med* 1987; **317**: 1237–45.

22. de Faire U, Ericsson CG, Grip L, Nilsson J, Svane B, Hamsten A. Secondary preventive potential of lipid-lowering drugs. The Bezafibrate Coronary Atherosclerosis Intervention Trial (BECAIT). *Eur Heart J* 1996; **17**(Suppl F): 37–42.

23. Effect of fenofibrate on progression of coronary-artery disease in type 2 diabetes: the Diabetes Atherosclerosis Intervention Study, a randomised study. *Lancet* 2001; **357**: 905–10.

24. Grundy SM. Hypertriglyceridemia, insulin resistance, and the metabolic syndrome. *Am J Cardiol* 1999; **83**: 25–9F.

25. Elwood PC, Burr ML, Sweetnam PM. Fish, fibre and heart disease. In: Marmot M, Elliott P (eds). Coronary Heart Disease Epidemiology. From Aetiology to Public Health, 1st edn. Oxford: Oxford University Press, 1992: 203–16.

26. Burr ML, Gilbert JF, Holliday RM, Elwood PC, Fehily AM, Rogers S, Sweetnam PM, Deadman NM. Effects of changes in fat, fish, and fibre intakes on death and myocardial reinfarction: diet and reinfarction trial (DART). *Lancet* 1989; **ii**: 757–61.

27. Dietary supplementation with n–3 polyunsaturated fatty acids and vitamin E after myocardial infarction: results of the GISSI-Prevenzione trial. Gruppo Italiano per lo Studio della Sopravvivenza nell'Infarto miocardico. *Lancet* 1999; **354**: 447–55.

28. Leaf A, Kang JX. Omega–3 fatty acids and cardiovascular disease. *World Rev Nutr Diet* 2001; **89**: 161–72.

29. Mori TA, Beilin LJ. Long-chain omega–3 fatty acids, blood lipids and cardiovascular risk reduction. *Curr Opin Lipidol* 2001; **12**: 11–17.

30. Leaf A. Diet and sudden cardiac death. *J Nutr Health Aging* 2001; **5**: 173–8.

31. Stein EA, Lane M, Laskarzewski P. Comparison of statins in hypertriglyceridemia. *Am J Cardiol* 1998; **81**: 66–9B.

32. Townend J, O'Sullivan J, Wilde JT. Hyperhomocysteinaemia and vascular disease. *Blood Rev* 1998; **12**: 23–34.

33. de Lorgeril M, Salen P, Paillard F, Lacan P, Richard G. Lipid-lowering drugs and homocysteine. *Lancet* 1999; **353**: 209–10.

34. Olszewski AJ, McCully KS. Fish oil decreases serum homocysteine in hyperlipemic men. *Coron Artery Dis* 1993; **4**: 53–60.

35. Jacob N, Bruckert E, Giral P, Foglietti MJ, Turpin G. Cysteine is a cardiovascular risk factor in hyperlipidemic patients. *Atherosclerosis* 1999; **146**: 53–9.

3

Diet, exercise and lipoprotein metabolism

Introduction

For the general population, and for those with primary hyperlipidaemia whether or not they require drug treatment, the avoidance of smoking, dietary modification and increased aerobic exercise are important factors in the reduction of cardiovascular risk. There remains much to be learnt in relation to these measures – not least how best to motivate people to make and maintain the necessary changes. Moreover there are still uncertainties about the nature of the optimum diet.

Diet

Low fat or modified fat?

The Nutrition Committee of the American Heart Association (AHA) recently published revised dietary guidelines |**1**|. These are not prescriptive in terms of targets for specific foodstuffs but emphasize patterns of eating. In particular the emphasis is on reduction of saturated and trans fatty acids, the principal cholesterol-raising fats. It is acknowledged that stearic acid may be less cholesterol raising but inadequate labelling of foods and its association with other saturates makes it difficult to use this fact in choosing foods.

The AHA recommends the consumption of a wide variety of fruits and vegetables, both as meals and snacks. These are rich in nutrients and fibre and relatively low in calories. Foods rich in starch (e.g. bread, pasta, cereal, potatoes) are recommended over sugars. Soluble fibre from some cereals, fruit and vegetables can produce a modest reduction in low-density lipoprotein cholesterol (LDL-C).

The AHA observes that despite the strong association between low high-density lipoprotein cholesterol (HDL-C) and risk of coronary heart disease (CHD) it remains to be established that increasing HDL-C by lifestyle changes reduces CHD events. Nevertheless the lifestyle changes which are likely to raise HDL-C (weight loss, increased exercise, giving up smoking) are all desirable on other grounds. The effect of moderate alcohol consumption in raising HDL-C may well have some impact in reducing CHD risk, but is unlikely to be emphasized in official recommendations given the potential for harm if such advice were given to the general public. They note the ability of low fat, high carbohydrate diets to reduce HDL-C in some subjects. This effect may be less marked if complex carbohydrate is substituted for fat rather than monosaccharides, particularly fructose. They acknowledge that substitution of unsaturated fatty acids for saturates may be a valuable alternative particularly in subjects with the atherogenic lipoprotein profile, who have high triglyceride (TG) levels and low HDL-C which may be exacerbated by high carbohydrate intake.

Atherogenic lipoprotein phenotype and diet–gene interactions.

Krauss RM. *J Nutr* 2001; **131**: 340–3S.

BACKGROUND. Studies employing analysis of LDL subclasses have demonstrated heterogeneity of the LDL response to low fat, high carbohydrate diets in healthy non-obese subjects. In individuals with a genetically influenced atherogenic lipoprotein phenotype, characterized by a predominance of small dense LDL (LDL subclass pattern B), lowering of plasma LDL-C levels by diets with ≤24% fat has been found to represent a reduction in numbers of circulating mid-sized and small LDL particles, and hence an expected lowering of cardiovascular disease risk. In contrast, in the majority of healthy individuals with larger LDL (pattern A, found in approximately 70% of men and a larger percentage of women), a significant proportion of the low fat diet-induced reduction in plasma LDL-C is made by depletion of the cholesterol content of LDL particles. This change in LDL composition is accompanied by a shift from larger to smaller LDL particle diameters. Moreover, with progressive reduction of dietary fat and isocaloric substitution of carbohydrate, an increasing number of subjects with pattern A convert to the pattern B phenotype. Studies in families have indicated that susceptibility to induction of pattern B by low fat diets is under genetic influence.

INTERPRETATION. Diet–gene interactions affecting LDL subclass patterns may contribute to substantial inter-individual variability in the effects of low fat diets on CHD risk.

Comment

Not all subjects respond to a low fat diet in which calories are replaced by carbohydrate with a reduction in LDL-C, some respond with an increase |**2**|. Moreover there may be adverse qualitative changes in the composition of the LDL. This paper gives a useful summary of studies that have examined the effects of changes in dietary fat on LDL composition. Those who have an adverse LDL pattern with an excess of small dense particles (pattern B) seem to benefit from reduction in fat intake and the concentration of the small LDL particles is reduced. Some of those with pattern A however achieve a reduction in LDL-C mainly by delipidation of LDL to form small dense particles without reduction in LDL particle number as judged by apoprotein B (apoB) concentrations. Thus they convert from pattern A to pattern B, which on present understanding would be undesirable. Conversion to

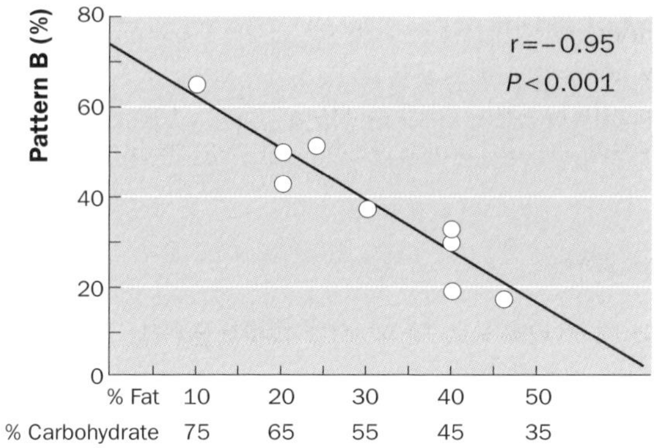

Fig 3.1 Relation between the percentages of dietary fat and carbohydrate and the prevalence of pattern B in men. Source: Krauss (2001).

pattern B depends on the degree of dietary change (see Fig. 3.1) and upon genetic factors. Subjects with parents who have LDL pattern B are more likely to transform from pattern A to pattern B when they take a low fat diet.

Impact of hydrogenated fat on high-density lipoprotein subfractions and metabolism.

Lichtenstein AH, Jauhiainen M, McGladdery S, *et al. J Lipid Res* 2001; **42**: 597–604.

Background. Relative to saturated fatty acids, trans-fatty acids/hydrogenated fat-enriched diets have been reported to increase LDL-C levels and either decrease or have no effect on HDL-C levels. To better understand the effect of trans-fatty acids/hydrogenated fat on HDL-C levels and metabolism, 36 subjects (female, $n = 18$; male, $n = 18$) were provided with each of three diets containing, as the major sources of fat, vegetable oil-based semiliquid margarine, traditional stick margarine, or butter for 35-day periods. LDL-C levels were 155 ± 27, 168 ± 30, and 177 ± 32 mg/dl after subjects followed the semiliquid margarine, stick margarine, and butter-enriched diets, respectively. HDL-C levels were 43 ± 10, 42 ± 9, and 45 ± 10 mg/dl, respectively. Dietary response in apolipoprotein (apo) A-I levels was similar to that in HDL-C levels. HDL_2 cholesterol levels were 12 ± 7, 11 ± 6, and 14 ± 7 mg/dl, respectively. There was virtually no effect of dietary fat on HDL_3 cholesterol levels. The dietary perturbations had a larger effect on particles containing apoA-I only (Lp A-I) than apoA-I and A-II (Lp A-I/A-II). Cholesterol ester transfer protein (CETP) activity was 13.28

CETP 5.76, 15.74 CETP 5.41, and 14.35 CETP 4.77 mmol/h/ml, respectively. Differences in CETP, phospholipid transfer protein activity, or the fractional esterification rate of cholesterol in HDL did not account for the differences observed in HDL-C levels.

Interpretation. These data suggest that the saturated fatty acid component, rather than the trans- or polyunsaturated fatty acid component, of the diets was the putative factor in modulating HDL-C response.

Comment

Trans-fatty acids occur naturally in small amounts in dairy products and meat and scarcely at all in plants. They are however generated in the partial hydrogenation of vegetable oils during the commercial preparation of fried and baked foods and some margarines. Several studies in the first half of the last decade suggested that substitution of oleic acid (cis 18:1) with its trans-isomer elaidic acid not only increased LDL-C but also reduced HDL-C in a dose-dependent manner |**3**|. The effect on HDL-C at levels of consumption seen in the diets of normal subjects (2–5% of dietary calories) was however very small and subsequent work has yielded contradictory results.

All three diets in the present study contained about 30% of calories from fat. The two margarine diets contained approximately 8.5% of calories from saturated fatty acids compared with 16.7% in the butter diet. The semiliquid margarine diet (and the butter diet) contained about 1% of calories from trans-fatty acids compared with 6.7% in the hard (stick) margarine diet. There was no statistical difference in HDL-C concentrations on the two margarine diets, both of which were associated with significantly lower HDL-C than the butter diet.

Although there was little difference in HDL-C between the soft and hard (high trans) margarine diets it should be noted that the latter did promote a significant increase in LDL-C and an adverse change therefore in total:HDL-C ratio. Overall, the findings support current advice to eat diets which are low in both saturated and trans-fatty acids.

Does dietary modification influence clinical events?

Much of the standard dietary advice is based upon epidemiological evidence and the effects of lipoprotein changes in small relatively short-term clinical trials. The most relevant question however, and the most difficult to answer, is whether the recommendations actually reduce cardiovascular events in large-scale clinical trials.

The following paper, based upon a Cochrane Review, suggests that dietary treatment to lower cholesterol over a prolonged period does have a modest impact on cardiovascular events.

Dietary fat intake and prevention of cardiovascular disease: systematic review.

Hooper L, Summerbell CD, Higgins JP, *et al. Br Med J* 2001; **322**: 757–63.

BACKGROUND. The study was carried out to assess the effect of reduction or modification of dietary fat intake on total and cardiovascular mortality and cardiovascular morbidity. It was designed as a sytematic review. The Cochrane Library, Medline, Embase, CAB abstracts, SIGLE, CVRCT registry, and biographies were searched; trials known to experts were included. Randomized controlled trials stating intention to reduce or modify fat or cholesterol intake in healthy adult participants over at least six months were studied. Inclusion decisions, validity, and data extraction were duplicated. Meta-analysis (random effects methodology), meta-regression, and funnel plots were performed. 27 studies (30 902 person years of observation) were included. Alteration of dietary fat intake had small effects on total mortality (rate ratio 0.98; 95% confidence interval [CI] 0.86–1.12). Cardiovascular mortality was reduced by 9% (0.91; 0.77–1.07) and cardiovascular events by 16% (0.84; 0.72–0.99), which was attenuated (0.86; 0.72–1.03) in a sensitivity analysis that excluded a trial using oily fish. Trials with at least two years' follow-up provided stronger evidence of protection from cardiovascular events (0.76; 0.65–0.90).

INTERPRETATION. There is a small but potentially important reduction in cardiovascular risk with reduction or modification of dietary fat intake, seen particularly in trials of longer duration.

Comment

It seems that reduction in saturated fat in the diet is likely to be effective in reducing cardiovascular events, though it takes two years for this to become apparent. There was a weak and insignificant trend to reduction in total mortality in trials lasting more than two years (rate ratio 0.93). This may be because the available studies lack the power to address the question. The relatively weak effect of diet on total and LDL-C makes this a difficult topic to study, needing very large numbers of individuals over long periods. The average reduction in total cholesterol in the diet studies was 11.1%, only half of that seen in the statin trials. If one trial (Minnesota Coronary Trial) is omitted the average cholesterol reduction with diet was only 7.3%. Inevitably, because of a lack of suitable trials, this review cannot distinguish between low saturated fat diets, where calories are replaced by carbohydrate, and modified fat diets, where calories from saturated fat are replaced from mono- and polyunsaturated fats. This is a key public health question that is likely to be unanswered for a considerable time.

The review above examines the effects of diet in subjects broadly representative of the general population, whereas that which follows looks at the same question in those with severe genetic hypercholesterolaemia, present from birth. There is a lack of adequate data.

Dietary treatment for familial hypercholesterolaemia (Cochrane Review).

Poustie VJ, Rutherford P. *Cochrane Database Syst Rev* 2001; **2**: CD001918.

BACKGROUND. Familial hypercholesterolaemia (FH) is an inherited disorder characterized by a raised blood cholesterol, the presence of xanthomatosis and premature ischaemic heart disease. The aim of treatment is the reduction of blood LDL-C concentrations in order to reduce the risk of ischaemic heart disease. Current treatment is based on a cholesterol-lowering diet alone or in combination with drug therapy. Many of the drugs found to be effective in treating adults with this disease are not licensed for use in children, therefore diet is the main treatment of children with FH. In addition to the cholesterol-lowering diet, several other dietary interventions have been suggested and consensus has yet to be reached on the most appropriate dietary treatment for children and adults with FH. The objective of the study was to examine the evidence that in children and adults with FH, a cholesterol-lowering diet is more effective at lowering cholesterol and reducing incidence of ischaemic heart disease than no intervention or than other dietary interventions. A search was carried out of the Cochrane Cystic Fibrosis and Genetic Disorders Trials Register, a specialist trials register which comprises references identified from comprehensive electronic database searches, handsearching relevant journals and handsearching abstract books of conference proceedings. Additional studies were identified from handsearching the Journal of Inherited Metabolic Disease (from inception, 1978 to 2000) and from the reference lists of identified studies. Selection criteria were randomized controlled trials (RCTs), both published and unpublished, where a cholesterol-lowering diet in children and adults with FH has been compared to other forms of dietary treatment or to no dietary intervention. Trials which include patients with FH alongside patients with non-familial hypercholesterolaemia were only included if the group of familial patients was well defined and the results for these patients were available. Two reviewers independently assessed the trial eligibility and methodological quality and one reviewer extracted the data, with independent verification of data extraction by a colleague. Only short-term outcomes could be assessed in this review due to the length of the five eligible studies. Compliance to treatment, quality of life, mortality and evidence of ischaemic or atheromatous disease were not assessed in the studies identified. No differences were found between the cholesterol-lowering diet and all other diets for all of the short-term outcomes assessed.

INTERPRETATION. No conclusions can be made about the effectiveness of the cholesterol-lowering diet, or any of the other dietary interventions suggested for FH, due to the lack of adequate data. A large, parallel, RCT is needed to investigate the effectiveness of the cholesterol-lowering diet and other dietary interventions for FH. It is also possible that data from trials including subjects with both familial and non-familial hypercholesterolaemia could alter the results of future updates of this review and until further evidence is available current dietary treatment of FH should continue to be observed and monitored with care.

Comment

Treatment with statins is the mainstay of the management of FH and can reduce total and LDL-C by the order of 50% though individual responses vary greatly. By comparison the reduction produced by dietary treatment of a few per cent is relatively trivial, though not to be ignored. Most adult males and post-menopausal females with FH should be on statins. In practice many younger women will also be treated though the drugs are not known to be safe for the fetus and should be avoided in pregnancy. It seems improbable that a large RCT of dietary treatments of sufficient duration to investigate clinical end-points will in fact be performed in FH. All the subjects would need to be treated with statins and the diets randomized.

Functional foods and dietary supplements

The use of so-called functional foods represents a stage between dietary treatment and intervention with drugs. Their role remains somewhat uncertain as far as the general population is concerned and is the subject of ongoing research |**4**|. The wealth of laboratory and epidemiological data suggesting benefit from intake of pharmacological doses of antioxidant vitamins has not been confirmed in clinical trials |**5,6**|, including the recently completed Heart Protection Study (HPS) |**7**|. It will be some time before it is known if folic acid supplements to reduce homocysteine levels are useful. The use of margarines enriched with plant sterol or stanol esters to lower cholesterol is another approach |**8,9**| and is the subject of a recent statement from the AHA.

AHA Science Advisory. Stanol/sterol ester-containing foods and blood cholesterol levels. A statement for healthcare professionals from the Nutrition Committee of the Council on Nutrition, Physical Activity, and Metabolism of the American Heart Association.

Lichtenstein AH, Deckelbaum RJ. *Circulation* 2001; **103**: 1177–9.

BACKGROUND. Considerable attention in the recent past has focused on the potential benefits or adverse effects of butter *versus* different types of margarines, usually with respect to their relative content of polyunsaturated, saturated, and trans-fatty acids, and the impact of these on LDL-C levels. Recently, a new class of margarines and other fat-derived products (e.g. salad dressings, mayonnaise) containing plant-derived sterols that are intended for use to lower blood cholesterol levels have been introduced into the food supply. These products are being marketed as adjuncts to low saturated-fat and low cholesterol diets to maximize reductions in LDL-C levels achievable by dietary means.

INTERPRETATION. Until long-term studies are performed to ensure the absence of adverse effects in all individuals ingesting plant sterol esters, these products should be reserved for adults requiring lowering of total and LDL-C levels because of hypercholesterolaemia or the need for secondary prevention after an atherosclerotic event.

Comment

The AHA is taking a cautious stance because of the observation that ingestion of plant sterol and stanol esters can be associated with reductions in plasma α- plus β-carotene, α-tocopherol and lycopene. With the exception of β-carotene the reductions tend to parallel those in LDL-C. In addition some individuals, such as those heterozygous for sitosterolaemia, may absorb greater quantities of sterol esters than normal with uncertain consequences. The AHA Committee do however concede that use of stanol/sterol esters could be considered in moderate to severely hypercholesterolaemic children, with monitoring of fat-soluble vitamins. They believe it would be premature to introduce these compounds routinely into the food supply by fortification of margarines and food oils. There would be economic implications and although they look promising adequate long-term safety data are not yet available.

Randomized controlled trial of use by hypercholesterolaemic patients of a vegetable oil sterol-enriched fat spread.

Neil HAW, Meijer GW, Roe LS. *Atherosclerosis* 2001; **156**: 329–37.

BACKGROUND. Plant sterols may be a useful additive therapy in the treatment of hypercholesterolaemic patients. The purpose of this study was to determine the effect of a fat spread enriched with vegetable oil sterols on plasma lipid, lipoprotein and apolipoprotein concentrations. A randomized double-blind placebo-controlled cross-over trial with two consecutive periods of 8 weeks was conducted. Thirty patients with heterozygous FH treated concurrently with an HMG-CoA reductase inhibitor (statin) and 32 patients with type IIa primary hypercholesterolaemia with a total cholesterol concentration >6.5 mmol/l not taking lipid-lowering drug therapy were recruited from a hospital lipid clinic. The active treatment was a fortified fat spread (25 g/day) providing 2.5 g of plant sterols. The control spread was indistinguishable in taste and appearance. Comparison at the end of the two 8-week trial periods showed a statistically significant reduction in total and LDL-C with use of the fortified spread but the results were confounded by a carry-over effect, which was partly explained by changes in the background diet. Because a carry-over effect was present, further analyses were restricted to the parallel arms of the first treatment period and were conducted on an intention-to-treat basis. After 4 weeks, LDL-C had decreased by 0.04 mmol/l (0.8%; 95% CI –0.44 to +0.37; not significant [NS]) in the placebo group and decreased by 0.76 mmol/l (15.0%; 95% CI –1.03 to –0.48, *P* <0.0001) in the active treatment group. After 8 weeks, the corresponding results were 0.0 mmol/l (0.0%; 95% CI –0.26 to +0.24 NS) and 0.51 mmol/l ([10.0%] 95% CI

–0.73 to –0.29 *P* <0.0001). There were no significant changes in apolipoprotein AI or B concentrations in the placebo group, but there was a small but statistically significant increase in apolipoprotein AI and a decrease in apolipoprotein B in the active treatment group. HDL-C and TG concentrations were unchanged. There was no difference in response between patients with statin-treated FH and patients with type IIa hyperlipoproteinaemia.

Interpretation. It was concluded that a fortified fat spread enriched with vegetable oil sterols reduces LDL-C by 10–15% with no difference in response between hypercholesterolaemic patients prescribed statins and those not taking lipid-lowering drug therapy.

Comment

The results of this study are similar to those obtained in previously published trials of spreads fortified with plant sterol esters and stanols |**9**|. In the general population it is suggested that reductions of LDL-C of this magnitude might reduce CHD events by up to 25%. As with drug therapy the effects of the plant sterol supplement were quite variable between individuals with some showing little or no response and others showing more than double the average reduction in LDL-C. It is not known if these individual responses would be consistent if retested.

It is at present unclear whether stanol or phytosterol ester-containing spreads are to be preferred. Their efficacy in reducing LDL-C is similar. Fat-soluble vitamin absorption is not affected, but both sterol and stanol margarines reduce β-carotene levels and the clinical relevance of this, if any, in the presence of reduced LDL concentration, is uncertain. Plant stanols (hydrogenated phytosterols) are present in only trace amounts in normal diets and very little is absorbed from stanol-fortified spreads. Phytosterols are present in larger amounts in normal diets and some are absorbed from spreads which are supplemented with them. This is a theoretical cause of concern since phytosterolaemia (also known as sitosterolaemia), a very rare inborn error of metabolism, leads to premature CHD |**10,11**|. In practice this is unlikely to be a problem. Despite there being a ten-fold increase in plant sterol intake in the present study plasma levels were less than doubled and remained in the reference range. The average β-sitosterol concentration in plasma in the treatment group in this study was only 0.9 mmol per mol of cholesterol. In normal subjects β-sitosterol levels in plasma average about 0.006 mmol/l while homozygotes for phytosterolaemia have levels in the range 0.2–1.4 mmol/l |**11**|.

The 25 g of spread (70% fat) consumed daily in this study contained about 160 Calories or some 8% of the total Calorie intake of the subjects. This could lead to a significant extra energy intake in individuals who have become used to low-fat diets and to taking little in the way of spreads. Both the stanol (Benecol®) and phytosterol-containing (Flora pro.activ®) spreads available in the UK are however available in low-fat formulations (32–35% fat), without reduction in the content of active ingredient.

Exercise

Aerobic exercise tends to reduce plasma TG and increase HDL-C. Beneficial effects on total and LDL-C are much less predictable. Resistance exercise seems less likely to be associated with favourable lipoprotein changes, particularly if there is no change in lean and fat body mass, though they may occur |**12**|.

Dose–response: variation with age, sex, and health status.

Wilmore JH. *Med Sci Sports Exerc* 2001; **33**: S622–34.

Background. The purpose of this study was to summarize the literature on the influence of age, sex, and health status on the changes in systolic (SBP) and diastolic blood pressure (DBP), triglycerides (TG), and HDL-C consequent to exercise training. A MEDLINE search was performed from January 1966 through to August 2000 to identify studies that have investigated the effects of age, sex, and health status on the changes in the outcome variables with exercise training. References from these studies and from review and meta-analysis studies were also reviewed.

Interpretation. The results indicate that age has little or no influence on the changes in SBP, DBP, TG, and HDL-C in response to exercise training. When looking at gender, females appear to have an attenuated response to exercise training compared with males with respect to SBP, DBP, and HDL-C, but the data for TG are equivocal. Finally, there appear to be more favourable changes in resting SBP and DBP, TG, and HDL-C in unhealthy subjects (hypertensive and post-myocardial infarction patients) when compared with healthy subjects.

Comment

This meta-analysis suggests that those who need it most are more likely to show beneficial changes in cardiovascular risk factors in response to exercise than healthy subjects. The question of dose–response and threshold to see favourable changes is however not addressed. Current US recommendations are for every adult to accumulate 30 minutes or more of moderate-intensity physical activity on most, preferably all, days of the week |**13**|.

This and the following paper are part of a symposium on dose–response issues related to physical activity and health and are accompanied by a consensus statement |**14**|.

The acute *versus* the chronic response to exercise.

Thompson PD, Crouse SF, Goodpaster B, Kelley D, Moyna N, Pescatello L. *Med Sci Sports Exerc* 2001; **33**: S438–45.

BACKGROUND. There is strong and consistent evidence that a single exercise session can acutely reduce TGs and increase HDL-C, reduce blood pressure, and improve insulin sensitivity and glucose homeostasis. Such observations suggest that at least some of the effects on atherosclerotic cardiovascular disease (ASCVD) risk factors attributed to exercise training may be the result of recent exercise. These acute and chronic exercise effects cannot be considered in isolation. Exercise training increases the capacity for exercise, thereby permitting more vigorous and/or more prolonged individual exercise sessions and a more significant acute effect. The intensity, duration, and energy expenditure required to produce these acute exercise effects are not clearly defined. The acute effect of exercise on TGs and HDL-C appears to increase with overall energy expenditure possibly because the effect may be mediated by reductions in intramuscular TGs. Prolonged exercise appears necessary for an acute effect of exercise on LDL-C levels. The acute effect of exercise on blood pressure is a low threshold phenomenon and has been observed after energy expenditures requiring only 40% maximal capacity. The acute effect of exercise on glucose metabolism appears to require exercise near 70% maximal, but this issue has not been carefully examined.

INTERPRETATION. Exercise has definite acute effects on blood lipids, blood pressure, and glucose homeostasis. Exercise also has acute effects on other factors related to atherosclerosis such as immunological function, vascular reactivity, and haemostasis. Considerable additional research is required to define the threshold of exercise required to produce these putatively beneficial effects.

Comment

As with chronic exercise, the acute effects are most apparent on plasma TG and HDL-C. They have most commonly been demonstrated with prolonged severe exercise and may persist for up to three days. The effect is said to be dose-related, depending upon energy expenditure, so there may be no biological threshold, simply a level of energy expenditure at which statistical changes are found in the experimental situation. Expenditures of 400 and 1000 kcal in a single bout of exercise have been reported as necessary to see measurable change in HDL-C, but it is not clear if these figures represent a true threshold. The acute effects of a single episode of exercise probably explain part of the effects seen in many longer-term studies of regular exercise, dependent on the relationship between the time of sampling and the last exercise session, but they are probably not the whole story.

Blood lipid and lipoprotein adaptations to exercise: a quantitative analysis.

Durstine JL, Grandjean PW, Davis PG, Ferguson MA, Alderson NL, DuBose KD. *Sports Med* 2001; **31**: 1033–62.

Background. Dose–response relationships between exercise training volume and blood lipid changes suggest that exercise can favourably alter blood lipids at low training volumes, although the effects may not be observable until certain exercise thresholds are met. The thresholds established from cross-sectional literature occur at training volumes of 24–32 km (15–20 miles) per week of brisk walking or jogging and elicit between 1200 and 2200 kcal/week. This range of weekly energy expenditure is associated with 2–3 mg/dl increases in HDL-C and TG reductions of 8–20 mg/dl. Evidence from cross-sectional studies indicates that greater changes n HDL-C levels can be expected with additional increases in exercise training volume. HDL-C and TG changes are often observed after training regimens requiring energy expenditures similar to those characterized from cross-sectional data. Training programmes that elicit 1200–2200 kcal/week in exercise are often effective at elevating HDL-C levels from 2 to 8 mg/dl, and lowering TG levels by 5–38 mg/dl. Exercise training seldom alters total cholesterol (TC) and LDL-C. However, this range of weekly exercise energy expenditure is also associated with TC and LDL-C reductions when they are reported. The frequency and extent to which most of these lipid changes are reported are similar in both genders, with the exception of TG.

Interpretation. For most individuals, the positive effects of regular exercise are exerted on blood lipids at low training volumes and accrue so that noticeable differences frequently occur with weekly energy expenditures of 1200–2200 kcal/week. It appears that weekly exercise caloric expenditures that meet or exceed the higher end of this range are more likely to produce the desired lipid changes. This amount of physical activity, performed at moderate intensities, is reasonable and attainable for most individuals and is within the American College of Sports Medicine's currently recommended range for healthy adults.

Comment

This extensive and detailed review supports the view that regular moderate aerobic exercise is most likely to exert a beneficial influence on TG and HDL-C concentrations, though the effects on TG in women are more contentious. Contrary to some early reports it seems that men with low initial HDL-C are less likely to see a favourable change. Also it appears that a reduction in weight or body fat is not essential to see changes in TG and HDL-C with exercise.

There remains uncertainty about the amount of exercise needed to see benefit, though it appears that energy expenditure is the determinant of change rather than exercise intensity. The thresholds emerging in this review of an energy expenditure of 1200–2200 kcal/week by walking or jogging 15–20 miles/week are achievable but will be regarded as substantial by most sedentary individuals. The energy equivalent of walking or jogging depends on body weight and for most men will be in excess of 100 kcal/mile. Thus walking 9–10 miles/week will expend 1200 kcal in many individuals and may be enough to produce measurable changes in some. It is questionable whether these energy thresholds have true biological validity or whether they are simply crude estimates of the amount of exercise needed to produce significant changes against the background of analytical and biological variability. Lesser energy expenditure may well be beneficial but effects on laboratory variables may not be apparent. To persuade sedentary individuals to take exercise at the upper end of the threshold range of 20 miles/week will often be difficult. Those of average fitness who wish to achieve this by walking would need to devote an hour a day on most days of the week.

Koutsari *et al.*, in the paper which follows, have investigated the possibility that exercise might mitigate some of the unwanted effects on lipoproteins of a low fat, high carbohydrate diet.

Exercise prevents the accumulation of triglyceride-rich lipoproteins and their remnants seen when changing to a high-carbohydrate diet.

Koutsari C, Karpe F, Humphreys SM, Frayn KN, Hardman AE. *Arterioscler Thromb Vasc Biol* 2001; **21**: 1520–5.

Background. This study tested the hypothesis that daily aerobic exercise opposes the fasting hypertriglyceridaemia and exaggerated post-prandial lipaemia observed after substituting dietary fat with carbohydrate. Eight healthy post-menopausal women aged 51 to 66 years consumed the same high-fat mixed meal on three occasions: (1) after 3 days on a low-carbohydrate diet (35%, 50%, and 15% energy from carbohydrate, fat, and protein, respectively); (2) after 3 days on an isoenergetic high-carbohydrate diet (corresponding values 70%, 15%, and 15%); and (3) after 3 days on the same high-carbohydrate diet with 60 minutes of brisk walking daily. Plasma TGs were higher after the high-carbohydrate diet than after the low-carbohydrate diet: fasting, 1.58 ± 0.19 *vs* 0.96 ± 0.12 mmol/l, respectively; 6-hour post-prandial area under concentration *vs* time curve, 13.74 ± 1.57 *vs* 10.12 ± 1.15 mmol/l × hour, respectively (both *P* <0.01). In the fasted and post-prandial states, concentrations of apolipoproteins B-48 and B-100 in the TG-rich lipoprotein fraction were significantly higher after the high-carbohydrate diet, as was the concentration of remnant-like lipoprotein particle cholesterol (a measure of lipoprotein remnants).

Interpretation. The carbohydrate-induced increases in the number of circulating TG-rich particles and their remnants were abolished when subjects had exercised daily during the high-carbohydrate diet.

Comment

This study is of considerable interest since it shows the potential for aerobic exercise to prevent the fasting and post-prandial hypertriglyceridaemia provoked by a high carbohydrate diet. The limitations of the study are that the diets used were very extreme and unlikely to be consumed by large populations of free living individuals in Westernized societies at least. This does not negate the findings however which might well be qualitatively similar but smaller in magnitude with lesser changes in diet. The different diets were consumed for very short periods (three

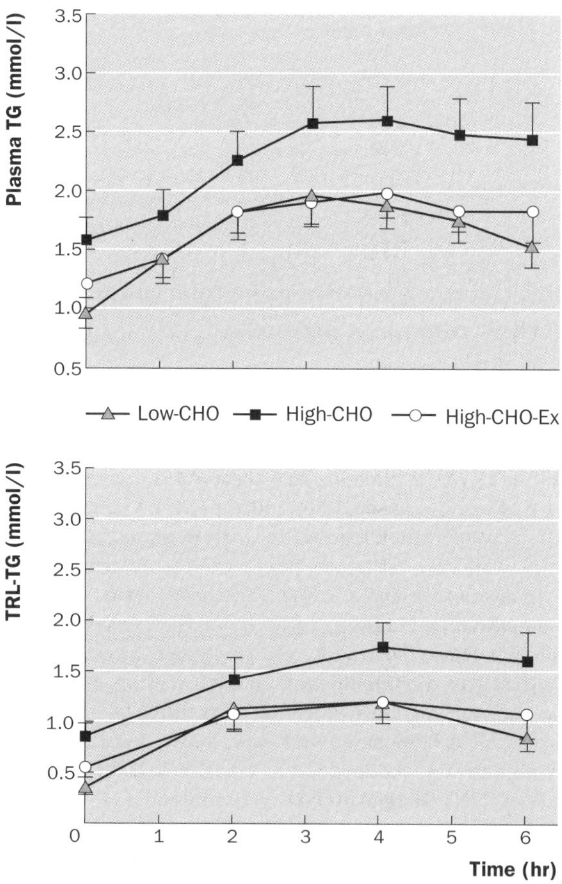

Fig 3.2 Plasma TG and TRL-TG concentrations in the fasted state (0 hours) and for 6 hours after consumption of a high-fat mixed meal after 3 days on a low-cholesterol (CHO) diet, after 3 days on a high-CHO diet, and after 3 days on a high-CHO Diet Ex. Values are mean ± SEM for 8 women. Source: Koutsari *et al.* (2001).

days) so that the effects on plasma lipoproteins were unlikely to be complete and would not represent the new steady state patterns that would be achieved in the long term. The exercise programme used (60 minutes brisk walking daily) is potentially achievable by a large proportion of the population, but as most clinicians who run clinics concerned with cardiovascular risk factors know, getting patients to change their lifestyles, whether in relation to diet or exercise, is not easy.

The same authors have done a similar study in men, using 30 minutes of moderate exercise daily |**15**|. The high carbohydrate diet was associated with higher fasting and post-prandial TG concentrations. Thirty minutes of moderate exercise essentially abolished the increased post-prandial lipaemia but did not over the short time frame of the study return fasting TG to control levels.

Effects of endurance exercise training on plasma HDL cholesterol levels depend on levels of triglycerides: evidence from men of the Health, Risk Factors, Exercise Training and Genetics (HERITAGE) Family Study.

Couillard C, Despres JP, Lamarche B, *et al. Arterioscler Thromb Vasc Biol* 2001; **21**: 1226–32.

BACKGROUND. HDL-C concentrations have been shown to increase with regular endurance exercise and, therefore, can contribute to a lower risk of CHD in physically active individuals compared with sedentary subjects. Although low HDL-C levels are frequently observed in combination with hypertriglyceridemia, some individuals may be characterized by isolated hypoalphalipoproteinemia, i.e. low HDL-C levels in the absence of elevated TG concentrations. The present study compared the responses of numerous lipoprotein-lipid variables to a 20-week endurance exercise training program in men categorized on the basis of baseline TG and HDL-C concentrations: (1) low TG and high HDL-C (normolipidaemia), (2) low TG and low HDL-C (isolated low HDL-C), (3) high TG and high HDL-C (isolated high TGs), and (4) high TGs and low HDL-C (high TG/low HDL-C). A series of physical and metabolic variables was measured before and after the training programme in a sample of 200 men enrolled in the Health, Risk Factors, Exercise Training and Genetics (HERITAGE) Family Study. At baseline, men with high TG/low HDL-C had more visceral adipose tissue than did men with isolated low HDL-C and men with normolipidemia. The 0.4% (NS) exercise-induced increase in HDL-C levels in men with isolated low HDL-C suggests that they did not benefit from the 'HDL-raising' effect of exercise. In contrast, men with high TG/low HDL-C showed a significant increase in HDL-C levels (4.9%, *P* <0.005). Whereas both subgroups of men with elevated TG levels showed reductions in plasma TGs (approximately –15.0%, *P* <0.005), only those with high TG/low HDL-C showed significantly reduced apolipoprotein B levels at the end of the study (–6.0%, *P* <0.005). Multiple regression analyses revealed that the exercise-induced change in abdominal subcutaneous adipose tissue (10.6%, *P* <0.01) was the only significant correlate of the increase in plasma HDL-C with training in men with high TG/low HDL-C.

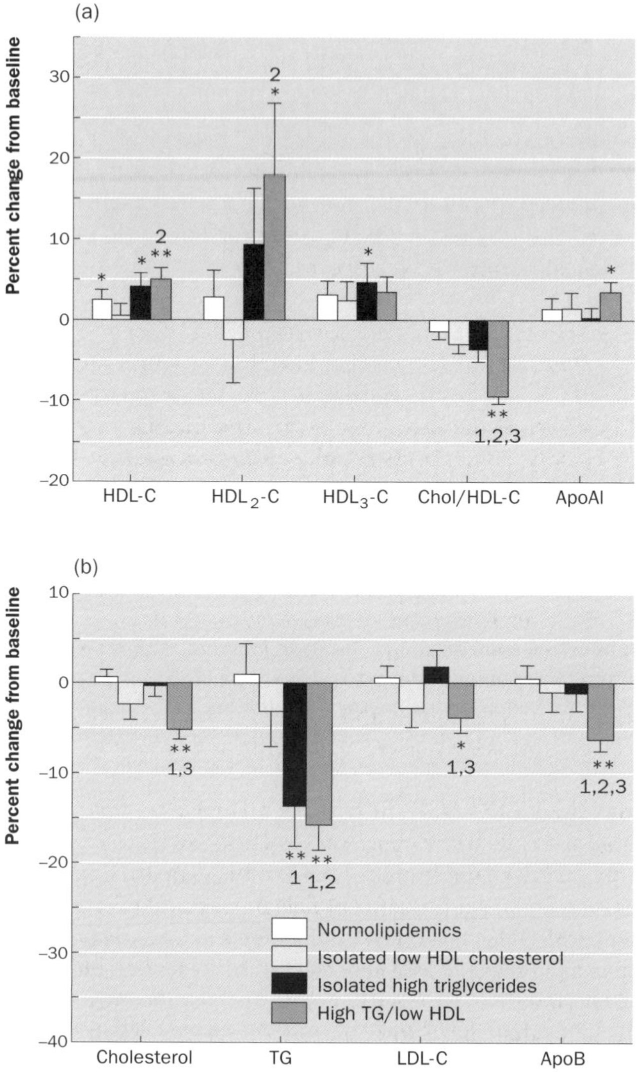

Fig 3.3 Exercise-induced changes (expressed in percentage from baseline) in HDL cholesterol (HDL-C) concentrations and related variables (a) and other lipid profile parameters (b) induced by the exercise training programme according to the different fasting plasma lipoprotein phenotypes at baseline. $*P<0.05$ and $**P<0.005$ for statistically significant change. Numbers indicate the folllowing: 1, significantly different from the normolipidaemic men; 2, significantly different from men with isolated low HDL-C; and 3, significantly different from men with isolated high TGs.
Source: Couillard *et al.* (2001).

Interpretation. Results of the present study suggest that regular endurance exercise training may be particularly helpful in men with low HDL-C, elevated TGs, and abdominal obesity.

Comment

This study involved sedentary white men up to the age of 65 years. The amount of exercise should be achievable by most reasonably healthy adults (with a little time to spare), and totalled 60 sessions on a bicycle ergometer over 20 weeks. Initially subjects worked at 55% of maximum oxygen consumption for 30 min increasing to 75% for 50 min. In practice only those who enjoy exercise are likely to sustain this level of activity in the long term.

There is a well known reciprocal relationship between plasma TG concentration and HDL-C, and those with elevated TG levels and low HDL-C tend also to have hypertension, abdominal obesity and insulin resistance – in other words an atherogenic lipoprotein profile as part of the metabolic syndrome. Regular endurance exercise has favourable metabolic effects in these subjects including an increase in HDL-C, which seems to correlate with decrease in body fat. Those who have isolated low HDL-C (i.e. without hypertriglyceridaemia) are less likely to be overweight and they did not benefit in terms of HDL-C with endurance training. An increase in HDL_3 was offset by a fall in HDL_2. Increase in post-heparin lipoprotein lipase, which is believed to be an important causal component of the exercise-induced increase in HDL-C, increased in all groups and does not therefore explain the failure of response in the isolated low HDL-C group.

In an accompanying editorial Thompson and Rader ask if exercise increases HDL-C in those who need it most – the answer seems to be only in some of them |**16**|.

In another report from the HERITAGE study using a similar protocol |**17**| 69 black men, when compared with 200 white men, were less likely to show changes in lipoprotein profile with the exercise programme. They exhibited an increase only in HDL_2-C, whereas the white subjects had significant falls in plasma cholesterol, TG and apoB and increases in total HDL-C and HDL_3-C. Exercise was associated with an increase in post-heparin lipoprotein lipase activity but this reached significance only in the white subjects. Hepatic lipase activity declined significantly in both groups. The authors speculate that the better baseline lipoprotein profile of the black subjects reduces the scope for improvement with exercise training.

Moderate exercise, post-prandial lipaemia and triacylglycerol clearance.

Gill JM, Mees GP, Frayn KN, Hardman AE. *Eur J Clin Invest* 2001; **31**: 201–7.

Background. Moderate intensity exercise reduces post-prandial TG concentrations. This study tested whether this reflects increased TG clearance. Eight

normotriglyceridaemic men, aged 48.3 ± 7.3 years (mean ± SD), performed two oral fat tolerance tests (blood samples taken in the fasted state and for six hours after a high-fat meal containing 1.00 g fat, 0.97 g carbohydrate, 58 kJ energy/kg fat-free body mass) and two intravenous fat tolerance tests (blood samples in the fasted state and after a bolus injection of intralipid, 0.1 g fat/kg body mass). The afternoon before one oral and one intravenous test, subjects walked briskly for 90 min; no exercise was performed before the control tests. Prior exercise reduced fasting TG concentration similarly in the oral (16 ± 7 %) (mean ± SEM) and intravenous (18 ± 7%) tests, and reduced post-prandial TG concentrations in the oral test by 18 ± 6% (all *P* <0.05). However, prior exercise did not increase intralipid clearance (disappearance curve slopes: control, 4.69 ± 0.49%/min; exercise, 4.85 ± 0.40%/min).

INTERPRETATION. These data suggest that mechanisms other than increased TG clearance mediate the lower post-prandial TG concentrations seen after moderate exercise.

Comment

It is generally thought that the attenuation in post-prandial lipaemia after exercise is the result of increased clearance resulting from increased activity of muscle lipoprotein lipase. The evidence from prolonged and vigorous exercise supports this. The mechanism is less clear with more moderate exercise. In the present study 90 min brisk walking the previous afternoon reduced basal and post-prandial TG

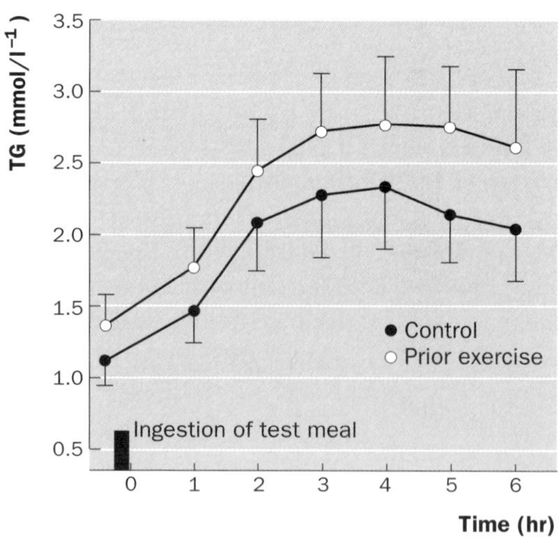

Fig 3.4 Plasma TG concentrations in the fasted state and for six hours after consumption of a high-fat mixed meal in control and prior exercise tests. Values are mean ± SEM, $n = 8$. Source: Gill *et al.* (2001).

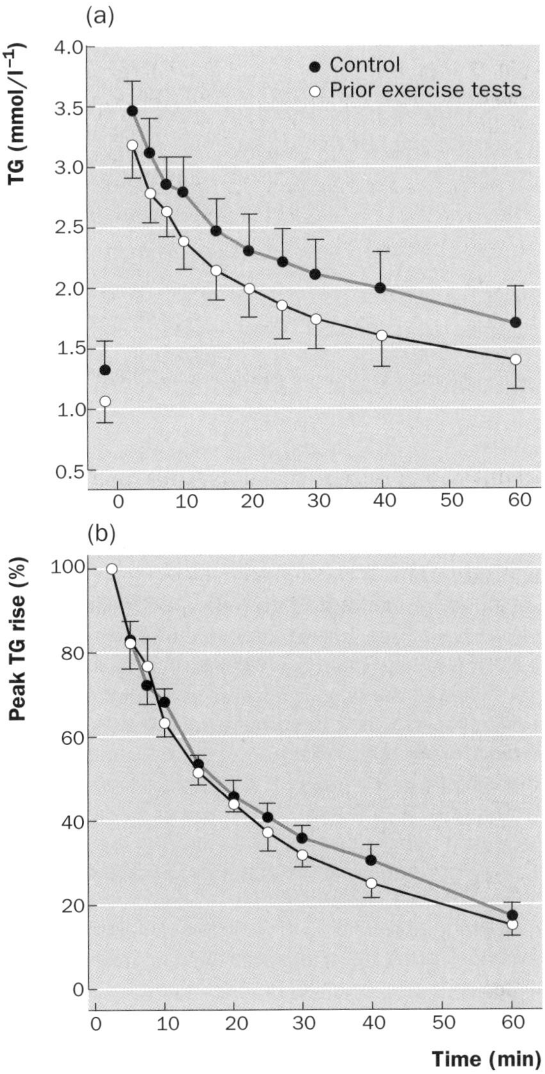

Fig 3.5 (a) Plasma TG concentrations prior to and for 60 min after an intravenous injection of a bolus of Intralipid (0.1 g kg $^{-1}$ body mass) in control and prior exercise tests. Values are mean ± SEM, n = 8. (b) Changes in plasma TG concentration in control and prior exercise tests are Intralipid injection, with values expressed as a percentage of the maximum rise in TG concentration. Values are mean ± SEM, n = 8. Source: Gill *et al.* (2001).

concentrations but had no effect on the clearance of exogenous TG given intravenously. The authors propose that under these circumstances other mechanisms must be involved in the hypotriglyceridaemic effects of exercise. Plasma paracetamol concentrations peaked at the same time after control and post-exercise oral fat tolerance tests suggesting that there was no difference in the rate of gastric emptying. This militates against a reduction in the rate of fat absorption after exercise but does not exclude it. The authors therefore suggest that the attenuated TG response to oral fat after exercise is due to reduced hepatic very low-density lipoprotein (VLDL)-TG secretion. There is some support for this in the following paper from the same group.

Effects of prior moderate exercise on exogenous and endogenous lipid metabolism and plasma factor VII activity.

Gill JM, Frayn KN, Wootton SA, Miller GJ, Hardman AE. *Clin Sci* 2001; **100**: 517–27.

BACKGROUND. Moderate exercise reduces post-prandial TG concentrations, which are a risk marker for CHD. The present study sought to determine the qualitative nature of exercise-induced changes in lipid metabolism and their association (if any) with changes in factor VII activation. Eleven normotriglyceridaemic men, aged 51.7 ± 6.1 years (mean ± SD), participated in two oral fat tolerance tests after different pre-conditions: control (no exercise), and exercise (90 min of brisk walking the day before). Venous blood samples were obtained in the fasted state and for 8 h after ingestion of a high-fat meal (1.32 g of fat, 1.36 g of carbohydrate, 0.30 g of protein and 10 mg of [1,1,1-^{13}C] tripalmitin/kg body mass). Prior exercise reduced post-prandial plasma TG concentrations by 25 ± 3% (mean ± SEM), with lower concentrations in the Svedberg flotation rate (Sf) 20–400 (VLDL) fraction accounting for 79 ± 10% of this reduction. There was no effect on plasma factor VII coagulant activity or on the concentration of the active form of factor VIIa. Prior exercise increased post-prandial serum 3-hydroxybutyrate and plasma fatty acid concentrations, decreased serum post-prandial insulin concentrations and increased exogenous (8 h ^{13}C breath excretion of 15.1 ± 0.9% of ingested dose compared with 11.9 ± 0.8%; *P* = 0.00001) and endogenous post-prandial fat oxidation.

INTERPRETATION. These data raise the possibility that reduced hepatic secretion of VLDL plays a role in the attenuation of plasma triacylglycerol concentrations seen after exercise, although it is possible that increased triacylglycerol clearance also contributes to this effect.

Comment

The design of this study is similar to that in the paper discussed immediately above. Normotriglyceridaemic men underwent oral, though not intravenous, fat tolerance tests with and without 90 minutes of brisk walking on the afternoon of the previous day. About 80% of the reduction in post-prandial lipaemia induced by exercise was in the VLDL flotation range, supporting the notion that after moderate exercise the reduction in TG concentration after a fatty meal is more dependent on reduced secretion of VLDL-TG than on increased clearance of dietary fat. There was a small reduction in particles in the chylomicron flotation range and moreover particles in the flotation range of VLDL will include some small chylomicrons and partially lipolyzed chylomicron remnants. The increase in 3-hydroxybutyrate concentrations during the post-exercise test would fit with diversion of fatty acids to the oxidation pathway as opposed to re-esterification and secretion in VLDL. This study, in accord with previously published work, demonstrated an increase in whole body fat oxidation, and the current work shows that this was true of the oxidation of both endogenous and exogenous fat.

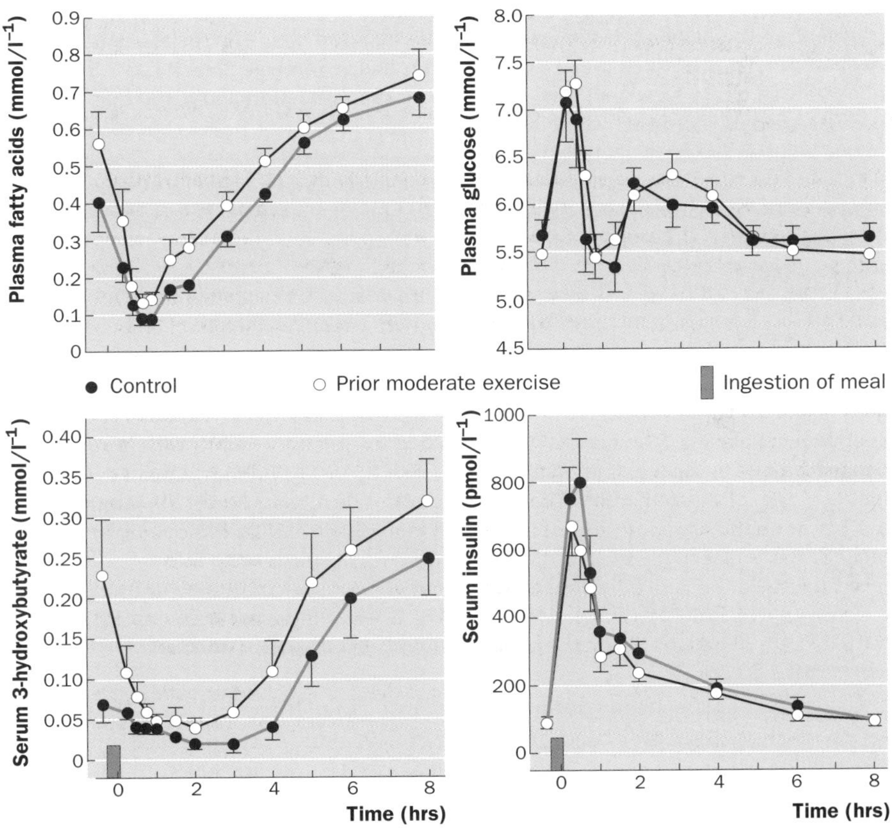

Fig 3.6 Plasma fatty acid, plasma glucose, serum 3-hydroxybutyrate and serum insulin concentrations in the fasted state and for 8 h after consumption of a mixed high-fat meal in control and prior moderate exercise trials. Values are mean ± SEM, $n = 11$. Source: Gill *et al.* (2001).

Resistance exercise training reduces hypertriglyceridemia in HIV-infected men treated with antiviral therapy.

Yarasheski KE, Tebas P, Stanerson B, Marin D, *et al. J Appl Physiol* 2001; **90**: 133–8.

BACKGROUND. Hypertriglyceridaemia, peripheral insulin resistance, and trunk adiposity are metabolic complications recently recognized in people infected with human immunodeficiency virus (HIV) and treated with highly active antiretroviral therapy (HAART). These complications may respond favourably to exercise training. Using a paired design, this study determined whether 16 weeks of weight-lifting exercise increased muscle mass and strength and decreased fasting serum triglycerides and adipose tissue mass in 18 HIV-infected men. The resistance exercise regimen consisted of three upper and four lower body exercises done for 1–1.5 h/day, 4 days/week for 64 sessions. Dual-energy X-ray absorptiometry indicated that exercise training increased whole body lean mass 1.4 kg (*P* = 0.005) but did not reduce adipose tissue mass (*P* = NS). Axial proton-magnetic resonance imaging indicated that thigh muscle cross-sectional area increased 5–7 cm² (*P* <0.005). Muscle strength increased 23–38% (*P* <0.0001) on all exercises. Fasting serum TG were decreased at the end of training (281–204 mg/dl; *P* = 0.02).

INTERPRETATION. These findings imply that resistance exercise training-induced muscle hypertrophy may promote TG clearance from the circulation of hypertriglyceridaemic HIV-infected men treated with antiviral therapy.

Comment

Protease inhibitors used for the treatment of HIV infection are well known to increase serum lipids with hypertriglyceridaemia being the dominant abnormality |**18,19**|. Ritonavir is more likely to provoke hypertriglyceridaemia though other protease inhibitors will have some adverse effect on LDL-C |**20**|. Ritonavir has featured in isolated case reports of chylomicronaemia sufficient to cause pancreatitis |**21,22**|.

In general benefits from resistance exercise on circulating lipids and lipoproteins have been more difficult to demonstrate than from aerobic training |**12**|. The literature suggests that resistance and endurance exercise are not harmful to subjects with HIV infection and this study demonstrated reduction in serum TG in 11 of 18 subjects, with the most marked reductions being seen in those with the highest baseline values. The final blood sample was taken about 16 h after the last bout of exercise so it is not clear to what extent the improvement of TG is a result of the last exercise session as opposed to a more sustained reduction from the whole exercise programme. The acute and chronic effects of exercise on lipids and lipoproteins were discussed above (Thompson *et al.* 2001).

References

1. Krauss RM, Eckel RH, Howard B, Appel LJ, Daniels SR, Deckelbaum RJ, Erdman JW Jr, Kris-Etherton P, Goldberg IJ, Kotchen TA, Lichtenstein AH, Mitch WE, Mullis R, Robinson K, Wylie-Rosett J, St Jeor S, Suttie J, Tribble DL, Bazzarre TL. AHA Dietary Guidelines: revision 2000: A statement for healthcare professionals from the Nutrition Committee of the American Heart Association. *Circulation* 2000; **102**: 2284–99.
2. Dreon DM, Fernstrom HA, Miller B, Krauss RM. Low-density lipoprotein subclass patterns and lipoprotein response to a reduced-fat diet in men. *FASEB J* 1994; **8**: 121–6.
3. Zock PL, Mensink RP. Dietary trans-fatty acids and serum lipoproteins in humans. *Curr Opin Lipidol* 1996; **7**: 34–7.
4. Lichtenstein AH. Approaches to reducing cardiovascular disease risk: food or pills? *Curr Opin Lipidol* 2001; **12**: 1–3.
5. Gruppo Italiano per lo Studio della Sopravvivenza nell'Infarto miocardico. Dietary supplementation with n-3 polyunsaturated fatty acids and vitamin E after myocardial infarction: results of the GISSI-Prevenzione trial. *Lancet* 1999; **354**: 447–55.
6. Yusuf S, Dagenais G, Pogue J, Bosch J, Sleight P. Vitamin E supplementation and cardiovascular events in high-risk patients. The Heart Outcomes Prevention Evaluation Study Investigators. *N Engl J Med* 2000; **342**: 154–60.
7. MRC/BHF Heart Protection Study of antioxidant vitamin supplementation in 20 536 high-risk individuals: a randomised placebo-controlled trial. *Lancet* 2002; **360**: 23–33.
8. Cater NB. Plant stanol ester: review of cholesterol-lowering efficacy and implications for coronary heart disease risk reduction. *Prev Cardiol* 2000; **3**: 121–30.
9. Law M. Plant sterol and stanol margarines and health. *Br Med J* 2000; **320**: 861–4.
10. Lutjohann D, von Bergmann K. Phytosterolaemia: diagnosis, characterization and therapeutical approaches. *Ann Med* 1997; **29**: 181–4.
11. Salen G, Shefer S, Nguyen LB, Ness GC, Tint GS. Sitosterolaemia. In: Betteridge DJ, Illingworth DR, Shepherd J (eds). *Lipoproteins in Health and Disease,* 1st edn. London: Arnold, 1999: 815–27.
12. Durstine JL, Thompson PD. Exercise in the treatment of lipid disorders. *Cardiol Clin* 2001; **19**: 471–88.
13. Pate RR, Pratt M, Blair SN, Haskell WL, Macera CA, Bouchard C, Buchner D, Ettinger W, Heath GW, King AC, *et al.* Physical activity and public health. A recommendation from the Centers for Disease Control and Prevention and the American College of Sports Medicine. *JAMA* 1995; **273**: 402–7.
14. Kesaniemi YK, Danforth E Jr, Jensen MD, Kopelman PG, Lefebvre P, Reeder BA. Dose-response issues concerning physical activity and health: an evidence-based symposium. *Med Sci Sports Exerc* 2001; **33**: S351–8.
15. Koutsari C, Hardman AE. Exercise prevents the augmentation of post-prandial lipaemia attributable to a low-fat high-carbohydrate diet. *Br J Nutr* 2001; **86**: 197–205.

16. Thompson PD, Rader DJ. Does exercise increase HDL cholesterol in those who need it the most? *Arterioscler Thromb Vasc Biol* 2001; **21**: 1097–8.

17. Bergeron J, Couillard C, Despres JP, Gagnon J, Leon AS, Rao DC, Skinner JS, Wilmore JH, Bouchard C. Race differences in the response of post-heparin plasma lipoprotein lipase and hepatic lipase activities to endurance exercise training in men: results from the HERITAGE Family Study. *Atherosclerosis* 2001; **159**: 399–406.

18. Koppel K, Bratt G, Eriksson M, Sandstrom E. Serum lipid levels associated with increased risk for cardiovascular disease is associated with highly active antiretroviral therapy (HAART) in HIV-1 infection. *Int J STD AIDS* 2000; **11**: 451–5.

19. Thiebaut R, Dabis F, Malvy D, Jacqmin-Gadda H, Mercie P, Valentin VD. Serum triglycerides, HIV infection, and highly active antiretroviral therapy, Aquitaine Cohort, France, 1996 to 1998. *J Acquir Immune Defic Syndr* 2000; **23**: 261–5.

20. Periard D, Telenti A, Sudre P, Cheseaux JJ, Halfon P, Reymond MJ, Marcovina SM, Glauser MP, Nicod P, Darioli R, Mooser V. Atherogenic dyslipidemia in HIV-infected individuals treated with protease inhibitors. The Swiss HIV Cohort Study. *Circulation* 1999; **100**: 700–5.

21. Echevarria KL, Hardin TC, Smith JC. Hyperlipidemia associated with protease inhibitor therapy. *Ann Pharmacother* 1999; **33**: 859–63.

22. Clark Perry R, Cushing HE, Deeg MA, Prince MJ. Ritonavir, triglycerides and pancreatitis. *Clin Infect Dis* 1999; **28**: 161–2.

4

Disorders which cause hypolipidaemia

Abetalipoproteinaemia, Anderson's disease and familial hypobetalipoproteinaemia

Abetalipoproteinaemia (ABL) and familial hypobetalipoproteinaemia (FHBL) are both characterized by hypolipidaemia as a result of deficiency of apolipoprotein B (apoB)-containing lipoproteins, fat malabsorption, acanthocytes, retinitis pigmentosa and spinocerebellar ataxia. Anderson's disease (chylomicron retention disease, CRD) can have similar manifestations, but the defect in lipoprotein metabolism seems to be confined to the small intestine |**1**|. The three disorders are however quite distinct genetically and in terms of their pathogenesis. Fat soluble vitamin supplementation is the key to management. Very high doses of vitamin E initiated early can prevent or at least delay the neurological and ophthalmological problems.

ABL, which behaves as an autosomal recessive condition, was first described in 1950 by Bassen and Kornzweig |**2**|, whose names became attached to the condition. They were principally interested in the haematological abnormalities.

A decade ago it was reported that ABL results not from defects in the production of apoB, but from abnormalities in the gene responsible for production of microsomal triglyceride (TG) transfer protein (MTP) |**3**|. MTP is essential for the normal lipidation of newly synthesized apoB and in its absence the apoB is rapidly degraded leading to a defect in the secretion of apoB-containing lipoproteins |**4**|.

FHBL is an autosomal codominant disorder in which the heterozygotes have intermediate concentrations of apoB-containing lipoproteins, while the homozygotes can be as severely hypolipidaemic as patients with ABL, from whom they may be clinically indistiguishable. Most cases seem to result from mutations in the apoB gene producing proteins which lead to the synthesis of apoB molecules truncated from the carboxyl end. These are designated using the same centile system that is used to describe apoB in normal individuals. Thus normal hepatic apoB is apoB-100 and apoB normally secreted by the intestine is apoB-48 and comprises the N-terminal 48% of hepatic apoB. Patients with mutations which lead to the production of molecules shorter than apoB-27 usually do not have detectable apoB in plasma.

CRD |**1**| was first described about 40 years ago in an infant with malabsorption of fat and accumulation of lipid in small intestinal epithelial cells. Serum cholesterol is usually below 2.5 mmol/l and apoB and low-density lipoprotein cholesterol (LDL-C) levels are about half normal. ApoB-100 is present in plasma but despite a small increase in plasma TG after fat feeding no chylomicrons are found. Vitamin E levels may be undetectable in plasma and the patients can develop neurological features. Enterocytes are intensely immunoreactive for apoB-48, but for reasons which remain uncertain there is a defect in secretion of apoB-48-containing lipoproteins (chylomicrons) into the intestinal lymph, hence these particles are absent from post-prandial plasma.

Long-term assessment of combined vitamin A and E treatment for the prevention of retinal degeneration in abetalipoproteinaemia and hypobetalipoproteinaemia patients.

Chowers I, Banin E, Merin S, Cooper M, Granot E. *Eye* 2001; **15**: 525–30.

BACKGROUND. The aim of the study was to assess the long-term efficacy of combined vitamin A and E treatment in preventing retinal degeneration in patients with ABL or homozygous hypobetalipoproteinaemia (HBL). Ten patients with ABL and three with homozygous HBL who were treated with oral supplements of vitamins A and E were studied. Systemic, ophthalmological and electroretinographic follow-up for a mean of 11.7 years (range 4–20 years) after onset of treatment was evaluated. Despite vitamin A and E treatment, seven of ten patients who began treatment prior to two years of age and all three patients who began treatment later in life manifested unusual fundoscopic pigmentary changes over time. At the end of follow-up, 11 of 13 patients had subnormal mixed cone–rod electroretinogram amplitudes. Seven of ten patients for whom perimetry was available had mild to severe constriction of the visual fields.

INTERPRETATION. Combined oral vitamin A and E supplementation that is initiated prior to two years of age can markedly attenuate the severe retinal degeneration that is associated with untreated ABL or homozygous HBL. Yet, fundoscopic and functional retinal changes do occur despite early initiation of vitamin treatment. Therefore, the adequacy of the present treatment protocol for ABL and homozygous HBL should be re-evaluated.

Comment

Fat-soluble vitamin supplementation is critical to the management of these disorders but even when initiated early it can be less than completely effective.

Rickets as an unusual initial presentation of abetalipoproteinemia and hypobetalipoproteinemia.

Narchi H, Amr SS, Mathew PM, El Jamil MR. *J Pediatr Endocrinol Metab* 2001; **14**: 329–33.

BACKGROUND. The paper gives a description of rickets as an unexpected initial manifestation in two children with ABL and HBL, and elucidation of its pathophysiology in these conditions. Two infants aged two and six months with ABL and HBL respectively had clinical rickets at presentation, confirmed radiologically and biochemically. Vitamin D intake and serum levels were measured and other causes of rickets were looked for. Vitamin D intake and laboratory studies levels were suggestive of rickets due to calcium deficiency instead of vitamin D deficiency. Healing of rickets occurred with dietary treatment of the malabsorption, without any dietary calcium or significant vitamin D supplementation.

INTERPRETATION. Steatorrhoea-induced calcium malabsorption seems to be the most likely cause of rickets in this entity.

Comment

Vitamin D is not dependent upon lipoproteins for absorption or transport and vitamin D deficiency is usually said not to be a feature of ABL and FHBL. Serum concentrations of vitamin D metabolites were not suggestive of vitamin D deficiency in these two infants. Their rickets healed rapidly with reduction of long-chain fatty acids in the diet and the substitution of medium-chain TGs leading the authors to postulate that the rickets resulted from calcium malabsorption secondary to steatorrhoea.

Phenotypic expression of familial hypobetalipoproteinaemia in three kindreds with mutations of apolipoprotein B gene.

Tarugi P, Lonardo A, Gabelli C, *et al.* *J Lipid Res* 2001; **42**: 1552–61.

BACKGROUND. This paper reports the clinical phenotype in three kindreds with FHBL carrying novel truncated apoBs of different sizes (apoB-8.15, apoB-33.4 and apoB-75.7). In D.A. kindred, three carriers of a C-deletion in exon 10 were found leading to the synthesis of apoB-8.15 not detectable in plasma. They showed steatorrhea and fatty liver. In N.L. kindred, the proband is heterozygous for a nonsense mutation in exon 26, leading to the formation of apoB-33.4. He had premature cerebrovascular disease and fatty liver; two apoB-33.4 carriers in this kindred showed only fatty liver. In B.E. kindred, the proband is heterozygous for a T-deletion in exon 26, which converts tyrosine at codon 3435 into a stop codon,

resulting in apoB-75.7. The proband, a heavy alcohol drinker, had steatohepatitis, whereas his teetotaller daughter, an apoB-75.7 carrier, had no detectable fatty liver.

INTERPRETATION. This study suggests that: (i) fatty liver invariably develops in FHBL carriers of short and medium-size truncated apoBs (<apoB-48), but its occurrence needs additional environmental factors in carriers of longer apoB forms; (ii) intestinal lipid malabsorption develops only in carriers of short truncated apoBs, which are not secreted into the plasma; and (iii) cerebrovascular disease due to premature atherosclerosis may occur even in FHBL subjects.

Comment

As expected, the family with the most severely truncated apoB (apoB-8.5) had no detectable apoB in plasma and had both steatorrhoea and fatty liver as judged by ultrasound scanning. None consented to liver biopsy so no comment can be made about the severity of any associated liver disease. The presence of steatorrhoea did not prevent two members of this kindred from becoming significantly overweight. Curiously, the concentration of apoB-100 in this family of FHBL heterozygotes was less than 50% of that predicted and this has been noted previously in other cases producing very short truncated versions of apoB. The synthetic rate of normal apoB-100 is disproportionately reduced for some reason.

A male patient in his seventies with apoB-33.4 had evidence of atherosclerotic vascular disease and had had a stroke at the age of 41 years. It has been suggested that patients with ABL and FHBL are protected against atherosclerotic disease, but unsurprisingly, protection is not absolute.

The finding of fatty liver, with some mild fibrosis on biopsy, was unexpected in the proband of the third family. ApoB-75.7 contains most of the lipid-associating domains and should therefore be able to allow secretion of triglyceride-rich lipoproteins from the liver and prevent their accumulation. The steatosis may have been the result of alcohol in this patient. It was not present in his 38-year-old daughter.

Apolipoprotein B48 glycosylation in abetalipoproteinemia and Anderson's disease.

Berriot-Varoqueaux N, Dannoura AH, Moreau A, *et al. Gastroenterol* 2001; **121**: 1101–8.

BACKGROUND. ABL and Anderson's disease are hereditary lipid malabsorption syndromes. In ABL, lipoprotein assembly is defective because of mutations in the microsomal triglyceride transfer protein. Here, we evaluated the intracellular transport of apolipoprotein B-48 to localize the defect in Anderson's disease. Asparagine-linked oligosaccharide processing of apoB-48 in normal and affected individuals was determined by the endoglycosidase H and F sensitivities of the protein after metabolic labelling of intestinal explants in organ culture. Cell ultrastructure was evaluated with electron microscopy. In Anderson's disease as in normal individuals, there was a time-dependent transformation of high mannose endoglycosidase H-sensitive

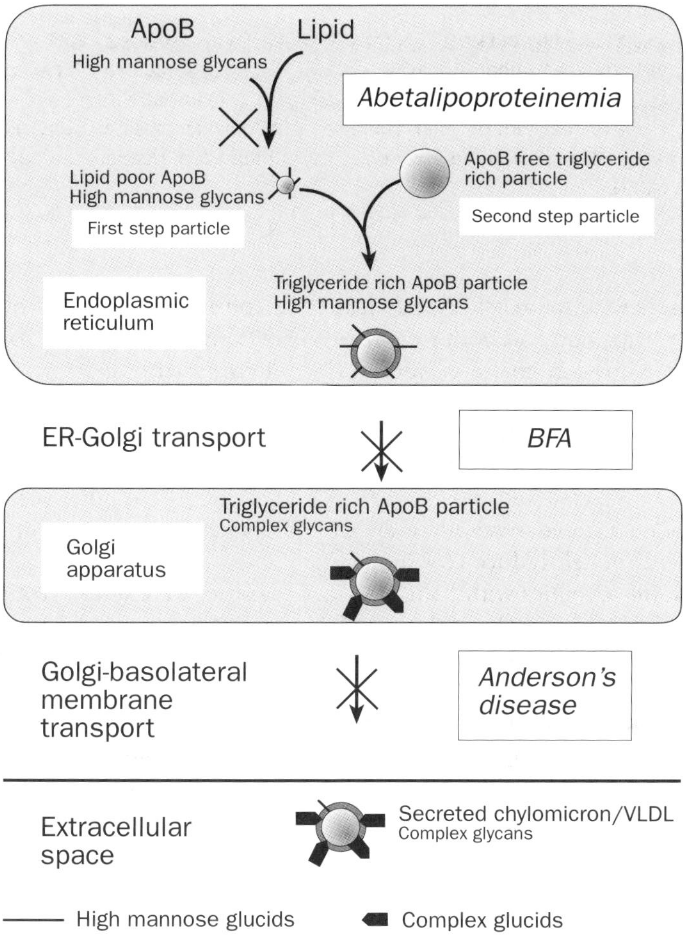

Fig. 4.1 Intracellular processing of intestinal ApoB. In the endoplasmic reticulum of normal individuals and of patients with Anderson's disease, apoB containing only high mannose glycans binds to lipid in an MTP-dependent step to form the first step particle. In a second step, in which the role of the MTP is not firmly established, the first step particle fuses with the apoB-free triglyceride-rich particle (second step particle) to form the nascent lipoprotein. The apoB acquires complex glycans in the Golgi apparatus. In contrast, in abetalipoproteinaemia and in BFA-treated intestinal biopsies, apoB does not reach the Golgi apparatus, as shown by the absence of complex glycans. In Anderson's disease, apoB-48 does reach the Golgi because it contains complex glycans. The disease is probably caused by a defect in the Golgi-basolateral transport mechanism. Source: Berriot-Varoqueaux *et al.* (2001).

oligosaccharides, of endoplasmic reticulum origin, to complex endoglycosidase H-resistant oligosaccharides, added in the Golgi network. In contrast, despite the translocation of apolipoprotein B-48 into the endoplasmic reticulum in patients with ABL and in biopsies treated with Brefeldin A, which blocks anterograde transport between the endoplasmic reticulum and the Golgi network, there was no transformation of endoglycosidase H-sensitive oligosaccharides.

INTERPRETATION. In ABL and Anderson's disease, apoB-48 is completely translocated into the endoplasmic reticulum, but only in Anderson's disease is the protein transported to the Golgi apparatus. This suggests that Anderson's disease is caused by a post-Golgi cargo-specific secretion defect.

Comment

Anderson's disease or CRD has clinical features in common with ABL and FHBL. In CRD however only apoB-48 is absent from plasma, even in the fed state. Previous work from the current authors has shown that the genetic defect in CRD does not appear to reside in the genes for the major apolipoproteins, MTP or fatty acid binding proteins 1 and 2 |**5**|. Chylomicron-like particles can be isolated from intestinal biopsies, but with a decreased mannose content of the apoB-48, suggesting a glycosylation defect. It has been unclear whether the defect occurs between the endosplasmic reticulum and the Golgi apparatus or in transport from the Golgi to the basolateral enterocyte membrane. The present work suggests that the defect is in the latter pathway in CRD and, once identified, it might provide a target to interfere with fat and cholesterol absorption for therapeutic purposes.

Hepatitis C virus induced hypobetalipoproteinemia: a possible mechanism for steatosis in chronic hepatitis C.

Serfaty L, Andreani T, Giral P, Carbonell N, Chazouilleres O, Poupon R. *J Hepatol* 2001; **34**: 428–34.

BACKGROUND. **Steatosis could be the result of hepatitis C virus (HCV)-induced hypobetalipoproteinaemia in patients with chronic hepatitis C. The aim of this study was to assess serum levels of main constituents of betalipoproteins and their relationship with steatosis in patients with chronic hepatitis C without known risk factors for steatosis. One hundred male patients with untreated biopsy proven non-cirrhotic chronic hepatitis C were included. Twenty-nine of these patients were further treated with interferon. Cholesterol concentration was significantly lower in patients compared to three control groups: reference male population, patients with chronic hepatitis B or individuals with non-alcoholic fatty liver. In multivariate analysis, low apoB concentration was an independent factor related to the degree of steatosis. HBL and degree of steatosis were significantly associated with infection with genotype 3. Among treated patients, only sustained virological responders had a significant increase of cholesterol (5.6 ± 1 *vs* 4.7 ± 1.3 mmol/l; $P = 0.03$) and apoB concentrations (113 ± 19 *vs* 75 ± 14 mg/dl; $P = 0.05$).**

Interpretation. In chronic hepatitis C, HBL is prevalent and associated with steatosis, especially in patients infected with genotype 3. The correction of HBL following HCV eradication suggests that HCV itself could induce HBL and steatosis.

Comment

HCV infection is of interest because it is a common cause of hepatic steatosis and hypolipidaemia, and the virus has been shown to be intimately associated with apoB-containing lipoproteins. It has been suggested that the LDL-receptor provides a means for the virus to enter hepatocytes. HCV core protein can induce steatosis and inhibit very low-density lipoprotein (VLDL) secretion in *in vitro* and animal studies.

In multivariate analysis serum apoB levels were inversely correlated with degree of steatosis and hepatic fibrosis was positively correlated. The authors were careful in this study to exclude patients who might have other common causes of hepatic steatosis such as obesity, diabetes and alcohol. It seems probable that HCV infection is a causal factor in fatty liver with secondary effects on plasma lipoproteins.

Tangier disease and ABCA1

ATP-binding cassette (ABC) transporters are present in all known organisms. About 1100 have been identified and about 50 in man [**6**]. Several genetic diseases are linked to mutations in ABC transporters including cystic fibrosis, age-related macular degeneration, Dubin-Johnson syndrome, Tangier disease [**7**], and phytosterolaemia (sitosterolaemia) [**8**].

Mutation of ATP-binding cassette transporter A1 (ABCA1) as the cause of Tangier disease was reported by three groups in 1999 [**9–11**], explaining the basis for the clinical syndrome which had been identified almost 40 years previously [**12,13**]. The original cases were a young boy and his sister, who lived on Tangier Island in Chesapeake Bay. They had large orange tonsils, the result of engorgement with cholesteryl esters, and barely detectable high-density lipoprotein (HDL).

The condition is rare. In view of the well established reciprocal relationship between HDL-C and the incidence of coronary heart disease (CHD) it is not surprising that Tangier cases are at increased risk, though the risk is probably attenuated by concomitant low concentrations of LDL-C.

Cellular cholesterol efflux.

Fielding CJ, Fielding PE. *Biochim Biophys Acta* 2001; **1533**: 175–89.

Background. Efflux of free cholesterol (FC) continues even when cellular FC mass is unchanged. This reflects a recirculation of preformed FC between cells and extracellular fluids which has multiple functions in cell biology including receptor

recycling and signalling as well as cellular FC homeostasis. Total FC efflux is heterogeneous. Simple diffusion to mature HDL, mainly via albumin as intermediate, initiates FC net transport driven by plasma lecithin:cholesterol acyltransferase activity. A second major efflux component reflects protein-facilitated transport from cell surface domains (caveolae, rafts) driven by FC binding to lipid-poor, pre-β-migrating HDL (pre-β-HDL). Facilitated efflux from caveolae, unlike simple diffusion, is highly regulated.

Interpretation. Neither ABC1 (the protein defective in Tangier disease) nor other ATP-dependent transporters now appear likely to contribute directly to FC efflux. Their role is limited to the initial formation of a particle precursor to circulating pre-β-HDL, which recycles without further lipid input from ATP-dependent transporter proteins. Lipid-free apoA-I, previously considered a surrogate for pre-β-HDL, has a reactivity much lower than that of native lipoprotein FC acceptors.

Comment

This thoughtful review emphasizes the complex and heterogeneous nature of the multiple mechanisms involved in efflux of cholesterol from cells. Important and exciting as the discovery that mutations in ABCA1 underlie Tangier disease and familial hypoalphaliproteinaemia are, it appears that the role of ABCA1 in transporting free cholesterol out of cells may be more restricted than was at first assumed. This was suggested by the observation that cholesterol efflux is normal in Tangier fibroblasts equilibrated with ^{3}H-unesterified cholesterol when incubated with mature normal HDL. It is proposed that the function of ABCA1 is to generate an initial apoAI-phopholipid complex (preβ_0-HDL) which takes up caveolar unesterified cholesterol to form preβ_1-HDL. Thus the direct role of ABCA1 would be to promote active efflux of phospholipid rather than cholesterol from cells.

ATP-binding cassette transporter A1 (ABCA1) functions as a cholesterol efflux regulatory protein.

Wang N, Silver DL, Thiele C, Tall AR. *J Biol Chem* 2001; **276**: 23742–7.

Background. ABCA1, an ATP-binding cassette transporter mutated in Tangier disease, promotes cellular phospholipid and cholesterol efflux by loading free apoA-I with these lipids. This process involves binding of apoA-I to the cell surface and phospholipid translocation by ABCA1. The goals of this study were to examine the relationship between ABCA1-mediated lipid efflux and apolipoprotein binding and to determine whether phospholipid and cholesterol efflux are coupled. Inhibition of lipid efflux by glibenclamide treatment or by mutation of the ATP-binding cassette of ABCA1 showed a close correlation between lipid efflux, the binding of apoA-I to cells, and cross-linking of apoA-I to ABCA1. The data suggest that a functionally important apoA-I binding site exists on ABCA1 and that the binding site could also involve lipids. After using cyclodextrin preincubation to deplete cellular cholesterol, ABCA1-mediated cholesterol efflux was abolished but phospholipid efflux and the binding of apoA-I were unaffected. The conditioned media from

cyclodextrin-pretreated, ABCA1-expressing cells readily promoted cholesterol efflux when added to fresh cells not expressing ABCA1, indicating that cholesterol efflux can be dissociated from phospholipid efflux. Further, using a photoactivatable cholesterol analogue, we showed that ABCA1 did not bind cholesterol directly, even though several other cholesterol-binding proteins specifically bound the cholesterol analogue.

INTERPRETATION. The data suggest that the binding of apoA-I to ABCA1 leads to the formation of phospholipid–apoA-I complexes, which subsequently promote cholesterol efflux in an autocrine or paracrine fashion.

Comment

This work provides evidence for the thesis of the review by the Fieldings discussed above in that ABCA1 seems to act primarily to translocate phospholipid to form apoA-I-phospholipid complexes which secondarily and indirectly promote cellular efflux of cholesterol.

Pivotal role of ABCA1 in reverse cholesterol transport influencing HDL levels and susceptibility to atherosclerosis.

Attie AD, Kastelein JP, Hayden MR. *J Lipid Res* 2001; **42**: 1717–26.

BACKGROUND. The identification of mutations in ABCA1 in patients with Tangier disease and familial HDL deficiency demonstrated that inadequate transport of phospholipid and cholesterol to the extracellular space results in the hypercatabolism of lipid-poor nascent HDL particles. However, the relationship between changes in ABCA1 activity and HDL levels is not clear. To address this question directly *in vivo*, bacterial artificial chromosome transgenic approaches were used, which allow for appropriate developmental and cellular localization of human ABCA1 in mouse tissues. Increased expression of ABCA1 is directly associated with an increase in HDL levels, and the relationship between the increase in efflux and HDL is completely linear ($r^2 = 0.87$). Preliminary data have suggested that CAD is increased in heterozygotes for ABCA1 deficiency. These results may have been biased by clinical sampling, and CAD end-points are insensitive markers. We have now used surrogate end-points of intima-media complex thickness (IMT) and have shown that mean IMT in ABCA1 heterozygotes is indeed increased. A strong correlation between adjusted IMT and HDL cholesterol values and apoA-I-driven efflux has been established.

INTERPRETATION. These studies suggest that compromised ABCA1 activity leads to accelerated and early atherogenesis and provides a link between the cholesterol deposition in macrophages within the arterial wall and cholesterol efflux in humans.

Comment

This is a detailed review which examines the process of reverse cholesterol transport and the key role played by ABCA1 in the initial stages of the process, thereby

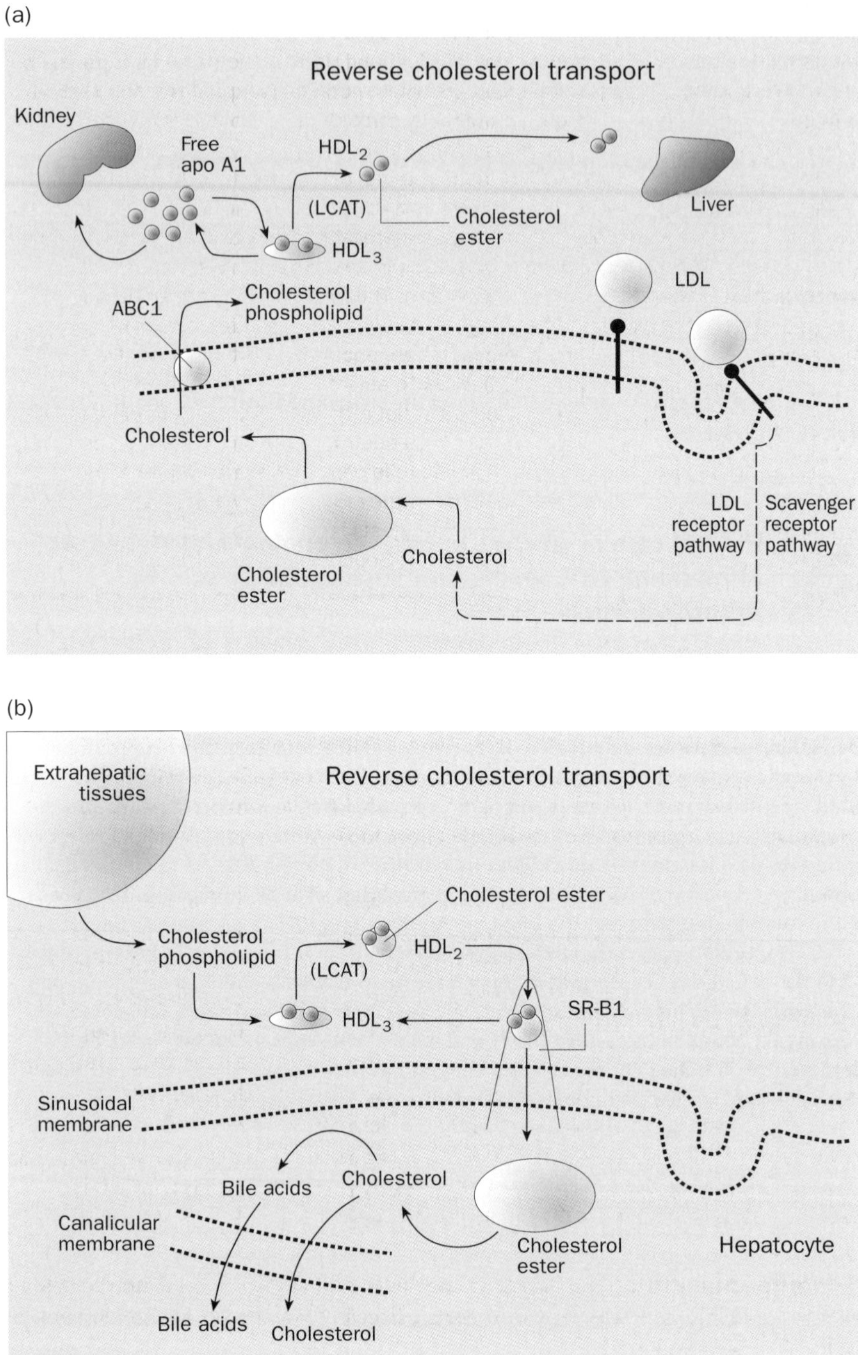
(a)
Reverse cholesterol transport
Kidney
Free
apo A1
HDL$_2$
(LCAT)
Cholesterol
ester
HDL$_3$
Liver
LDL
ABC1
Cholesterol
phospholipid
Cholesterol
Cholesterol
ester
Cholesterol
LDL
receptor
pathway
Scavenger
receptor
pathway
(b)
Extrahepatic
tissues
Reverse cholesterol transport
Cholesterol ester
Cholesterol
phospholipid
HDL$_2$
(LCAT)
HDL$_3$
SR-B1
Sinusoidal
membrane
Bile acids
Cholesterol
Canalicular
membrane
Cholesterol
ester
Hepatocyte
Bile acids
Cholesterol

Fig. 4.2 (a) Reverse cholesterol transport. Extrahepatic tissues synthesize cholesterol and also derive cholesterol through the uptake of lipoproteins via the LDL-receptor and scavenger receptors. The cholesteryl ester is in a dynamic equilibrium with free cholesterol, through the opposing actions of ACAT and neutral cholesterol esterase. Free cholesterol effluxes to extracellular acceptors, most notably phospholipid/apoA-I disks (pre-β-HDL). This process is directly (or indirectly through phospholipid efflux) dependent on functional ABCA1. Proper lipidation is essential for the stability of HDL. In the absence of sufficient cholesterol efflux, apoA-1 is rapidly cleared from the circulation by the kidneys. Cholesterol that associates with apoA-1/phospholipid disks is substrate for LCAT. LCAT transfers a fatty acyl chain from phosphatidylcholine to cholesterol, forming cholesteryl ester. The cholesteryl ester partitions into the hydrophobic core of the lipoprotein, thus forming spherical HDL particles. These particles can then deliver cholesteryl ester to the liver and steroidogenic tissues. (b) Selective uptake of cholesteryl esters from HDL. The interaction of spherical HDL particles with the scavenger receptor class B type I (SR-BI) leads to selective delivery of cholesteryl esters. SR-BI interacts with spherical HDL particles but not with apoA-I or poorly lipidated HDL disks. The cholesteryl esters are hydrolyzed by a neutral cholesterol esterase, providing free cholesterol for secretion across the apical (bile canalicular) membrane of the hepatocyte and for bile acid synthesis. Although the diagram shows cholesterol coming from extrahepatic tissues, growing evidence suggests that a major source of cholesterol for ABCA1-mediated transport to HDL is the liver. Source: Attie *et al.* (2001).

exerting a controlling influence on HDL-C concentrations and the well-established inverse relationship with incidence of CHD. Consonant with the conclusions of the two papers discussed above, the authors point out that there is no evidence that cholesterol is the primary substrate of ABCA1 and that the direct role of ABCA1 is to generate sufficient apoA-I–phospholipid complexes to act as a sink for cellular cholesterol.

Although overall in epidemiological studies high HDL-C is a powerful predictor of low CHD risk, the authors point out that at the individual level the impact on risk is likely to depend upon whether the increase in HDL-C is the result of active reverse cholesterol transport or impaired delivery of HDL-C to the liver. They go on to demonstrate, in a large cohort of individuals with hypoalphalipoproteinaemia heterozygous for mutated ABCA1, a threefold increase in symptomatic vascular disease compared to unaffected relatives. The residual activity of ABCA1 in these subjects was closely related to HDL-C level and risk of CHD. Using IMT of peripheral arteries as a surrogate for developing atherosclerosis showed that individuals heterozygous for a defect in ABCA1 activity developed an IMT of 0.75 mm on average at the age of 55 years, about 25 years earlier than unaffected family members.

The question remains as to whether the increased risk in ABCA1 heterozygotes relates to low HDL concentrations or more directly to defective ABCA1 activity in vessel wall macrophages.

Subpopulations of high density lipoproteins in homozygous and heterozygous Tangier disease.

Asztalos BF, Brousseau ME, McNamara JR, Horvath KV, Roheim PS, Schaefer EJ. *Atherosclerosis* 2001; **156**: 217–25.

BACKGROUND. Tangier disease (TD) is characterized by severe HDL deficiency, hypercatabolism of HDL constituents, impaired cellular cholesterol efflux, and mutations in the gene of ABC1. In the present study, the authors determined plasma lipid and apolipoprotein levels, and HDL subpopulations, in 110 subjects from a large TD kindred in which the proband was homozygous for an A→C missense mutation at nucleotide 5338 of the ABC1 transcript. In the proband HDL-C, apoA-I, and apoA-II concentrations were 2, 1, and 2 mg/dl, respectively, apoA-I was present only in preβ_1, while apoA-II was found free of apoA-I in two distinct α mobility subpopulations with different sizes. The smaller size particles contained only apoA-II while the larger one contained apoA-II and apo(a). Relative to unaffected male relatives (*n* = 30), male heterozygotes (*n* = 21) had significant reductions (*P* <0.001) in plasma HDL-C (–45%), apoA-I (–34%), apoA-II (–59%), apoA-IV (–40%), Lp(a) (–62%), and apoB (–55%) concentrations, and a significant increase (*P* <0.05, +33%) in plasma apoC-III levels. Female heterozygotes (*n* = 11) similarly had significant reductions (*P* <0.001) in the concentrations of plasma HDL-C (–42%), apoA-I (–27%), apoA-II (–52%), Lp(a) (–27%), and (*P* <0.01) apoA-IV (–28%), apoB (–13%), and a significant increase (*P* <0.05) in plasma apoE levels (+29%) as compared to unaffected female relatives (*n* = 41). Large size HDL subpopulations, especially the two LpA-I particles: α_1 and preα_1 were dramatically reduced in both male and female heterozygotes relative to their unaffected family members. Since apoA-II decreased more than apoA-I in both male and female heterozygotes, the ratios of apoA-I/apoA-II were significantly (*P* <0.01) increased. The prevalence of CHD was 60% higher in the 32 heterozygotes than in the 71 unaffected relatives even though the latter group was on average 7 years older.

INTERPRETATION. The authors conclude that TD homozygotes have only preβ_1 apoA-I-containing HDL subpopulations, while heterozygotes have HDL that is selectively depleted in the large α_1, preα_1, and α_2, preα_2 subpopulations, resulting in HDL particles that are small in size, poor in cholesterol, but relatively enriched in apoA-I compared to those of their unaffected relatives. These abnormalities appear to result in a higher risk of CHD in heterozygotes than in unaffected controls.

Comment

It has generally been thought that the defect in ABCA1 in Tangier homozygotes prevents the translocation of cholesterol to the outer layer of cell membranes making it unavailable for acceptors such as preβ_1-HDL (lipid-poor apoA-I). Current thinking (see Fielding and Fielding, and Wang *et al.*, above) suggests that the direct defect as a result of non-functional ABCA1 is a failure to form phospholipids–apoA-I complexes, which are the acceptors of cholesterol. In any event HDL precursors are not converted into mature spherical HDL of α-mobility, and are rapidly catabolized. Heterozygotes in this study had disproportionate reductions in the largest HDL particles (α_1 and preα_1). α_1 is an LpA-I particle which resembles HDL_2 and its selective reduction may be implicated in the increased prevalence of CHD in Tangier heterozygotes. The increase in risk, which is modest, may be attenuated by the marked reduction in apoB-containing lipoproteins, which in turn could result in part from a lack of available cholesteryl ester for transfer from HDL. As discussed in the next paper, the LDL which are present in Tangier disease however have an unfavourable composition even though present in small quantities.

Cholesterol and apolipoprotein B metabolism in Tangier disease.

Schaefer EJ, Brousseau ME, Diffenderfer MR, *et al. Atherosclerosis* 2001; **159**: 231–6.

BACKGROUND. TD, caused by mutations in the gene encoding ABCA1, is a rare genetic disorder in which homozygotes have a marked deficiency of HDL, as well as concentrations of LDL that are typically 40% of normal. Although it is well known that the reduced levels of HDL in TD are due to hypercatabolism, the mechanism responsible for the low LDL levels has not been defined. Recently, it has been reported that intestinal cholesterol absorption is altered in ABCA1-deficient mice, suggesting that aberrant cholesterol metabolism may contribute to the LDL reductions in TD. In order to explore this possibility, as well as to define the role that ABCA1 plays in the metabolism of apoB-containing lipoproteins, the kinetics of apoB-100 within lipoproteins, and cholesterol absorption, biosynthesis, and turnover, were determined in a compound heterozygote for TD. The levels of HDL cholesterol, LDL cholesterol and LDL apoB-100 in this subject were 7, 27 and 69% of normal, respectively, the latter of which was due to a two-fold increase in LDL catabolism (0.54 *vs* 0.26 ± 0.07 pools/day relative to controls (*n* = 11). Nuclear magnetic resonance (NMR) analysis of plasma lipoproteins revealed that 91% of the LDL cholesterol in the TD subject was contained within small, dense LDL, as compared with only 20% for controls (*n* = 70). Cholesterol absorption was 97% of the value for controls (*n* = 15) in the TD subject, at 45%, with cholesterol synthesis and turnover increased modestly by 17 and 25%, respectively.

INTERPRETATION. These data are consistent with the concept that the reductions of LDL observed in TD are due to enhanced catabolism, secondary to changes in LDL composition and size, with neither cholesterol absorption nor metabolism being significantly influenced by mutations in ABCA1.

Comment

Mutations in ABCA1, which underlie the metabolic effects on HDL in Tangier disease, do not appear to influence intestinal cholesterol absorption in man though there is conflicting information about this in mice.

LDL-C levels in Tangier patients are often less than half those in controls. This may help to protect them against adverse consequences of low HDL-C, but the situation is not as favourable as it might seem. In this patient LDL-C was reduced to 27% of normal, but apoB concentration, which reflects the number of circulating LDL particles, was only reduced to 69% of normal, as a result of increased catabolism. Thus the LDL particles were depleted in lipid and the particle distribution was skewed towards small dense LDL, which are believed to be more atherogenic than larger lipid-rich LDL. This patient carried 91% of his LDL-C in small, dense particles compared with about 20% in controls.

It is suggested that the severe deficiency of HDL-C in Tangier patients results in defective transfer of cholesteryl esters to LDL under the influence of cholesteryl ester transfer protein. This leaves a TG-rich LDL from which the TG is removed by hepatic lipase resulting in a small lipid-poor LDL. This alteration in composition is proposed as the reason for the unusually rapid catabolism of LDL. This contrasts with the situation in non-Tangier subjects with an excess of small, dense LDL. These particles are believed to be atherogenic in part because of their slower removal by the LDL-receptor |**14**|.

A point mutation in ABC1 gene in a patient with severe premature coronary heart disease and mild clinical phenotype of Tangier disease.

Bertolini S, Pisciotta L, Seri M, *et al. Atherosclerosis* 2001; **154**: 599–605.

BACKGROUND. The proband is a 50-year-old woman born from a consanguineous marriage. She has been suffering from angina pectoris since the age of 38 and underwent coronary bypass surgery for three-vessel disease at 48. The presence of low plasma levels of total cholesterol and HDL cholesterol (2.4 and 0.1 mmol/l) and apoA-I (<15 mg/dl), associated with corneal lesions and a mild splenomegaly suggested the diagnosis of Tangier disease. However, none of the other features of Tangier disease, including hepatomegaly, anaemia and peripheral neuropathy, were present. The analysis of the dinucleotide microsatellites located in chromosome 9q31 region demonstrated that the proband was homozygous for the alleles of D9S53, D9S1784 and D9S1832. The mother and son of the proband, both with low levels of HDL cholesterol, shared one of the proband's haplotypes, whereas neither of these haplotypes was present in the normolipidaemic proband's sister. The sequence of ATP-binding cassette transporter 1 (ABC1-1) cDNA obtained by reverse transcription-PCR (RT-PCR) of total RNA isolated from cultured fibroblasts showed that the proband was homozygous for a C>T transition in exon 13, which caused a tryptophan for arginine substitution (R527W). This mutation was confirmed by direct sequencing of exon 13 amplified from genomic DNA. It can be easily screened, as the nucleotide change introduces a restriction site for the enzyme Afl III. R527W substitution occurs in a highly conserved region of the NH2 cytoplasmic domain of ABC1 protein. R527W co-segregated with the low HDL phenotype in the family and was not found in 200 chromosomes from normolipidaemic individuals.

Interpretation The index case in this study had a relatively mild phenotype for Tangier disease in so far as she did not have a peripheral neuropathy which is found in more than half of cases |**15**|, nor did she exhibit anaemia or an enlarged liver. She did however have premature cardiovascular disease which has been reported in 44% of Tangier homozygotes over the age of 35 years |**15**|. There is a wide spectrum of clinical manifestations in patients who are homozygous for mutations in ABCA1 leading to Tangier disease and familial HDL deficiency.

Comment

Discovery of ABCA1 and its role in promoting cholesterol efflux from tissues has provided an understanding of the pathophysiology of Tangier disease and of the low HDL-C concentrations in Tangier heterozygotes. It is likely in the future to provide information about variability in HDL-C in the general population and its contribution to CHD risk and may be a target for pharmacological modulation in an attempt to modify that risk |**16**|.

The following two studies, and there are likely to be more of a similar nature, have attempted to determine if common polymorphisms in ABCA1 underlie variability in CHD risk and HDL-C in the general population.

Common variants in the gene encoding ATP-binding cassette transporter 1 in men with low HDL cholesterol levels and coronary heart disease.

Brousseau ME, Bodzioch M, Schaefer EJ, *et al. Atherosclerosis* 2001; **154**: 607–11.

Background. HDL-C deficiency is the most common lipid abnormality observed in patients with premature CHD. Recently, our laboratory and others demonstrated that mutations in the ABCA1 gene are responsible for Tangier disease, a rare genetic disorder characterized by severely diminished plasma HDL-C concentrations and a predisposition for CHD. To address the question of whether common variants within the coding sequence of ABCA1 may affect plasma HDL-C levels and CHD risk in the general population, the frequencies of three common ABCA1 variants (G596A, A2589G and G3456C) were determined in men participating in the Veterans Affairs Cooperative HDL Cholesterol Intervention Trial (VA-HIT), a study designed to examine the benefits of HDL-raising in men having low HDL-C (≤40 mg/dl) and established CHD, as well as in CHD-free men from the Framingham Offspring Study (FOS). Allele frequencies (%) in VA-HIT were 31, 16, and 4 for the G596A, A2589G, and G3456C variants, respectively, *versus* 27, 12, and 2 in FOS (*P* <0.03).

Interpretation. None of the variants were significantly associated with plasma HDL-C concentrations in either population; however, in VA-HIT, the G3456C variant was associated with a significantly increased risk for CHD end-points, suggesting a role for this variant in the premature CHD observed in this population.

Comment

ABCA1 is involved in several functions, which include ingestion of apoptotic cells and secretion of interleukin-1β by macrophages, and transport of lipids from the Golgi apparatus to the plasma membrane in addition to its role in promoting cholesterol efflux from cells. Not all of its effects on the atherosclerotic process are likely therefore to be readily explained simply by changes in HDL metabolism.

No relationship was found between the three polymorphisms studied and HDL-C concentration, though the least common of the variants in ABCA1 was associated with an increased risk of further CHD end-points in the VA-HIT and, paradoxically, a small reduction in plasma apoB levels. The significance of these observations is however unclear and they were not replicated in CHD-free men in the FOS.

Novel polymorphisms in promoter region of ATP binding cassette transporter gene and plasma lipids, severity, progression, and regression of coronary atherosclerosis and response to therapy.

Lutucuta S, Ballantyne CM, Elghannam H, Gotto AM Jr, Marian AJ. *Circ Res* 2001; **88**: 969–73.

BACKGROUND. Identification of mutations in the ABCA1 gene in patients with TD disease, who exhibit reduced HDL-C and apoA-I levels and premature coronary atherosclerosis, has led to the hypothesis that common polymorphisms in the ABCA1 gene could determine HDL-C and apoA-I levels and the risk of coronary atherosclerosis in the general population. We sequenced a 660-bp 5′ fragment of the ABCA1 gene in 24 subjects and identified three novel polymorphisms: -477C/T, -419A/C, and -320G/C. Assays were developed and 372 participants were genotyped in the prospective Lipoprotein Coronary Atherosclerosis Study (LCAS). The association of the variants with fasting plasma lipids and indices of quantitative coronary angiograms obtained at baseline and 2.5 years after randomization to fluvastatin or placebo were determined. Distribution of -477C/T and -320G/C genotypes were 127 CC, 171 CT, and 74 TT and 130 GG, 168 GC, and 75 CC, respectively, and were in complete linkage disequilibrium (*P* <0.0001). Data for -477C/T are presented. The -419A/C variant was uncommon (present in 1 of 63 subjects). Heterozygous subjects had a modest reduction in HDL-C (*P* = 0.09) and apoA-I (*P* = 0.05) levels and a lesser response of apoA-I to treatment with fluvastatin (*P* = 0.04). The mean number of coronary lesions causing 30% to 75% diameter stenosis was greater in subjects with the TT genotype (3.1 ± 2.1) or CT genotype (2.9 ± 1.9) than in subjects with the CC genotype (2.2 ± 1.8; *P* = 0.002). Similarly, compared with subjects with the CC genotype, greater numbers of subjects with the TT or CT genotype had ≥1 coronary lesion (*P* = 0.001). No association between the genotypes and progression of coronary atherosclerosis or clinical events was detected.

INTERPRETATION. It was concluded that ABCA1 genotypes are potential risk factors for coronary atherosclerosis in the general population.

Comment

Three new polymorphisms in the promoter region of the ABCA1 gene were identified one of which proved to be rare and was not further pursued. The association of the -477C/T variant with HDL-C and apoA-I are of questionable significance but it was strongly related to severity of CHD as judged angiographically in this LCAS population. It is perhaps surprising then that this polymorphism did not predict coronary events or change in angiographic appearances during the study. The authors suggest that this is because events and angiographic progression are more likely to be the result of plaque rupture and thrombosis than gradual accretion of lipid in macrophages, which is more likely to be influenced by ABCA1 activity. Moreover the numbers involved were relatively small and the duration of the study only 2.5 years.

References

1. Malloy MJ, Kane JP. Disorders involving deficiencies of lipoproteins that contain B apolipoproteins. In: Betteridge DJ, Illingworth DR, Shepherd J (eds). *Lipoproteins in Health and Disease*, 1st edn. London: Arnold, 1999: 863–77.
2. Bassen FA, Kornzweig AL. Malformation of the erythrocytes in a case of atypical retinitis pigmentosa. *Blood* 1950; **5**: 381–7.
3. Wetterau JR, Aggerbeck LP, Bouma M-E, Eisenberg C, Munck A, Hermier M, Schmitz J, Gay G, Rader DJ, Gregg RE. Absence of microsomal triglyceride transfer protein in individuals with abetalipoproteinemia. *Science* 1992; **258**: 999–1001.
4. Gordon DA. Recent advances in elucidating the role of the microsomal triglyceride transfer protein in apolipoprotein B lipoprotein assembly. *Curr Opin Lipidol* 1997; **8**: 131–7.
5. Dannoura AH, Berriot-Varoqueaux N, Amati P, Abadie V, Verthier N, Schmitz J, Wetterau JR, Samson-Bouma ME, Aggerbeck LP. Anderson's disease: exclusion of apolipoprotein and intracellular lipid transport genes. *Arterioscler Thromb Vasc Biol* 1999; **19**: 2494–508.
6. Gottesman MM, Ambudkar SV. Overview: ABC transporters and human disease. *J Bioenerg Biomembr* 2001; **33**: 453–8.
7. Schmitz G, Langmann T. Structure, function and regulation of the ABC1 gene product. *Curr Opin Lipidol* 2001; **12**: 129–40.
8. Lee MH, Lu K, Patel SB. Genetic basis of sitosterolemia. *Curr Opin Lipidol* 2001; **12**: 141–9.
9. Bodzioch M, Orso E, Klucken J, Langmann T, Bottcher A, Diederich W, Drobnik W, Barlage S, Buchler C, Porsch-Ozcurumez M, Kaminski WE, Hahmann HW, Oette K, Rothe G, Aslanidis C, Lackner KJ, Schmitz G. The gene encoding ATP-binding cassette transporter 1 is mutated in Tangier disease. *Nat Genet* 1999; **22**: 347–51.

10. Brooks-Wilson A, Marcil M, Clee SM, Zhang LH, Roomp K, van Dam M, Yu L, Brewer C, Collins JA, Molhuizen HO, Loubser O, Ouelette BF, Fichter K, Ashbourne-Excoffon KJ, Sensen CW, Scherer S, Mott S, Denis M, Martindale D, Frohlich J, Morgan K, Koop B, Pimstone S, Kastelein JJ, Hayden MR, *et al.* Mutations in ABC1 in Tangier disease and familial high-density lipoprotein deficiency. *Nat Genet* 1999; **22**: 336–45.

11. Rust S, Rosier M, Funke H, Real J, Amoura Z, Piette JC, Deleuze JF, Brewer HB, Duverger N, Denefle P, Assmann G. Tangier disease is caused by mutations in the gene encoding ATP-binding cassette transporter 1. *Nat Genet* 1999; **22**: 352–5.

12. Fredrickson DS, Altrocchi PH, Avioli LV, Goodman DS, Goodman HC. Tangier disease – combined clinical staff conference at the National Institutes of Health. *Ann Intern Med* 1961; **55**: 1016–31.

13. Assmann G, Von Eckardstein A. Tangier disease. In: Betteridge DJ, Illingworth DR, Shepherd J (eds). *Lipoproteins in Health and Disease,* 1st edn. London: Arnold, 1999: 767–81.

14. Campos H, Arnold KS, Balestra ME, Innerarity TL, Krauss RM. Differences in receptor binding of LDL subfractions. *Arterioscler Thromb Vasc Biol* 1996; **16**: 794–801.

15. Serfaty-Lacrosniere C, Civeira F, Lanzberg A, Isaia P, Berg J, Janus ED, Smith MP Jr, Pritchard PH, Frohlich J, Lees RS, *et al.* Homozygous Tangier disease and cardiovascular disease. *Atherosclerosis* 1994; **107**: 85–98.

16. Oram JF. Novel approaches to treating cardiovascular disease: lessons from Tangier disease. *Expert Opin Investig Drugs* 2001; **10**: 427–38.

Part II

Diabetes and dyslipidaemia

5

The rising tide of type 2 diabetes – pathogenesis and epidemiology

Introduction

The prevalence of type 2 diabetes (T2D) is increasing rapidly and varies widely between populations, reflecting potential differences in environmental and genetic factors |**1**|. Currently it is estimated that 150 million people worldwide have diabetes |**2**|. This number is due to increase to over 300 million by 2025 |**2**|, the majority of this increase being in the form of T2D. This huge increase in the global prevalence of T2D affects both developed and developing nations alike, with particularly high rates of T2D in the Pacific and Asia. Data from the United States estimate that T2D affects approximately 8% of the adult population |**3**|, while recent data from Australia estimates an overall prevalence of T2D of 7.4% in Australian adults, with 5.7% of younger individuals (25–34 years) demonstrating abnormal glucose tolerance (Dunstan *et al.*, p. 109). In the United Kingdom a similar picture exists with the recent audit commission report projecting an increase in the incidence of T2D to over 3 million in the UK by the year 2010 |**1**|.

T2D, previously considered a disease of old age, is increasingly common in younger subjects and is a serious and costly disease associated with a significant disease burden resulting in an increased morbidity and mortality, primarily related to cardiovascular disease. Treatment may aid in minimizing the complications associated with T2D, but does not eliminate all the adverse consequences. The diagnosis of T2D is also often delayed, and up to 20% of patients at diagnosis will have evidence of diabetes-related complications |**4**|. Because of the macro- and microvascular injury typical of T2D, the economic and functional burdens associated with the condition are greatest during mid to late life, an issue further compounded by the all too common delay in confirming the diagnosis and instituting appropriate therapy. Since the current methods of managing T2D remain suboptimal, prevention is a preferable approach. T2D thus represents a major public health issue and in order to develop optimal prevention strategies for T2D, it is important that we have a clear understanding of the epidemiology, aetiology and risk factors for T2D.

T2D is a heterogeneous disorder, characterized by hyperglycaemia and resulting in multiple metabolic abnormalities, in particular perturbations of lipid metabolism.

The aetiology of T2D is multifactorial, with insulin resistance associated with beta cell dysfunction appearing to be the primary defect in the majority of patients with T2D |**5**|. Although the main physiological abnormalities associated with T2D are well recognized, the specific determinants of these factors remain uncertain.

In the forthcoming discussion we present and evaluate some of the most recent data relating to our current understanding of both the epidemiology of T2D and of the factors which may be important in the aetiology and pathogenesis of T2D.

Prevalence of impaired glucose tolerance among children and adolescents with marked obesity.

Sinha R, Fisch G, Teague B, *et al.* *N Engl J Med* 2002; **346**: 802–10.

BACKGROUND. Childhood obesity, epidemic in the United States, has been accompanied by an increase in the prevalence of T2D among children and adolescents. This investigation determined the prevalence of impaired glucose tolerance in a multiethnic cohort of 167 obese children and adolescents. All subjects underwent a two-hour oral glucose-tolerance test (1.75 mg of glucose per kilogram of body weight), and glucose, insulin, and C-peptide levels were measured. Fasting levels of proinsulin were obtained, and the ratio of proinsulin to insulin was calculated. Insulin resistance was estimated by homeostatic model assessment, and beta cell function was estimated by calculating the ratio between the changes in the insulin level and the glucose level during the first 30 minutes after the ingestion of glucose. Impaired glucose tolerance was detected in 25% of the 55 obese children (4–10 years of age) and 21% of the 112 obese adolescents (11–18 years of age); silent T2D was identified in 4% of the obese adolescents. Insulin and C-peptide levels were markedly elevated after the glucose-tolerance test in subjects with impaired glucose tolerance but not in adolescents with diabetes, who had a reduced ratio of the 30-minute change in the insulin level to the 30-minute change in the glucose level. After the body-mass index (BMI) had been controlled for, insulin resistance was greater in the affected cohort and was the best predictor of impaired glucose tolerance.

INTERPRETATION. Impaired glucose tolerance is highly prevalent among children and adolescents with severe obesity, irrespective of ethnic group. Impaired oral glucose tolerance was associated with insulin resistance while beta cell function was still relatively preserved. Overt T2D was linked to beta cell failure.

Comment

Childhood obesity is reaching epidemic proportions in the western world and represents a major factor in the projected increase in the prevalence of T2D. In adults T2D develops over a long period and most patients initially have impaired glucose tolerance, representing an intermediate stage in the natural history of T2D which predicts the risk of the development of both T2D and cardiovascular disease |**6**|. With appropriate interventions, progression from impaired glucose tolerance to T2D may be delayed or even prevented |**7**|. Abnormal beta cell function, manifested

Table 5.1 Clinical and metabolic phenotype of obese children and adolescents with normal glucose tolerance, impaired glucose tolerance, or type 2 diabetes

Variable	Children (n = 55)			Adolescents (n = 112)				
	Normal glucose tolerance (n = 41)	Impaired glucose tolerance (n = 14)	*P* value	Normal glucose tolerance (n = 85)	Impaired glucose tolerance (n = 23)	*P* value	Type 2 diabetes (n = 4)	*P* value
Race or ethnic group (no.)								
Non-Hispanic white	25	9		53	10		0	
Non-Hispanic black	8	4		17	7		2	
Hispanic	8	1		15	6		2	
Sex (no.)								
Male	15	2		40	12		2	
Female	26	12		45	11		2	
Age (yr)	7.5 ± 0.3	8 ± 0.5		14 ± 1	15 ± 0.4		15 ± 1	
Range	4–10	5–10		11–18	12–18		11–18	
Body mass index	30 ± 1	32 ± 1	<0.05	37 ± 0.9	41 ± 1	<0.001	41 ± 5	<0.001
Fasting insulin level (μU/ml)	20 ± 5	34 ± 5	<0.001	30 ± 2	56 ± 7	<0.001	50 ± 10	<0.001
Fasting C-peptide level (ng/ml)	2.6 ± 0.3	3.2 ± 0.2	<0.001	3.6 ± 0.1	4.7 ± 0.3	<0.001	4.5 ± 1	<0.001
Insulin resistance index	5 ± 0.6	7.2 ± 1	<0.001	6.3 ± 0.4	13 ± 7	<0.001	14 ± 3	<0.001

Data is expressed as mean ± SE. *P* values are for comparison with the group with normal glucose tolerance. Insulin resistance index was determined by HOMA model assessment and calculated as the product of fasting plasma insulin and fasting plasma glucose divided by 22.5. Lower insulin-resistance values indicate a higher insulin sensitivity, whereas higher insulin-resistance values indicate a lower insulin sensitivity.

Source: Sinha *et al.* (2002).

as relative hyperproinsulinaemia is common in patients with overt T2D, while disproportionate hyperproinsulinaemia represents impending beta cell failure. The earlier an increase in the ratio of proinsulin to insulin occurs, the more likely it is that abnormal processing of insulin by beta cells is fundamental to the pathogenesis of T2D |**8**|. This study examined both the prevalence of glucose intolerance in obese children and adolescents as well as evaluating the processing of proinsulin to assess whether alterations are present early in the development of glucose intolerance in obese children and adolescents.

In order to assess these issues 55 children (4–10 years of age) and 112 adolescents (11–18 years of age) were studied. All subjects had a BMI that was greater than the 95th centile age and sex and were thus classed as obese. All subjects followed a weight maintenance diet of at least 250 g of carbohydrate for seven days prior to the study. Baseline measurements of plasma glucose, insulin, C-peptide, proinsulin and lipid profiles were made. Thereafter, all subjects underwent a two-hour oral glucose-tolerance test (1.75 mg of glucose/kg body weight), with measurements made at 30 minute intervals for plasma glucose, insulin and C-peptide levels. Calculating the ratio of the increment in plasma insulin level to plasma glucose over the first 30 minutes after the oral glucose load assessed beta cell function. Insulin resistance was calculated by a HOMA model assessment (fasting plasma glucose × fasting plasma insulin/22.5). Impaired glucose tolerance was detected in 25% of the obese children and in 21% of the obese adolescents, while frank T2D was identified in 4% of the obese adolescents. A total of 30% of the combined group with impaired glucose tolerance and frank T2D had a parental history of T2D, the rate being 25% in those with normal glucose tolerance. There were no significant differences in ethnicity between subjects with normal and abnormal glucose tolerance. In keeping with recent observations, six girls with apparent polycystic ovarian syndrome had impaired glucose tolerance and two had frank diabetes |**9**|. Insulin and C-peptide levels were markedly elevated after the glucose-tolerance test in subjects with impaired glucose tolerance but not in adolescents with diabetes, who had a reduced ratio of the 30-minute change in the insulin level to the 30-minute change in the glucose level. After the BMI had been allowed for, insulin resistance was greater in the affected cohort and was the best predictor of impaired glucose tolerance.

Impaired glucose tolerance appeared to be highly prevalent among children and adolescents with severe obesity irrespective of ethnicity or parental history. The risk factors associated with impaired glucose tolerance included insulin resistance, marked hyperinsulinaemia, both fasting and post-glucose load, and fasting hyperproinsulinaemia. Insulin resistance was identified as a strong predictor of elevated two-hour plasma glucose levels and may thus play an important role in the transition to impaired glucose tolerance. Family history of T2D or the degree of obesity were not found to be significant risk factors for glucose intolerance, although the effects of obesity are likely to be related to its association with insulin resistance. Beta cell function, measured by insulinogenic index, was partially preserved in adolescents with impaired glucose tolerance whereas it was significantly reduced in subjects with frank T2D. The vigorous hyperinsulinaemic response to glucose seen in both the children and adolescents with impaired glucose tolerance may reflect chronic upregulation of beta cell function as a result of insulin resistance [8].

This study thus suggests that insulin resistance, initially associated with hyperinsulinaemia and hyperproinsulinaemia, is the most important linked to the development of impaired glucose tolerance in children and adolescents with severe obesity. Although beta cell function appears preserved in the setting of impaired glucose tolerance, in the presence of established T2D beta cell failure becomes manifest.

Dairy consumption, obesity, and the insulin resistance syndrome in young adults: the CARDIA study.

Pereira MA, Jacobs DR Jr, Van Horn L, Slattery ML, Kartashov AI, Ludwig DS. *JAMA* 2002; **287**: 2081–9.

BACKGROUND. Components of the insulin resistance syndrome (IRS), including obesity, glucose intolerance, hypertension, and dyslipidaemia, are major risk factors for T2D and heart disease. Although diet has been postulated to influence IRS, the independent effects of dairy consumption on development of this syndrome have not been investigated. The aim of this study was to examine associations between dairy intake and incidence of IRS, adjusting for confounding lifestyle and dietary factors. The Coronary Artery Risk Development in Young Adults (CARDIA) study was used, a population-based prospective study. Participants were from a general community sample from 4 US metropolitan areas consisting of 3157 black and white adults aged 18–30 years who were followed up from 1985–1986 to 1995–1996. The main outcome measure was the ten-year cumulative incidence of IRS and its association with dairy consumption, measured by diet history interview.

INTERPRETATION. Dietary patterns characterized by increased dairy consumption have a strong inverse association with IRS among overweight adults and may reduce risk of T2D and cardiovascular disease.

Comment

The risk of developing impaired glucose tolerance and T2D in adolescents and young adults appears closely linked to insulin resistance (Sinha *et al.*, p. 104). Insulin resistance is also associated with multiple factors, which contribute to an increased risk of cardiovascular disease [10]. The so-called IRS, includes glucose intolerance, obesity, hypertension, a complex dyslipidaemia characterized by low high-density lipoprotein cholestolerol (HDL-C) levels and hypertriglycerideamia, and multiple other recognized vascular risk factors such as enhanced inflammation, vascular dysfunction and hyperhomocystaemia. The IRS is increasing in incidence in all ethnic, racial and social classes and has been described in children, adolescents and young adults. Age-adjusted prevalence of the IRS among adults has recently been estimated at 24% [11]. Various environmental factors such as physical inactivity and smoking are known to promote insulin resistance; the effect of dietary composition on the IRS is poorly understood. Epidemiological and experimental data suggest that dairy products may have favourable effects on body mass in children and adults and may reduce the risk for hypertension, coronary heart disease (CHD) and stroke |**12,13**|. The purpose of this study was to prospectively examine independent associations between dairy consumption and the IRS, after correcting for potentially confounding variables such as physical activity, macronutrient and fibre intake.

The CARDIA study was a multicentre population-based study. A total of 3157 adults aged 18–30 were followed up over a 10-year period. Standard questionnaires were used to assess dietary history as well as other variables such as smoking history and physical activity. Various features of the IRS were defined based on standard classifications. Abnormal glucose metabolism was defined as fasting plasma glucose >6.1 mmol/l and a fasting plasma insulin concentration >90th centile for distribution of insulin. Obesity was defined by a BMI >30 kg/m^2 or waist:hip ratio of 0.85 for women and 0.9 for men. Hypertension was defined as a blood pressure BP >130/85 mm/Hg or by the use of antihypertensive medication. Dyslipidaemia was defined by a low HDL-C <0.9 mmol/l or high plasma triglyceride (TG) >2.25 mmol/l. Presence of the IRS was defined by the presence of two or more of the above components. Dairy consumption was inversely associated with the incidence of all components of the IRS among individuals who were overweight (BMI ≥25 kg/m^2) at baseline but not among leaner individuals (BMI <25 kg/m^2). The adjusted odds of developing IRS (two or more components) were 72% lower (odds ratio [OR], 0.28; 95% confidence interval [CI] 0.14–0.58) among overweight individuals in the highest (≥35 times per week, 24/102 individuals) compared with the lowest (<10 times per week, 85/190 individuals) category of dairy consumption. Each daily occasion of dairy consumption was associated with a 21% lower odds of insulin resistance (OR, 0.79; 95% CI 0.70–0.88). These associations were similar for black and white subjects and for men and women. Other factors, including physical activity and dietary macronutrients and micronutrient content, did not appear to explain the association between dairy intake and insulin resistance.

Frequent dairy product consumption appeared to reduce the risk of developing abnormal glucose metabolism, dyslipidaemia, hypertension and obesity in young overweight subjects over a 10-year period, the incidence of the IRS being reduced by almost two-thirds among those overweight individuals with the highest dairy product consumption (>5 times daily). These associations did not appear to be confounded by any other dietary or lifestyle factors. This study is limited by its observational nature and it is thus impossible to conclude that increased dairy product intake reduces the incidence of the IRS in a causal manner. The longitudinal nature of the study however enables a better estimate of both dietary and lifestyle factors and also limits the tendency for bias. Although saturated fat in dairy products may result in increased LDL-C concentrations there are several mechanisms by which dairy intake may protect against insulin resistance. These include potential effects on satiety and the effects of magnesium, calcium and potassium on intracellular mechanisms involved in body weight regulation |**14**|. Alternative explanations include alterations in dietary intake associated with dairy intake or the presence of factors in dairy products such as conjugated linoleic acid, which may influence adipocyte metabolism and insulin sensitivity through potential anti-inflammatory effects |**15**|. Irrespective of the mechanisms involved, this study suggests that dietary patterns characterized by high dairy product intake may protect overweight individuals against the development of insulin resistance and thus T2D and cardiovascular disease.

The rising prevalence of diabetes and impaired glucose tolerance: the Australian Diabetes, Obesity and Lifestyle study.

Dunstan DW, Zimmet PZ, Welborn TA, *et al. Diabetes Care* 2002; **25**: 829–34.

BACKGROUND. The aim of this study was to determine the population-based prevalence of diabetes and other categories of glucose intolerance (impaired glucose tolerance [IGT] and impaired fasting glucose [IFG]) in Australia and to compare the prevalence with previous Australian data. Diabetes prevalence has more than doubled since 1981, and this is only partially explained by changes in age profile and obesity.

INTERPRETATION. Australia has a rapidly rising prevalence of diabetes and other categories of abnormal glucose tolerance. The prevalence of abnormal glucose tolerance in Australia is one of the highest yet reported from a developed nation with a predominantly Europid background.

Comment

The ageing population combined with the effects of modernization on lifestyle and diet has led to a dramatic increase in the risk factors for developing T2D and in the estimated prevalence of T2D. Despite the increasing awareness of the growing problem of diabetes, and the recent publication of a number of predictions of current and future prevalence of T2D worldwide, the US remains the only developed nation with accurate data on the prevalence of T2D. This study assessed the population-based prevalence of T2D and glucose intolerance (IGT, IFG) in Australia. The results obtained were compared to estimates of obesity and diabetes made in 1981.

A national sample involving 11 247 participants aged ≥25 years living in 42 randomly selected areas were examined in a cross-sectional survey using the 75-g oral glucose-tolerance test to assess fasting and 2-h plasma glucose concentrations. The World Health Organization (WHO) diagnostic criteria were used to determine the prevalence of abnormal glucose tolerance. The prevalence of T2D in Australia was 8.0% in men and 6.8% in women, and an additional 17.4% of men and 15.4% of women had IGT or IFG. Even in the youngest age group (25–34 years), 5.7% of subjects had abnormal glucose tolerance. The overall diabetes prevalence in Australia was 7.4%, and an additional 16.4% had IGT or IFG.

Compared to 1981 estimates, the prevalence of T2D had more than doubled, although this was also associated with an increase in recognized risk factors for T2D, in particular obesity and an ageing population. However these factors did not fully account for the observed increase in T2D. This may be a reflection of the limitation of BMI as a measurement of obesity, since measures of central obesity, such as waist:hip ratio or waist circumference show better correlations with insulin resistance and cardiovascular mortality |**16**|. The role of physical activity is also likely to impact on the risk of developing T2D (Sinha *et al.*, p. 104), indeed in both epidemiological and interventional studies physical activity seems to provide strong protection from the development of obesity, insulin resistance and T2D. This study demonstrates that T2D is affecting increasing numbers of younger subjects and although there is an increasing awareness of T2D, this study demonstrated that up to 50% of subjects with T2D still remain undiagnosed. This study thus further confirms the major public health issues presented by T2D, with almost a doubling in prevalence of T2D and impaired glucose metabolism over 20 years in a developed nation with a predominantly Europid background. Particularly alarming is the rapid increase seen in young people.

Relation of impaired fasting and post-load glucose with incident type 2 diabetes in a Dutch population: the Hoorn Study.

de Vegt F, Dekker JM, Jager A, *et al. JAMA* 2001; **285**: 2109–13.

Background. Persons with IGT are known to have an elevated risk of developing diabetes mellitus. Less is known about diabetes risk among persons with IFG or with normal glucose levels. The aim of this study was to determine the incidence of diabetes in relation to baseline fasting and post-load glucose levels and other risk factors. It was designed as a population-based cohort study conducted from October 1989 to February 1992 among 1342 non-diabetic white residents of Hoorn, The Netherlands, aged 50–75 years at baseline, in whom fasting plasma glucose (FPG) levels and glucose levels 2 hours after a 75-g oral glucose-tolerance test were measured at baseline and at follow-up in 1996–1998. The main outcome measure was the cumulative incidence of diabetes, defined according to the diagnostic criteria of the World Health Organization (WHO, 1985 and WHO, 1999) and the American Diabetes Association (ADA, 1997), during a mean follow-up of 6.4 years, compared among participants with IFG, IGT, and normal glucose levels at baseline. The cumulative incidence of diabetes was 6.1%, 8.3%, and 9.9% according to the WHO–1985, ADA, and WHO–1999 criteria, respectively. The cumulative incidence of diabetes (WHO–1999 criteria) for participants with both IFG and IGT was 64.5% compared with 4.5% for those with normal glucose levels at baseline. The OR values for diabetes (WHO, 1999 criteria), adjusted for age, sex, and follow-up duration, were 10.0 (95% CI 6.1–16.5), 10.9 (95% CI 6.0–19.9), and 39.5 (95% CI 17.0–92.1),

Table 5.2 Cumulative incidence of diabetes (WHO-1999 criteria) for combinations of impaired fasting and impaired post-load glucose levels*

Category	**No.**	**Cutoff values, FPG/2hPG, mg/dl**	**Cumulative incidence, No. (%)**	**Mean follow-up duration†**	**Incidence density (No./1000 person-years)**	**OR (95% CI)‡**
Normal	1125	<110/140	51 (4.5)	6.47	7.0	1.0
IFG and NGT	106	110–126/<140	35 (33.0)	6.42	51.4	10.0 (6.1–16.5)
NFG and IGT	80	<110/140–200	27 (33.8)	5.83	57.9	10.9 (6.0–19.9)
IFG and IGT	31	110–126/140–200	20 (64.5)	5.75	112.2	39.5 (17.0–92.1)

*OR indicates odds ratio; CI, confidence interval; FPG, fasting plasma glucose; IFG, impaired fasting glucose; NGT, normal glucose tolerant; NFG, normal fasting glucose; IGT, impaired glucose tolerance. To convert mg/dl to mmol/l, multiply mg/dl by 0.05551.
†Years between baseline and follow-up measurements.
‡Calculated by logistic regression adjusted for follow-up duration, age, and sex.

Source: de Vegt *et al.* (2001).

respectively, for those having isolated IFG, isolated IGT, and both IFG and IGT. In addition to FPG and 2-hour post-load glucose levels (P <0.001 for both), the waist:hip ratio also was an important risk factor for developing diabetes (P = 0.002).

INTERPRETATION. In this study, the cumulative incidence of diabetes was strongly related to both IFG and IGT at baseline and, in particular, to the combined presence of IFG and IGT.

Comment

With the rising prevalence of T2D, there is great interest in developing optimal methods of identifying subjects at risk of developing the condition. This is particularly true in view of the recent evidence demonstrating the benefits of lifestyle intervention on reducing progression to T2D in susceptible individuals. Although impaired glucose tolerance is recognized as a pre-diabetic condition, little is however known about the progression from normal glucose tolerance or impaired FPG to frank diabetes. Previous studies on the incidence of T2D were mainly performed in subjects with IGT only or in high-risk populations. This prospective study conducted over a six-year follow-up period compared the incidence of T2D among subjects with normal glucose levels, IFG, IGT or both IFG and IGT at baseline. Other potential variables predicting the development of T2D were also determined.

In this study 64.5% of subjects with both IFG and IGT progressed to develop T2D. Of those subjects with normal glucose parameters 4.5% had T2D at follow-up while 28.5% of those with IGT only had T2D and 38% of those with IFG only progressed to develop T2D.

This was a population-based study (total baseline cohort 2484). Of the 2086 subjects invited for follow-up 1513 attended and these subjects tended to be younger, less hypertensive, have lower waist:hip ratio and lower glucose levels than the population as a whole. Thus the observed prevalence of T2D may be an underestimation. Additionally, glucose levels were only measured once at baseline and follow-up and due to the variability, particularly on post-load glucose, this may lead to an increased potential for misclassification.

Table 5.3 Variables predictive of the development of diabetes during the 6 year follow-up period following adjustment for age, sex, and follow-up duration

Variable	SD (n = 1342)	OR (95% CI), per SD Difference	P Value
FPG, mg/dl	9.36	2.32 (1.85–2.90)	<0.001
2hPG, mg/dl	29.33	1.97 (1.59–2.44)	<0.001
WHR × 100	8.46	1.57 (1.19–2.08)	0.002

OR indicates odds ratio; CI, confidence interval; FPG, fasting plasma glucose; 2hPG, 2-hour post-load glucose; and WHR, waist-hip ratio.

Source: de Vegt *et al.* (2001).

The observations of this study are generally in keeping with previous studies in which the cumulative incidence of T2D in subjects with IFG ranges from 24.9–39% and between 28 and 60% in subjects with IGT [**17,18**].

IFG and IGT represent different physiological abnormalities. Fasting hyperglycaemia is primarily related to elevated rate of hepatic glucose output, while IGT is characterized by impaired insulin secretion and sensitivity. The combined presence of these disorders thus significantly increases the risk of T2D as demonstrated in this study. Waist:hip ratio rather than body mass was an important predictor of the incidence of T2D, suggesting that an assessment of body fat distribution may be a superior measure of future risk of T2D.

This study thus supports the notion that subjects at the highest risk of future T2D are those with both IFG and IGT, with elevated waist:hip ratio. These individuals thus represent an important target group for intervention.

Genome-wide search for type 2 diabetes mellitus susceptibility loci in Finnish families: the Botnia study.

Lindgren CM, Mahtani MM, Widen E, *et al. Am J Hum Genet* 2002; **70**: 509–16.

BACKGROUND. Type 2 diabetes mellitus is a heterogeneous inherited disorder characterized by chronic hyperglycaemia resulting from pancreatic beta cell dysfunction and insulin resistance. Although the pathogenic mechanisms are not fully understood, manifestation of the disease most likely requires interaction between both environmental and genetic factors. In the search for such susceptibility genes, a genome-wide scan was performed in 58 multiplex families in Finland. Initially, linkage between chromosome 12q24 and impaired insulin secretion had been reported, by Mahtani *et al.* [19] in a subsample of 26 families. In the present study, the initial genome-wide scan was extended to include 32 additional families, update the affectation status, and fine map regions of interest, and an attempt was made to replicate the initial stratification analysis. In the analysis of all 58 families, suggestive linkage to one region, chromosome 9p13–q21 (non-parametric linkage (NPL) score 3.9; *P* <0.0002) was identified. When chromosome 12q24 was analyzed in only the 32 additional families, a nominal *P* value <0.04 was observed.

INTERPRETATION. Together with data from other published genome-wide scans, these findings lend support to the hypothesis that regions on chromosome 9p13-q21 and 12q24 may harbour susceptibility genes for T2D.

Comment

T2D is increasing in prevalence and although many environmental factors such as obesity, physical inactivity, smoking and diet contribute to this increase, the role of genetic susceptibility remains unclear. Insulin resistance and beta cell dysfunction are recognized factors in the aetiology of T2D. Although the precise mechanisms are unknown, the manifestation of these factors as the clinical disease state of T2D is likely to involve interactions between both genetic and environmental factors |**17**|. In the search for potential susceptibility genes for T2D this study performed a genome-wide scan in 58 muiltiplex families (comprising 440 individuals, 229 of whom were affected) from the Botnia region in Finland.

In this analysis, a suggestive linkage to one region, chromosome 9p13–q21 (NPL score 3.9; $P < 0.0002$) was identified. Regions with nominal P values <0.05 included chromosomes 2p11 (NPL score 2.0; $P < 0.03$), 3p24–p22 (NPL score 2.2; $P < 0.02$), 4q32–q33 (NPL score 2.5; P <0.01), 12q24 (NPL score 2.1; $P < 0.03$), 16p12–11 (NPL score 1.7; $P < 0.05$), and 17p12–p11 (NPL score 1.9; $P < 0.03$). Previous studies have suggested linkage between chromosome 12q24 and impaired insulin secretion |**18**|. When this chromosome was analyzed in this study a nominal P value <0.04 was observed. This thus lends support to the hypothesis that regions on chromosome 9p13–q21 and 12q24 may harbour susceptibility genes for T2D.

A number of previously published studies have suggested a link between chromosome 12q |**18**| and T2D, with several genome-wide scans supporting the hypothesis that chromosome 12q24 might harbour a gene increasing the susceptibility to T2D. Similar supportive evidence exists for a genetic association with other loci on chromosomes 3p, 4q and 17q |**20,21**|. This study provides an extension of these data, but further detailed analysis will be required to provide a comprehensive linkage association for T2D. With the availability of dense genetic maps such an extensive analysis is becoming increasingly feasible.

Endogenous sex hormones and the development of type 2 diabetes in older men and women: the Rancho Bernardo study.

Oh JY, Barrett-Connor E, Wedick NM, Wingard DL; the Rancho Bernardo Study. *Diabetes Care* 2002; **25**: 55–60.

BACKGROUND. The aim of this study was to determine the prospective association between endogenous sex hormones and the development of T2D in older men and women. A standardized medical history was obtained, an oral glucose-tolerance test was performed, and plasma samples for sex hormones and covariates were collected from ambulatory, community-dwelling men and women at baseline from 1984 to 1987. Approximately eight years later (1992–1996), another medical history was obtained, an oral glucose-tolerance test was performed, fasting and 2-h insulin levels were measured, and the homeostasis model assessment for insulin resistance (HOMA-IR) was evaluated. This report is based on the 294 men and 233 women, aged 55–89 years, who completed both visits and who did not have diabetes as determined by history or glucose-tolerance test at baseline, as well as women who were post-menopausal and not taking replacement oestrogen. In age-adjusted correlation analyses, total testosterone was inversely and significantly related to subsequent levels of fasting and post-challenge glucose and insulin in men, whereas bioavailable testosterone and bioavailable oestradiol were positively and significantly related to fasting and post-challenge glucose and insulin in women (all *P* <0.05). There was

similar significant association with HOMA-IR in unadjusted and multiply adjusted analyses (P <0.05). There were 26 men and 17 women with new (incident) diabetes. The odds for new diabetes were 2.7 (95% CI 1.1–6.6) for men in the lowest quartile of total testosterone and 2.9 (CI 1.1–8.4) for women in the highest quartile of bioavailable testosterone.

Interpretation. Low testosterone levels in men and high testosterone levels in women predict insulin resistance and incident T2D in older adults.

Comment

Various studies have demonstrated gender differences in the effects of endogenous sex hormones on insulin resistance |**22–24**|. In men low testosterone levels are associated with obesity and insulin resistance, whereas hyperandrogenicity is associated with an insulin resistance and an increased risk of cardiovascular disease in women. Furthermore, previous studies have suggested that low testosterone and sex-hormone binding globulin (SHBG) levels predict the development of T2D in elderly and middle-aged men |**25**|. In women by contrast some data suggest that high levels of testosterone and low SHBG levels are associated with an increased risk of T2D |**26**|. Controversy exists regarding the associations between sex hormones and the risk of T2D, with studies providing conflicting evidence regarding the association between SHBG levels and T2D in men and women. There are no large population-based studies of total and bioavailable (non-SHBG bound) testosterone as predictors of insulin resistance or T2D in men and women. This study addressed these issues and sought to determine whether sex hormones predicted future T2D and whether the direction was similar in men and women.

In men total and bioavailable testosterone inversely associated with baseline BMI and waist girth, total testosterone was also inversely associated with plasma TG levels and positively correlated with baseline HDL-C levels. In women total and bioavailable oestradiol were positively correlated with BMI, waist girth and triglyceride levels. Bioavailable testosterone was positively correlated with BMI, waist girth and diastolic BP, while total testosterone positively correlated with HDL-C levels.

In this prospective study, low levels of total testosterone in men and high levels of bioavailable testosterone in women predicted the development of T2D. This observation is in agreement with previous data in which men with impaired glucose tolerance had lower levels of total testosterone and women with either impaired glucose tolerance or T2D had higher levels of bioavailable testosterone and total and bioavailable oestrogen |**27**|. Hyperandrogenicity is associated with insulin resistance in women with polcystic ovarian syndrome and may decrease insulin sensitivity via an effect on muscle glycogen synthase. In males, experimental data suggest a dimorphic relationship between testosterone and insulin resistance, with both low and abnormally high levels being associated with reduced insulin sensitivity. The biological mechanism of the insulin resistance–testosterone relationship however remains uncertain. This study demonstrated an association between higher levels of oestrogens and the development of insulin resistance and T2D in these post-menopausal women. The main source of oestrogens in post-menopausal women is from the aromatization of androgens |**28**|. Therefore the higher oestrogen levels may be a further mechanism linking hyperandrogenaemia to insulin resistance and an increased risk of T2D. Irrespective of the mechanisms concerned, this study demonstrated gender differences in the effect of sex hormones on insulin resistance and the risk of developing T2D. The identification of low levels of testosterone and high levels of bioavailable testosterone in women may aid in identifying older subjects at higher risk of developing T2D. The demonstration of low levels of testosterone in men and high levels of bioavailable testosterone and oestrogen in women may aid in the identification of subjects with insulin resistance, which is now recognized as a therapeutic target in its own right.

C-reactive protein is an independent predictor of risk for the development of diabetes in the West of Scotland Coronary Prevention Study.

Freeman DJ, Norrie J, Caslake MJ, *et al.*; West of Scotland Coronary Prevention Study. *Diabetes* 2002; **51**: 1596–600.

BACKGROUND. Accumulating evidence implicates inflammation as a potential pathway in the pathogenesis of T2D. The objective of the present study was to assess the ability of C-reactive protein (CRP) to predict the development of diabetes in middle-aged men in the West of Scotland Coronary Prevention Study (WOSCOPS). Baseline plasma samples for CRP measurement were available for 5245 men of whom 127 were classified as having a transition from normal glucose control to overt diabetes during the study, based on ADA criteria. Baseline CRP was an important

Table 5.4 Univariate and multivariate analysis of CRP quintiles as a predictor of developing diabetes, the definition of diabetes being two fasting plasma glucose measurements >7 mmol/l

	Univariate	Multivariate	Multivariate
P†	<0.0001	0.041	0.045
CRP (antilog) (mg/l)			
Quintile I: ≤0.66	Referent level	Referent level	Referent level
Quintile II: >0.66 to ≤1.28	2.30 (0.94–5.58)	1.76 (0.72–4.30)	1.35 (0.62–2.94)
Quintile III: >1.28 to ≤2.27	3.60 (1.56–8.28)	2.35 (1.01–5.46)	2.01 (0.98–4.13)
Quintile IV: >2.27 to ≤4.18	4.64 (2.05–10.5)	2.75 (1.19–6.36)	2.27 (1.11–4.64)
Quintile V: >4.18	6.13 (2.76–13.6)	3.07 (1.33–7.10)	2.46 (1.20–5.04)

Data are hazard ratios (95% CIs). Multivariate analysis included all the covariates regardless of statistical significance on univariate analysis. †*P* for the χ^2 test on four degrees of freedom for the equality of the five quintiles.

Source: Freeman *et al.* (2002).

predictor of the development of diabetes in univariate analysis (hazard ratio [HR] for an increase of 1 SD = 1.55; 95% CI 1.32–1.82; *P* <0.0001). In multivariate analysis, CRP remained a predictor of diabetes development (HR 1.30; 95% CI 1.07–1.58; *P* = 0.0075) independent of other clinically employed predictors, including baseline BMI and fasting TG and glucose concentrations. Moreover, there was a graded increase in risk across CRP quintiles throughout the study, evident at even one year of follow-up. The highest quintile (CRP >4.18 mg/l) was associated with a greater than three-fold risk of developing diabetes (HR 3.07; 95% CI 1.33–7.10) in a multivariate analysis at five years. Thus, CRP predicts the development of T2D in middle-aged men independently of established risk factors.

Interpretation. Because CRP, the most commonly used acute-phase protein in clinical practice, is very stable in serum, these observations have clinical potential in helping to better predict individuals destined to develop T2D. They also add to the notion that low-grade inflammation is important in the pathogenesis of T2D.

Comment

Enhanced inflammation is a recognized association of the IRS |**29**|. Accumulating evidence implicates inflammation as an important factor in the pathogenesis of cardiovascular disease and T2D. CRP is the most commonly used acute phase reactant marker of inflammation. Cross-sectional studies have demonstrated correla-

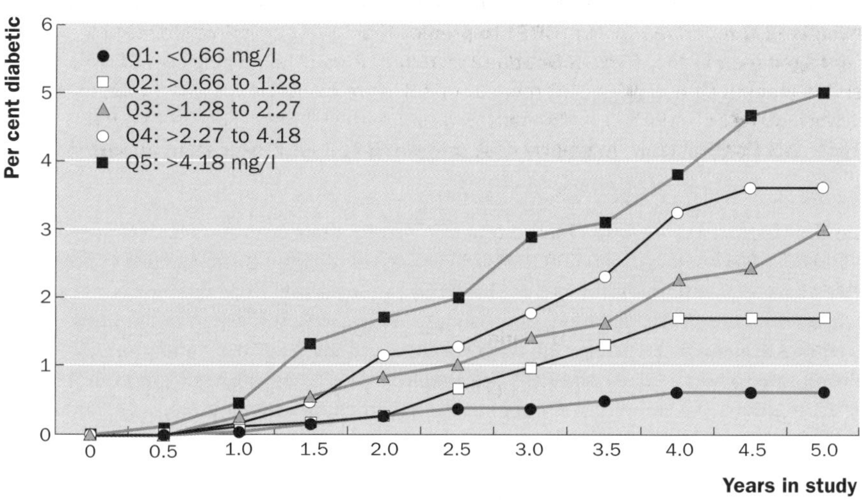

Fig. 5.1 Kaplan-Meier plot of time to diabetes by quintiles of $\log_e$ CRP. The time to becoming diabetic was taken as the semi-annual visit at which they had at least two post-randomization plasma glucose levels ⩾7 mmol/l and at least one plasma glucose >2 mmol/l above baseline glucose, or the post-randomization visit at which they first indicated taking hypoglycaemic agents. Source: Freeman *et al.* (2002).

tions between CRP levels and obesity, insulin resistance, hypertriglyceridaemia and low HDL-C levels |**29**|. Furthermore recent data from the Women's Health Study demonstrated CRP to predict the development of T2D in women (Pradhan *et al.*, see below). The objective of this study was to assess the role of a baseline CRP measurement in predicting the development of T2D in men, in conjunction with other established predictors such as hypertriglyceridaemia, over the five-year follow-up period of the WOSCOPS.

This study demonstrated for the first time that raised CRP levels may predict the development of T2D in middle-aged men, independent of other established risk factors such as hypertriglyceridaemia and obesity. In multivariate analysis a 1 SD change in CRP concentration was associated with an HR similar to those of BMI, plasma glucose and TG concentrations, factors known to predict diabetes. In addition men in the top quartile of CRP (>4 .18 mg/l) had a greater than three-fold risk of developing T2D compared to those in the lowest quartile (<0.66 mg/l). Although this study was a *post hoc* analysis of the WOSCOPS data, in which the primary end-point was cardiovascular disease and not the development of T2D, these findings support similar observations in women and the elderly regarding the association between CRP levels and risk of T2D. CRP levels thus appear to be a useful predictor of T2D. Because of its stability in serum and the availability of an easy and internationally standardized method of measurement, CRP may be of use in the assessment of patients at risk of developing T2D. Furthermore, this study also adds to the notion that enhanced inflammation is an important factor in the pathogenesis of T2D.

C-reactive protein, interleukin 6, and risk of developing type 2 diabetes mellitus.

Pradhan AD, Manson JE, Rifai N, Buring JE, Ridker PM. *JAMA* 2001; **286**: 327–34.

BACKGROUND. Inflammation is hypothesized to play a role in development of type 2 diabetes mellitus (DM); however, clinical data addressing this issue are limited. The aim of this study was to determine whether elevated levels of the inflammatory markers interleukin 6 (IL-6) and CRP are associated with development of type 2 DM in healthy middle-aged women. A prospective, nested case–control study was carried out within the Women's Health Study, an ongoing US primary prevention, randomized clinical trial initiated in 1992. From a nationwide cohort of 27 628 women free of diagnosed DM, cardiovascular disease, and cancer at baseline, 188 women who developed diagnosed DM over a 4-year follow-up period were defined as cases and matched by age and fasting status with 362 disease-free controls. The main outcome measures were incidence of confirmed clinically diagnosed type 2 DM assessed by baseline levels of IL-6 and CRP. Baseline levels of IL-6 ($P<0.001$) and CRP ($P<0.001$) were significantly higher among cases than among controls. The relative risks of future DM for women in the highest vs lowest quartile of these inflammatory markers were 7.5 for IL-6 (95% CI 3.7–15.4) and 15.7 for CRP (95% CI 6.5–37.9). Positive

associations persisted after adjustment for BMI, family history of diabetes, smoking, exercise, use of alcohol, and hormone replacement therapy; multivariate relative risks for the highest *vs* lowest quartiles were 2.3 for IL-6 (95% CI 0.9–5.6; *P* for trend = 0.07) and 4.2 for CRP (95% CI 1.5–12.0; P for trend = 0.001). Similar results were observed in analyses limited to women with a baseline haemoglobin A(1c) of 6.0% or less and after adjustment for fasting insulin level.

Interpretation. Elevated levels of CRP and IL-6 predict the development of type 2 DM. These data support a possible role for inflammation in diabetogenesis.

Comment

Although the main physiological abnormalities in T2D are insulin resistance and beta cell function, the underlying determinants of these metabolic defects remain uncertain. An accumulating body of evidence suggests that inflammation may play a crucial role in these processes |**29**|. CRP is a stable and most commonly used marker of inflammation. Interleukin-6 (IL-6) is a major pro-inflammatory cytokine, produced in a variety of tissues, including activated leucocytes, endothelial cells and adipocytes. CRP is the principal downstream mediator of inflammatory responses mediated by IL-6. Experimental models have demonstrated that IL-6-mediated inflammation may promote hepatic gluconeogenesis, hyperglycaemia and insulin resistance |**30**|. With these issues in mind this study sought to assess whether enhanced inflammation as measured by plasma levels of CRP and IL-6 predicated the future risk of developing T2D.

This study was conducted in the setting of the Women's Health Study, an ongoing US primary prevention, randomized clinical trial initiated in 1992. From a

Table 5.5 Relative risk of type 2 diabetes according to baseline concentration of plasma interleukin-6 and CRP following adjustment for insulin levels in addition to clinical risk factors*

Inflammatory marker	Quartile of plasma biomarker level				*P* for trend
	1	2	3	4	
Interleukin 6					
Relative risk (95% CI)	1.0	0.7 (0.2–2.6)	0.9 (0.3–2.9)	1.5 (0.5–4.8)	0.23
P value		0.56	0.89	0.51	
C-reactive protein					
Relative risk (95% CI)	1.0	0.9 (0.2–4.3)	3.1 (0.8–12.2)	4.3 (1.1–17.1)	0.01
P value		0.91	0.10	0.04	

*Matched on age and fasting status, controlled for BMI, family history of diabetes, smoking, alcohol consumption, physical activity, and hormone replacement therapy. Limited to subjects providing fasting blood specimens with haemoglobin A_{1c} of 6.5% or less: 126 cases, 225 controls. CI indicates confidence interval.

Source: Pradhan *et al.* (2001).

nationwide cohort of 27 628 women free of diagnosed DM, cardiovascular disease, and cancer at baseline, 188 women who developed diagnosed DM over a 4-year follow-up period were defined as cases and matched by age and fasting status with 362 disease-free controls. In addition randomized treatment assignment to vitamin E or low dose aspirin appeared to have no impact on the observed risk relationships during the 4-year observation period. This study was composed of primarily healthy middle-aged women and the results may thus not be generalizable to other patient groups. Undetected T2D at baseline may also have biased the results, this potential was minimized by the exclusion of individuals with a baseline HbA_{1c} >6.5%, whilst all women diagnosed with T2D in the first year of the study were also excluded. Adjustment was made for the common measures of obesity, namely BMI, waist:hip ratio and waist circumference, however statistical modelling using these parameters may not fully account for the metabolic effects of obesity.

In this prospective study of apparently healthy middle-aged women, two markers of systemic inflammation, CRP and IL-6 were found to determine the risk of developing T2D. CRP appeared to be a powerful independent predictor after adjustment for obesity, clinical risk factors and fasting insulin levels. Although this study further supports an aetiological role for inflammation and in particular CRP in T2D, potential mechanisms remain speculative. It is possible that these associations reflect underlying atherosclerosis; particularly since elevated CRP levels are a recognized feature of atherogeneis |**31**|. However the 4-year cardiovascular event rate was similar in both groups in this study, even in those with highest CRP levels. Obesity-mediated cytokine production is another potentially important mechanism. Visceral fat cells have been shown to produce 2–3 times more inflammatory cytokines than subcutaneous tissue |**32**|. In the present analysis BMI as a principal measure of obesity attenuated relative risk (RR) estimates for both CRP and IL-6. Adjustment for waist:hip ratio and waist circumference, indices of abdominal fat mass similarly weakened the risk relationships. However, significant associations for both IL-6 and CRP were maintained after adjusting for these factors and a stepwise RR gradient was evident even among non-obese subjects. Enhanced inflammation associates with endothelial dysfunction, which may be attenuated with aspirin |**33**|. Altered endothelial permeability and diminished peripheral blood flow may limit insulin delivery and promote insulin resistance in metabolically active tissues. Elevated IL-6 and CRP levels may represent adipocyte activation. Other potential mediators of insulin resistance arising from adipocyte metabolism include free fatty acids, tumour necrosis α, leptin and resistin |**34**|. This study provides further epidemiological evidence supporting a role for inflammation in the pathogenesis of T2D and further raises the possibility that measurement of CRP may provide an adjunctive method for early detection of subjects at risk of T2D.

Elevated levels of acute-phase proteins and plasminogen activator inhibitor-1 predict the development of type 2 diabetes: the insulin resistance atherosclerosis study.

Festa A, D'Agostino R Jr, Tracy RP, Haffner SM; the Insulin Resistance Atherosclerosis study. *Diabetes* 2002; **51**: 1131–7.

BACKGROUND AND INTERPRETATION. Elevated serum levels of acute-phase proteins, indicating chronic subclinical inflammation, have been associated with cardiovascular disease as well as the IRS. Chronic inflammation may also be a risk factor for developing T2D. The authors studied the concentrations of CRP, fibrinogen, and plasminogen activator inhibitor-1 (PAI-1) in 1047 non-diabetic subjects in relation to incident diabetes within five years in the Insulin Resistance Atherosclerosis Study. Subjects with diabetes at follow-up (n = 144) had higher baseline levels of fibrinogen (mean ± SD; 287.8 ± 58.8 *vs* 275.1 ± 56.0 mg/dl; P = 0.013) as well as of CRP (median [interquartile range]; 2.40 [1.29, 5.87] *vs* 1.67 mg/l [0.75, 3.41]; P = 0.0001) and PAI-1 (24 [15, 37.5] *vs* 16 ng/ml [9, 27]; P = 0.0001) than non-converters. The OR of converting to diabetes was significantly increased with increasing baseline concentrations of the inflammatory markers. In contrast to PAI-1, the association of CRP and fibrinogen with incident diabetes was significantly attenuated after adjustment for body fat (BMI or waist circumference) or insulin sensitivity (S(I)), as assessed by a frequently sampled intravenous glucose-tolerance test. In a logistic regression model that included age, gender, ethnicity, clinical centre, smoking, BMI, S(I), physical activity, and family history of diabetes, PAI-1 still remained significantly related to incident T2D (OR [95% CI] for 1 SD increase: 1.61 [1.20–2.16]; P = 0.002). Chronic inflammation emerges as a new risk factor for the development of T2D; PAI-1 predicts T2D independent of insulin resistance and other known risk factors for diabetes.

Comment

Various experimental and epidemiological studies have demonstrated an association between measures of inflammation, in particular CRP, with both cardiovascular disease and incident T2D. It has been hypothesized that T2D and cardiovascular disease represent manifestations of a 'common soil' pathophysiology. Inflammation and abnormal haemostasis relating to insulin resistance may represent such a process [**35**]. Inflammatory markers such as CRP, high fibrinogen, high white cell count and low albumin along with markers of altered haemostasis such as elevated factor VII and PAI-1 levels have all been linked to the development of both cardiovascular disease and T2D [**29**]. This study sought to investigate the relationship between CRP, PAI-1 and fibrinogen to insulin resistance and incident T2D during a five-year period.

Concentrations of CRP, fibrinogen, and PAI-1 were measured in 1047 non-diabetic subjects in relation to incident diabetes within five years in the Insulin Resistance Atherosclerosis Study. The results further support the role of chronic

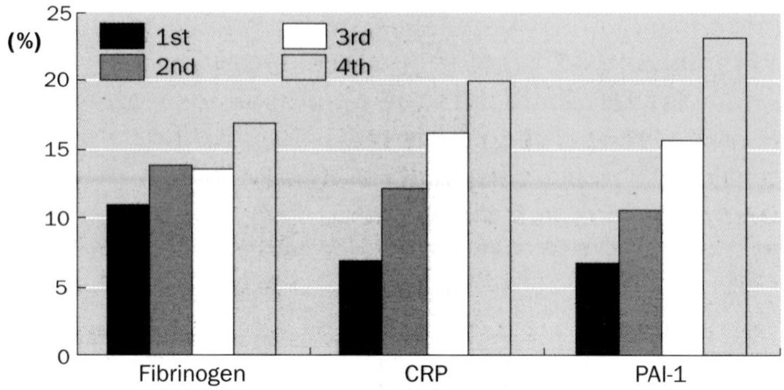

Fig. 5.2 Incidence of type 2 diabetes according to stratification by quartiles of inflammatory markers. *P* values for χ^2 were 0.001 for PAI-1 and CRP and 0.06 for fibrinogen. Source: Festa *et al.* (2002).

inflammation in the pathogenesis of T2D and suggest that PAI-1 may be a superior independent predictor of incident T2D.

This study demonstrated a significant relationship between CRP, fibrinogen and PAI-1 levels with the development of T2D, which was consistent among men and women, obese and non-obese subjects. The association of PAI-1 was particularly strong and independent of several other known risk factors for T2D. In particular this association was independent of both body fat and insulin resistance, unlike other studies of CRP and incident T2D, where the association was attenuated by measures of obesity |**36**|. The relation of PAI-1 to incident T2D was stronger in subjects with normal glucose tolerance. In subjects with IGT, the relation may be confounded by other features of insulin resistance, which associate with both PAI-levels and T2D. Glycaemia also attenuated the relation of PAI-1 to incident T2D, which may reflect the effect of glucose on PAI-1 gene expression. These observations may be of particular relevance by offering the potential to identify a population with a high risk of developing T2D (high PAI-1 levels) among subjects with an *a priori* low risk (normal glucose tolerance). Both PAI-1 and CRP levels predict the development of atherosclerosis. Based on the results of this study we may speculate that the common antecedent for atherosclerosis, insulin resistance and T2D may be chronic inflammation with PAI-1 over expression. If this were the case then PAI-1 levels would represent an important marker for targeting therapeutic intervention. Indeed a variety of studies have shown that interventions that reduce the risk of T2D and cardiovascular disease such as weight loss, statin therapy, acetylcholine esterase (ACE)-inhibitors and metformin |**37,38**| also reduce PAI-1 levels. Thus high PAI-1 levels in subjects with normal glucose tolerance may aid in identifying a high-risk population with the potential to prevent both cardiovascular disease and T2D.

High white blood cell count is associated with a worsening of insulin sensitivity and predicts the development of type 2 diabetes.

Vozarova B, Weyer C, Lindsay RS, Pratley RE, Bogardus C, Tataranni PA. *Diabetes* 2002; **51**: 455–61.

BACKGROUND AND INTERPRETATION. Chronic low-grade inflammation may be involved in the pathogenesis of insulin resistance and T2D. This study examined whether a high white blood cell count (WBC), a marker of inflammation, predicts a worsening of insulin action, insulin secretory function, and the development of T2D in Pima Indians. The investigators measured WBC in 352 non-diabetic Pima Indians (215 men and 137 women, aged 27 ± 6 years [means ± SD], body fat 32 ± 8%, WBC 8107 ± 2022 cells/mm^3) who were characterized for body composition (by hydrodensitometry or dual-energy X-ray absorptiometry), glucose tolerance (by 75-g oral glucose-tolerance test), insulin action (M; by hyperinsulinaemic clamp), and acute insulin secretory response (AIR; by 25-g intravenous glucose challenge). Among 272 subjects who were normal glucose tolerant (NGT) at baseline, 54 developed diabetes over an average follow-up of 5.5 ± 4.4 years. Among those who remained non-diabetic, 81 subjects had follow-up measurements of M and AIR. Cross-sectionally, WBC was related to per cent body fat (r = 0.32, P <0.0001) and M (r = –0.24, P <0.0001), but not to AIR (r = 0.06, P = 0.4). In a multivariate analysis, when adjusted for age and sex, both percentage body fat (P <0.0001) and M (P = 0.03) were independently associated with WBC. A high WBC value predicted diabetes (RR 90th *vs* 10th percentiles [95% CI] of 2.7 [1.3–5.4], P = 0.007) when adjusted for age and sex. The predictive effect of WBC persisted after additional adjustment for established predictors of diabetes, i.e. percentage body fat, M, and AIR (RR 2.6 [1.1-6.2], P = 0.03). After adjustment for follow-up duration, a high WBC at baseline was associated with a subsequent worsening of M (P = 0.003), but not a worsening of AIR. A high WBC predicts a worsening of M and the development of T2D in Pima Indians. These findings are consistent with the hypothesis that a chronic activation of the immune system may play a role in the pathogenesis of T2D.

Comment

Various studies have demonstrated the potential importance of low-grade inflammation in the pathogenesis of T2D. Elevated cytokine levels, C-reactive protein and PAI-1 levels have all been shown to predict incident T2D. Elevated WBC is also a marker of enhanced inflammation. This study examined the role of an elevated WBC in the development of T2D in Pima Indians, a population characterized by severe insulin resistance. This study specifically aimed to assess whether a high WBC was associated with later development of increasing insulin resistance and

T2D and if so whether these associations were independent of factors such as insulin action, secretion and adipocity.

This study demonstrated that among Pima Indians with normal glucose tolerance a high WBC predicted the future development of T2D and was associated with a decline in insulin sensitivity. This study supports previous data in which elevated WBC was seen to predict T2D, the predictive nature of leucocytosis in this study was however independent of recognized risk factors for T2D such as adiposity, insulin resistance and insulin secretion. The mechanisms underlying this apparent association between activation of the immune system as measured by an elevated WBC with insulin resistance and T2D remain speculative. It may be that this observation represents a manifestation of the 'thrifty gene hypothesis'. An insulin resistant geneotype associated with an activated immune system may have been historically advantageous when life span was short and infections and injury common. In modern times this selection advantage no longer exists, resulting in the expression of this genotype in the form of insulin resistance and T2D |**39**|. Another possible explanation is that higher WBC and insulin resistance may reflect underlying activation of the immune system, mediated by inflammatory cytokines such as IL-6 |**40**|. The findings of this study provide further support for the notion that inflammation and chronic activation of the immune system are important in the pathogenesis of T2D.

Relation between a diet with a high glycemic load and plasma concentrations of high-sensitivity C-reactive protein in middle-aged women.

Liu S, Manson JE, Buring JE, Stampfer MJ, Willett WC, Ridker PM. *Am J Clin Nutr* 2002; **75**: 492–8.

BACKGROUND. Recent prospective data suggest that intake of rapidly digested and absorbed carbohydrates with a high dietary glycaemic load is associated with an increased risk of ischaemic heart disease. This study examined whether a high dietary glycaemic load was associated with elevated high-sensitivity C-reactive protein (hs-CRP) concentrations and whether this association was modified by BMI (in kg/m^2). In 244 apparently healthy women, plasma hs-CRP concentrations were measured and average dietary glycaemic loads determined with a validated semiquantitative food-frequency questionnaire. Using multiple regression models, the association between dietary glycaemic load and plasma hs-CRP was evaluated after adjusting for age, treatment status, smoking status, BMI, physical activity level, parental history of myocardial infarction, history of hypertension, diabetes, and high cholesterol, post-menopausal hormone use, alcohol intake and other dietary variables. A strong and statistically significant positive association between dietary glycaemic load and plasma hs-CRP was found. The median hs-CRP concentration for the lowest quintile of dietary glycaemic load was 1.9 mg/l and for the highest quintile was 3.7 mg/l; corresponding multivariate-adjusted geometric means were 1.4 and 3.8 mg/l, respectively (*P* for trend <0.01). This association was significantly modified by BMI.

Among women with a BMI ≥25, the multivariate-adjusted geometric mean hs-CRP concentration in the lowest quintile was 1.6 mg/l and in the highest quintile was 5.0 mg/l; however, among women with a BMI <25, the corresponding means were 1.1 and 3.1 mg/l, respectively (*P* = 0.01 for interaction).

INTERPRETATION. Dietary glycaemic load is significantly and positively associated with plasma hs-CRP in healthy middle-aged women, independent of conventional risk factors for ischaemic heart disease. Exacerbation of the proinflammatory process may be a mechanism whereby a high intake of rapidly digested and absorbed carbohydrates increases the risk of ischaemic heart disease, especially in overweight women prone to insulin resistance.

Comment

These data suggest that dietary glycaemic load is significantly and positively associated with plasma CRP concentrations in healthy middle-aged women, particularly in overweight subjects. Exacerbation of the proinflammatory process may thus be a mechanism whereby a high intake of rapidly digested and absorbed carbohydrates may increase the risk of insulin resistance, cardiovascular disease and T2D, especially in obese subjects that may already exhibit inappropriate inflammatory responses, predisposing to insulin resistance. Attenuated inflammatory responses consequent upon reduced intake of rapidly absorbed carbohydrates, may therefore provide a mechanism for the improved metabolic status observed in lifestyle intervention studies such as the Diabetes Prospective Prevention Program.

Pravastatin and the development of diabetes mellitus: evidence for a protective treatment effect in the West of Scotland Coronary Prevention Study.

Freeman DJ, Norrie J, Sattar N, *et al. Circulation* 2000; **103**: 357–62.

BACKGROUND. **This study examined the development of new DM in men aged 45 to 64 years during the WOSCOPS. The definition of DM was based on the ADA threshold of a blood glucose level of ≥7.0 mmol/l. Subjects who self-reported diabetes at baseline or had a baseline glucose level of ≥7.0 mmol/l were excluded from the analyses. A total of 5974 of the 6595 randomized subjects were included in the analysis, and 139 subjects became diabetic during the study. The baseline predictors of the transition from normal glucose control to diabetes were studied. In the univariate model, BMI, log TG, log WBC, systolic blood pressure, total and HDL cholesterol, glucose, and randomized treatment assignment to pravastatin were significant predictors. In a multivariate model, BMI, log TG, glucose, and pravastatin therapy were retained as predictors of diabetes in this cohort.**

INTERPRETATION. It was concluded that the assignment to pravastatin therapy resulted in a 30% reduction (*P* =0.042) in the risk of becoming diabetic. By lowering plasma TG levels, pravastatin therapy may favourably influence the development of diabetes, but other explanations, such as the anti-inflammatory properties of this drug in combination with its endothelial effects, cannot be excluded with these analyses.

Comment

Factors that influence the development of diabetes mellitus have been the subject of considerable research. Blood lipid and lipoprotein levels have been confirmed as important predictors of this pathology, and more recently, markers of low-grade inflammation have been shown to predict the occurrence of T2D. Despite these strong associations, there have been very few investigations into the effects of drugs that influence the blood lipid profile or that have anti-inflammatory actions on the development of diabetes. One such drug that exhibits potentially beneficial pleiotropic effects on both plasma lipids and the inflammatory response is pravastatin. In patients with T2D and impaired glucose tolerance pravastatin has been shown to reduce cardiovascular mortality, but no attempt had been made to examine the effects of pravastatin on glycaemia. This study used the WOSCOPS database to study prospectively the effects of pravastatin therapy on the risk of developing diabetes in subjects with follow-up ranging from 3.5 to 6.1 years.

T2D was defined based on the ADA threshold of a blood glucose level of ≥7.0 mmol/l. Subjects who self-reported diabetes at baseline or had a baseline glucose level of ≥7.0 mmol/l were excluded from the analyses. The baseline predictors of the transition from normal glucose control to diabetes were studied. The principal findings of this study were that with a definition based on serial fasting glucose measurements and the ADA criteria, the well-recognized predictors of loss of glucose control such as BMI and plasma TG level were further confirmed. In addition to establishing these variables as independent predictors, an association between pravastatin therapy and a reduced risk of developing diabetes was also identified.

The observed effects of pravastatin on incident T2D were generated from a *post hoc* analysis of WOSCOPS and as such, should be treated as hypothesis generating, rather than absolute clinical fact. However they raise important new clinical possibilities, which should now be explored in further prospective studies and trials, such as the ongoing Prospective Study of Pravastatin in the Elderly at Risk (PROSPER). There are however a variety of potential mechanisms, which may account for any effect of pravastatin on the risk of developing T2D. First, the TG-lowering effect of pravastatin therapy may be important over the long term in reducing insulin resistance. Secondly, the anti-inflammatory effects of pravastatin may be important. Pravastatin has been shown to reduce circulating levels of the cytokines interleukin-6 and TNF-α and C-reactive protein. Growing evidence supports the role of inflammatory cytokines, derived in part from adipose tissue, in the pathogenesis of insulin resistance and the development of T2D. The anti-inflammatory properties of pravastatin may therefore interrupt the natural progression from central obesity to insulin resistance and T2D mediated by the adipose tissue-derived cytokines. Finally, the effects of pravastatin on the risk of developing T2D may be related to its influence on vascular function. Pravastatin has been consistently shown to improve endothelial function; hence by restoring endothelial function, pravastatin may improve selective tissue perfusion and thereby beneficially affect glucose and insulin transport and thus insulin resistance.

Early differential defects of insulin secretion and action in 19-year-old Caucasian men who had low birth weight.

Jensen CB, Storgaard H, Dela F, Holst JJ, Madsbad S, Vaag AA. *Diabetes* 2002; **51**: 1271–80.

BACKGROUND AND INTERPRETATION. Several studies have linked low birth weight (LBW) and T2D. This study investigated hepatic and peripheral insulin action including intracellular glucose metabolism in 40 19-year-old men (20 LBW, 20 matched control subjects), using the hyperinsulinaemic-euglycaemic clamp technique at two physiological insulin levels (10 and 40 mU/m^2/min), indirect calorimetry, and [3-^{3}H]glucose. Insulin secretion was examined during an oral and intravenous glucose-tolerance test. Fasting plasma glucose was higher in the LBW group (5.6 $\pm$ 0.1 *vs* 5.4 $\pm$ 0.1; P <0.05). Basal plasma glycerol concentrations were significantly lower in the LBW group. Insulin-stimulated glycolytic flux was significantly reduced, and suppression of endogenous glucose production was enhanced in the LBW group. Nevertheless, basal and insulin-stimulated rates of whole-body peripheral glucose disposal, glucose oxidation, lipid oxidation, exogenous glucose storage, and non-oxidative glucose metabolism were similar in the two groups. Insulin secretion was reduced by 30% in the LBW group, when expressed relative to insulin sensitivity (disposition index = insulin secretion $\times$ insulin action). It was proposed that reduced insulin-stimulated glycolysis precedes overt insulin resistance in LBW men. A lower insulin secretion may contribute to impaired glucose tolerance and ultimately lead to diabetes.

Comment

Various studies have linked birth weight or indices of poor fetal growth with later development of IGT and T2D. Such observations have led to the so called 'Barker' hypothesis in which IGT and T2D may arise as a result of persistent changes in pancreatic structure or function as a result of adverse environmental exposure at critical periods of development. Indeed several studies have demonstrated that subjects with previous LBW demonstrate insulin resistance; there is however no evidence relating to the effects of LBW on insulin secretion or action. The aim of this study was to resolve whether a birth weight below the 10th centile was associated with impaired insulin secretion or action in young Caucasian men with normal glucose tolerance. The results suggest that LBW may result in reduced insulin secretion, which may thus predispose to the future development of both IGT and T2D.

Basal and insulin stimulated rates of glucose disposal were similar in both LBW and control groups, implying no difference in insulin sensitivity between the two groups. Fasting plasma and 2-hour post-load glucose was higher in the LBW group, while insulin stimulated glycolysis was reduced along with a reduction in glucose stimulated insulin secretion (when adjusted for insulin sensitivity). In addition, the LBW subjects demonstrated lower fasting and steady state glycerol concentrations, possibly reflecting a lower lipolytic rate and also exhibited enhanced suppression of glucose production during high physiological insulin concentrations. These observations suggest that in young men with previous LBW, insulin secretion may be impaired but insulin sensitivity is maintained as consequence of compensatory mechanisms. The effect of other factors such as westernised diet and sedentary life style may impact upon these mechanisms, programmed in fetal life, resulting in a tendency to develop T2D.

Effects of size at birth and childhood growth on the insulin resistance syndrome in elderly individuals.

Eriksson JG, Forsen T, Tuomilehto J, Jaddoe VW, Osmond C, Barker DJ. *Diabetologia* 2002; **45**: 342–8.

BACKGROUND. A study of 7086 men and women born in Helsinki, Finland, has shown that the development of type 2 (non-insulin-dependent) diabetes mellitus is associated with LBW followed by accelerated gain in height and weight during childhood and with high maternal BMI but the processes which underlie these associations are largely not known. These investigators carried out standard oral glucose-tolerance tests, and measured plasma insulin and proinsulin, serum lipid concentrations and blood pressure in 474 patients from the Helsinki cohort. They used four indices of insulin resistance: fasting and 2-h plasma insulin, and fasting proinsulin and 32-33 split proinsulin concentrations. These were associated with small body size at birth and during childhood, rapid growth in height and low maternal BMI.

INTERPRETATION: Insulin resistance and T2D share common associations with retarded fetal growth and accelerated growth during childhood. They are dissimilar, however, in that insulin resistance is associated with thinness in childhood and low maternal BMI, while T2D is associated with high BMI in childhood and high maternal BMI.

Comment

LBW with subsequent accelerated childhood growth has recently been identified as a potential factor predisposing to insulin resistance and the development of T2D in adult life |**41**|. The aim of this study was to assess whether the components of the IRS form the link between birth weight, childhood growth and the future development of T2D.

Indices of insulin resistance were associated with LBW and low BMI in childhood, after current BMI had been taken into consideration. Low childhood BMI may reflect a low lean body mass, which persists into adult life so that for any given BMI there will be a raised fat to muscle ratio predisposing to insulin resistance and

Table 5.6 Mean values of components of the insulin resistance syndrome, according to BMI at 7 years of age

	BMI (kg/m^2) at 7 years						
	<14.3	**−14.9**	**−15.5**	**−16.2**	**>17.5**	**All**	***P***
Patients (*n*)	98	89	96	91	100	474	
Fasting glucose (mmol/l)[a]	5.6	5.5	5.5	5.5	5.4	5.5	0.06
2-hour glucose (mmol/l)[a]	7.8	7.7	7.6	7.5	7.7	7.7	0.29
Fasting insulin (pmol/l)[a]	69	73	62	68	63	67	0.21
2-hour insulin (pmol/l)[a]	550	590	433	485	457	498	0.005
Acute insulin response (pmol/l)	484	487	445	481	473	474	0.922
Fasting proinsulin (pmol/l)[a]	3.3	3.4	2.7	3.1	2.7	3.0	0.01
Fasting 32–33 split proinsulin (pmol/l)[a]	8.6	8.7	7.3	7.6	7.7	8.0	0.12
Triglyceride (mmol/l)[a]	1.31	1.45	1.22	1.30	1.25	1.30	0.22
LDL cholesterol (mmol/l)	3.8	3.9	3.8	3.9	3.8	3.8	0.95
HDL cholesterol (mmol/l)[a]	1.43	1.40	1.47	1.42	1.45	1.43	0.97
Systolic pressure (mmHg)	162	159	159	158	154	158	0.005
Diastolic pressure (mmHg)	90	91	89	89	86	89	0.002

Values adjusted for age, sex and adult BMI.
[a]Values are geometric means.
P values are given for trends for each component of the metabolic syndrome.

Source: Eriksson *et al.* (2002).

Table 5.7 Mean values of components of the insulin resistance syndrome, according to change in standard deviation score for height between 7 and 15 years

	Change in standard deviation scores for height						
	>0.17 fall	**0.17 fall to 0.06 rise**	**0.06 to 0.31 rise**	**0.31 to 0.63 rise**	**>0.63 rise**	**All**	***P***
Patients (*n*)	94	95	95	95	95	474	
Fasting glucose (mmol/l)[a]	5.5	5.4	5.5	5.7	5.6	5.5	0.19
2-h glucose (mmol/l)[a]	7.7	7.4	7.8	7.6	7.7	7.7	0.80
Fasting insulin (pmol/l)[a]	61	71	62	69	72	67	0.04
2-h insulin (pmol/l)[a]	442	517	516	519	501	498	0.05
Acute insulin response (pmol/l)	445	455	471	518	479	474	0.28
Fasting proinsulin (pmol/l)[a]	2.8	2.9	2.6	3.6	3.3	3.0	0.04
Fasting 32–33 split proinsulin (pmol/l)[a]	6.9	7.6	7.1	10.2	8.6	8.0	0.001
Triglyceride (mmol/l)[a]	1.29	1.28	1.32	1.30	1.32	1.30	0.32
LDL cholesterol (mmol/l)[a]	3.8	4.0	3.9	3.8	3.8	3.8	0.34
HDL cholesterol (mmol/l)[a]	1.49	1.47	1.42	1.37	1.43	1.43	0.04
Systolic pressure (mmHg)	159	160	159	157	157	158	0.43
Diastolic pressure (mmHg)	88	90	89	89	88	89	0.75

Values are adjusted for age, sex and adult BMI.
[a]Values are geometric means.
P values are given for trends for each component of the metabolic syndrome.

Source: Eriksson *et al.* (2002).

T2D. Rapid childhood growth (7–15 years) was also associated with all indices of insulin resistance. This lends support to the hypothesis that insulin resistance may be the result of over activation of the GH-IGF-1 system in association with accelerated post-natal growth. This study also demonstrated that low maternal BMI in late pregnancy was associated with insulin resistance in the offspring. This association appeared to be independent of birth weight and childhood BMI. The mechanisms are unclear but low protein turnover may be one possibility. Consistent with the importance of insulin resistance in the aetiology of T2D, both insulin resistance and T2D were associated with the same childhood and maternal factors, although T2D was also associated with increased maternal BMI. This may be a reflection of relative maternal hyperglycaemia with subsequent effects on fetal beta cell development and function. Interestingly low childhood BMI also seemed to associate with adult systolic and diastolic BP suggesting that the fetal and childhood factors may account for the relationship between insulin resistance and hypertension.

Summary

T2D is becoming an increasingly common condition worldwide, with an alarming increase in the prevalence of impaired glucose tolerance and T2D in young adults, adolescents and children (Sinha *et al.*, p. 104). T2D is a polygenic condition, resulting from interactions between environmental and genetic factors |**41,42**| (Lindgren *et al.*, p. 113). Although the genetics of T2D remain unclear, data from linkage analysis and genome-wide scans are beginning to identify chromosomal regions, which may associate with T2D. In order to fully elucidate the genetics of T2D, comprehensive linkage analysis using very large data sources, combined with dense genetic mapping are required.

The fetal environment resulting in potential functional and structural abnormalities in insulin metabolism and vascular function |**43**| (Jensen *et al.*, p. 127), manifest as LBW are increasingly recognized as a potentially important factor in the aetiology and pathogenesis of both T2D and cardiovascular disease. Indeed a number of studies have demonstrated that young adults with LBW demonstrate endothelial dysfunction and insulin resistance |**44**| (Eriksson *et al.*, p. 128) both of which are recognized risk factors for the development of T2D. Recent evidence suggests that microvascular dysfunction and impaired insulin secretion are early events in the processes linking fetal environment to the future development of T2D and cardiovascular disease.

The alarming increase in the frequency of T2D, particularly in the young, is largely related to a huge increase in the prevalence of obesity and associated insulin resistance. Indeed, the importance of obesity and insulin resistance as a therapeutic target has been recently recognised by the National Cholesterol Education Programme Adult Treatment Panel III guidelines.

T2D and cardiovascular disease may be considered as two phenotypic manifestations of a 'common soil' pathogenesis. Insulin resistance associated with enhanced inflammation as measured by increased concentrations of pro-inflammatory cytokines and elevated WBC represents a potential mechanism linking cardiovascular disease and T2D. A growing body of evidence supports this concept, with various studies demonstrating that elevated levels of inflammatory cytokines such as CRP, IL-6 and PAI-1 correlate with hypertension, cardiovascular disease and incident T2D, independent of associated features of the IRS, such as hypertriglyceridaemia, reduced HDL-C, hyperglycaemia, insulin sensitivity, obesity and measures of insulin sensitivity. Thus the measurement of such factors in clinical practice may prove useful in identifying subjects at high risk of developing both T2D and cardiovascular disease. Similarly with increasing evidence demonstrating significant associations between endogenous sex hormone levels and incident T2D in older subjects, our ability to identify subjects at risk of developing the condition may become further refined.

Identifying high-risk subjects and developing strategies to prevent the development of T2D is an important goal, since although treatment may prevent some of the complications associated with T2D, it does not restore normal metabolism or eliminate all the adverse consequences. The hypothesis that T2D is preventable is supported by clinical trials and observation studies of diet, exercise or both in high-risk individuals. This concept was further supported by the recent Diabetes Prevention Programme |**8**| in which the incidence of diabetes was 58% in susceptible subjects following life-style intervention, effects which were similar across all sex, racial and ethnic groups. The important influence of diet on the aetiology of both T2D and vascular disease has also been recently demonstrated by studies in relating reduced dairy content to insulin resistance, high carbohydrate content to enhanced inflammation and high saturated fat content to vascular dysfunction |**45**|.

Although the potential epidemic of T2D represents a major public health problem, our increased understanding of the epidemiology and aetiology of the condition may enable the early identification of high-risk subjects and thus the more effective targeting of interventions aimed at preventing the onset of the condition.

References

1. Amos AF, McCarty DJ, Zimmet P. The rising global burden of diabetes and its complications: estimates and projections to the year 2010. *Diabet Med* 1997; **14**(Suppl 5): S1–85.

2. King H, Aubert RE, Herman WH. Global burden of diabetes, 1995-2025: prevalence, numerical estimates, and projections. *Diabetes Care* 1998; **21**: 1414–31.

3. Harris MI, Flegal KM, Cowie CC, Eberhardt MS, Goldstein DE, Little RR, Wiedmeyer HM, Byrd-Holt DD. Prevalence of diabetes, impaired fasting glucose, and impaired glucose tolerance in U.S. adults. The Third National Health and Nutrition Examination Survey, 1988–1994. *Diabetes Care* 1998; **21**: 518–24.

4. Harris MI, Eastman RC. Early detection of undiagnosed diabetes mellitus: a US perspective. *Diabetes Metab Res Rev* 2000; **16**: 230–6.

5. Porte D Jr. Insulin resistance in the treatment of type 2 diabetes mellitus and its complications. *Diabetes Metab Res Rev* 2001; **17**: 181–8.

6. Edelstein SL, Knowler WC, Bain RP, Andres R, Barrett-Connor EL, Dowse GK, Haffner SM, Pettitt DJ, Sorkin JD, Muller DC, Collins VR, Hamman RF. Predictors of progression from impaired glucose tolerance to NIDDM: an analysis of six prospective studies. *Diabetes* 1997; **46**: 701–10.

7. Knowler WC, Barrett-Connor E, Fowler SE, Hamman RF, Lachin JM, Walker EA, Nathan DM; Diabetes Prevention Program Research Group. Reduction in the incidence of type 2 diabetes with lifestyle intervention or metformin. *N Engl J Med* 2002; **346**: 393–403.

8. Porte D Jr, Kahn SE. Beta-cell dysfunction and failure in type 2 diabetes: potential mechanisms. *Diabetes* 2001; **50**(Suppl 1): S160–3.

9. Arslanian SA, Lewy VD, Danadian K. Glucose intolerance in obese adolescents with polycystic ovary syndrome: roles of insulin resistance and beta-cell dysfunction and risk of cardiovascular disease. *J Clin Endocrinol Metab* 2001; **86**: 66–71.

10. Reaven GM. Banting Lecture 1988. Role of insulin resistance in human disease. *Diabetes* 1988; **37**: 1595–607.

11. Ford ES, Giles WH, Dietz WH. Prevalence of the metabolic syndrome among US adults. Findings from the Third National Health and Nutrition Examination survey. *JAMA* 2002; **287**: 356–9.

12. Ascherio A, Hennekens C, Willett WC, Sacks F, Rosner B, Manson J, Witteman J, Stampfer MJ. Prospective study of nutritional factors, blood pressure, and hypertension among US women. *Hypertension* 1996; **27**: 1065–72.

13. Bostick RM, Kushi LH, Wu Y, Meyer KA, Sellers TA, Folsom AR. Relation of calcium, vitamin D and dairy food intake to ischemic heart disease mortality among postmenopausal women. *Am J Epidemiol* 1999; **149**: 151–61.

14. Zemel MB, Shi H, Greer B, Dirienzo D, Zemel PC Regulation of adiposity by dietary calcium. *FASEB* J 2000; **14**: 1132–8.

15. Kelly GS. Conjugated linoleic acid: a review. *Altern Med Rev* 2001; **6**: 367–82.

16. Haffner SM, Stern MP, Hazuda HP, Pugh J, Patterson JK. Do upper-body and centralized adiposity measure different aspects of regional body-fat distribution? Relationship to non-insulin-dependent diabetes mellitus, lipids, and lipoproteins. *Diabetes* 1987; **36**: 43–51.

17. Groop LC, Tuomi T. Non-insulin-dependent diabetes mellitus – a collision between thrifty genes and an affluent society. *Ann Med* 1997; **29**: 37–53.

18. Ehm MG, Karnoub MC, Sakul H, Gottschalk K, Holt DC, Weber JL, Vaske D, Briley D, Briley L, Kopf J, McMillen P, Nguyen Q, Reisman M, Lai EH, Joslyn G, Shepherd NS, Bell C, Wagner MJ, Burns DK; American Diabetes Association GENNID Study Group. Genetics of NIDDM. Genome-wide search for type 2 diabetes susceptibility genes in four American populations. *Am J Hum Genet* 2000; **66**: 1871–81.

19. Mahtani MM, Widen E, Lehto M, Thomas J, McCarthy M, Brayer J, Bryant B, Chan G, Daly M, Forsblom C, Kanninen T, Kirby A, Kruglyak L, Munnelly K, Parkkonen M, Reeve-Daly MP, Weaver A, Brettin T, Duyk G, Lander ES, Groop LC. Mapping of a gene for type 2 diabetes associated with an insulin secretion defect by a genome scan in Finnish families. *Nat Genet* 1996; **14**(1): 90–4.

20. Ghosh S, Watanabe RM, Valle TT, Hauser ER, Magnuson VL, Langefeld CD, Ally DS, Mohlke KL, Silander K, Kohtamaki K, Chines P, Balow Jr J, Birznieks G, Chang J, Eldridge W, Erdos MR, Karanjawala ZE, Knapp JI, Kudelko K, Martin C, Morales-Mena A, Musick A, Musick T, Pfahl C, Porter R, Rayman JB. The Finland-United States investigation of non-insulin-dependent diabetes mellitus genetics (FUSION) study. I. An autosomal genome scan for genes that predispose to type 2 diabetes. *Am J Hum Genet* 2000; **67**: 1174–85.

21. Mitchell BD, Cole SA, Hsueh WC, Comuzzie AG, Blangero J, MacCluer JW, Hixson JE. Linkage of serum insulin concentrations to chromosome 3p in Mexican Americans. *Diabetes* 2000; **49**: 513–6.

22. Haffner SM. Sex hormones, obesity, fat distribution, type 2 diabetes and insulin resistance: epidemiological and clinical correlation. *Int J Obes Relat Metab Disord* 2000; **24**(Suppl 2): S56–8.

23. Haffner SM, Dunn JF, Katz MS. Relationship of sex hormone-binding globulin to lipid, lipoprotein, glucose, and insulin concentrations in post-menopausal women. *Metabolism* 1992; **41**: 278–84.

24. Haffner SM, Katz MS, Stern MP, Dunn JF. The relationship of sex hormones to hyperinsulinemia and hyperglycemia. *Metabolism* 1988; **37**: 683–8.

25. Stellato RK, Feldman HA, Hamdy O, Horton ES, McKinlay JB. Testosterone, sex hormone-binding globulin and the development of type 2 diabetes in middle-aged men: prospective results from the Massachusetts Male Aging study. *Diabetes Care* 2000; **23**: 490–4.

26. Haffner SM, Valdez RA, Morales PA, Hazuda HP, Stern MP. Decreased sex hormone-binding globulin predicts non-insulin-dependent diabetes mellitus in women but not in men. *J Clin Endocrinol Metab* 1993; **77**: 56–60.

27. Goodman-Gruen D, Barrett-Connor E. Sex differences in the association of endogenous sex hormone levels and glucose tolerance status in older men and women. *Diabetes Care* 2000; **23**: 912–8.

28. Morley JE. Androgens and ageing. *Maturitas* 2001; **38**: 61–73.

29. Festa A, D'Agostino R Jr, Howard G, Mykkanen L, Tracy RP, Haffner SM. Chronic subclinical inflammation as part of the insulin resistance syndrome: the Insulin Resistance Atherosclerosis Study (IRAS). *Circulation* 2000; **102**(1): 42–7.

30. Tsigos C, Papanicolaou DA, Kyrou I, Defensor R, Mitsiadis CS, Chrousos GP. Dose-dependent effects of recombinant human interleukin-6 on glucose regulation. *J Clin Endocrinol Metab* 1997; **82**: 4167–70.

31. van der Meer IM, de Maat MP, Bots ML, Breteler MM, Meijer J, Kiliaan AJ, Hofman A, Witteman JC. Inflammatory mediators and cell adhesion molecules as indicators of severity of atherosclerosis: the Rotterdam study. *Arterioscler Thromb Vasc Biol* 2002; **22**: 838–42.

32. Visser M, Bouter LM, McQuillan GM, Wener MH, Harris TB. Elevated C-reactive protein levels in overweight and obese adults. *JAMA* 1999; **282**: 2131–5.

33. Kharbanda RK, Walton B, Allen M, Klein N, Hingorani AD, MacAllister RJ, Vallance P. Prevention of inflammation-induced endothelial dysfunction: a novel vasculo-protective action of aspirin. *Circulation* 2002; **105**: 2600–4.

34. Steppan CM, Bailey ST, Bhat S, Brown EJ, Banerjee RR, Wright CM, Patel HR, Ahima RS, Lazar MA. The hormone resistin links obesity to diabetes. *Nature* 2001; **409**: 307–12.

35. Ross R. Atherosclerosis and inflammatory disease. *N Engl J Med* 1999; **344**: 115–26.

36. Schmidt MI, Duncan BB, Sharrett AR, Lindberg G, Savage PJ, Offenbacher S, Azambuja MI, Tracy RP, Heiss G. Markers of inflammation and prediction of diabetes mellitus in adults (Atherosclerosis Risk in Communities study): a cohort study. *Lancet* 1999; **353**: 1649–52.

37. Mavri A, Alessi MC, Bastelica D, Geel-Georgelin O, Fina F, Sentocnik JT, Stegnar M, Juhan-Vague I. Subcutaneous abdominal, but not femoral fat expression of plasminogen activator inhibitor-1 (PAI-1) is related to plasma PAI-1 levels and insulin resistance and decreases after weight loss. *Diabetologia* 2001; **44**: 2025–31.

38. Chu NV, Kong AP, Kim DD, Armstrong D, Baxi S, Deutsch R, Caulfield M, Mudaliar SR, Reitz R, Henry RR, Reaven PD. Differential effects of metformin and troglitazone on cardiovascular risk factors in patients with type 2 diabetes. *Diabetes Care* 2002; **25**: 542–9.

39. Lindsay RS, Krakoff J, Hanson RL, Bennett PH, Knowler WC. Gamma globulin levels predict type 2 diabetes in the Pima Indian population. *Diabetes* 2001; **50**: 1598–603.

40. Pankow JS, Folsom AR, Cushman M, Borecki IB, Hopkins PN, Eckfeldt JH, Tracy RP. Familial and genetic determinants of systemic markers of inflammation: the NHLBI family heart study. *Atherosclerosis* 2001: **154**: 681–9.

41. Forsen T, Eriksson J, Tuomilehto J, Reunanen A, Osmond C, Barker D. The fetal and childhood growth of persons who develop type 2 diabetes. *Ann Intern Med* 2000; **133**: 176–82.

42. Hales CN, Barker DJ. The thrifty phenotype hypothesis. *Br Med Bull* 2001; **60**: 5–20.

43. Goh KL, Shore AC, Quinn M, Tooke JE. Impaired microvascular vasodilatory function in 3-month-old infants of low birth weight. *Diabetes Care* 2001; **24**: 1102–7.

44. Leeson CP, Kattenhorn M, Morley R, Lucas A, Deanfield JE. Impact of low birth weight and cardiovascular risk factors on endothelial function in early adult life. *Circulation* 2001; **103**: 1264–8.

45. Anderson RA, Evans ML, Ellis GR, Graham J, Morris K, Jackson SK, Lewis MJ, Rees A, Frenneaux MP. The relationships between post-prandial lipaemia, endothelial function and oxidative stress in healthy individuals and patients with type 2 diabetes. *Atherosclerosis* 2001; **154**(2): 475–83.

6

Dyslipidaemia of diabetes

The use of lipid-regulating therapy

Over 75% of people with type 2 diabetes will die of atherosclerotic macrovascular disease. The lipid abnormalities in type 2 diabetes essentially consist of above average levels of triglycerides (TGs), below average levels of high-density lipoprotein (HDL) cholesterol and the total and low-density lipoprotein (LDL) cholesterol levels are about typical of the population from which the patients are derived |**1**|. There are, in addition, qualitative differences in the lipoprotein particle in patients with diabetes which render them more atherogenic. These changes and the possible mechanisms by which these disturbances in lipid metabolism result in premature, accelerated atherosclerosis have been extensively reviewed elsewhere |**2,3**|. Some of the greater risk in diabetes may result from changes in the LDL particles, in particular – they are smaller, denser and more prone to oxidation and glycation. Quantitatively, the lipid abnormalities may be minor leading to clinicians being sanguine if not nihilistic about whether treatment and/or intervention is required.

The recent consensus statement by the Expert Panel on Detection, Evaluation and Treatment of High Blood Cholesterol in Adults (ATPIII) have recently revised the US guidelines and recognized, for the first time, that having a diagnosis of diabetes is equivalent in risk to having pre-existing coronary disease |**4**|. These guidelines will be reviewed later in this section.

Should an individual patient with type 2 diabetes require pharmacotherapy for the underlying lipoprotein abnormalities (after ensuring optimal lifestyle changes for that individual (diet, exercise, weight)), the two basic groups of drugs in clinical practice are the hydroxymethyl glutaryl coenzyme A reductase inhibitors (the statins) and the fibric acid derivatives (the fibrates).

The statins

The primary consequence of statin therapy is a reduction in the LDL cholesterol level. There is a wealth of evidence for their efficacy in preventing coronary heart disease (CHD), mainly in non-diabetic subjects. There are, however, four statin trials, from which, by *post hoc* subgroup analysis, data on diabetic subgroups can be analysed. The Air Force/Texas Coronary Atherosclerosis Prevention Study (AFCAPS) |**5**|, had only a very people with type 2 diabetes included (n = 155). This number is far too small for any meaningful inferences to be drawn from the data. The remaining three studies – the Simvastatin Scandinavian Survival Study (4S) study |**6**|, the Cholesterol and Recurrent Events (CARE) study |**7**| and the Long-term Intervention with Pravastatin in Ischaemic Disease (LIPID) study |**8**| all showed benefit in the diabetic sub group. In the 4S study, this was significant amongst those who had known diabetes at randomization |**9**| and those who had impaired fasting glucose levels |**10**|. Similar analyses were done on the CARE data set |**11**| but the definitions of the end-points considered in CARE had to be extended to also include those who had undergone coronary revascularization or intervention procedures to achieve significance. Recently, the data sets from the LIPID, CARE and West of Scotland Coronary Prevention Study (WOSCOPS) studies have been pooled into a meta-analysis and data from the subgroup with diabetes analyzed |**12**|. WOSCOPS, which is a primary prevention study from the West of Scotland, contributed only 76 of the 1444 people in this subgroup. Thus the same pattern as seen from CARE and LIPID inevitably emerged. The data depends very much on subgroup analyses each of which therefore has limitations. Much has been written about the caution that must be exercised in extrapolating any conclusions from these data sets to the typical patient with type II diabetes. This gap in reliable data from randomized controlled trials prompted a recent editorial in *Heart* (see later) to rationalize the argument for the need for a large-scale prospective randomized controlled trial to study the effect of LDL cholesterol reduction in people with diabetes mellitus (DM). The authors argue that, as total and LDL cholesterol levels in patients with diabetes are similar to those in the general population, this may have contributed to the belief (mistakenly) that LDL cholesterol is of little relevance to the cardiovascular risk in diabetes. This is an excellent and well argued editorial and provides the rationale for the inclusion of 5963 patients with diabetes randomly entered into the Heart Protection Study (HPS; see below)

The fibrates

Intervention studies in diabetes have been conducted with gemfibrozil, bezafibrate and fenofibrate. The earliest was a *post hoc* subgroup analysis of men with diabetes who participated in the Helsinki Heart Study |**13**|. Those receiving gemfibrozil in this primary prevention study had fewer coronary events compared to placebo but the number of individuals was so small that the difference failed to achieve statistical significance. The Veterans Affairs High-Density Lipoprotein Cholesterol Intervention Trial (VA-HIT) |**14**| was a secondary intervention trial also with gemfibrozil in which a pre specified *post hoc* analysis of the diabetic cohort was conducted. Those with diabetes also had fewer events ($P = 0.052$). The St. Mary's, Ealing, Northwick Park Diabetes Cardiovascular Disease prevention study (SENDCAP) |**15**| examined the effects of bezafibrate on carotid intimal-medial thickness in people with type 2 diabetes. No benefit was found but a subsequent re-analysis showed a significant decrease in CHD as defined by myocardial infarction (MI) and ischaemic ECG changes.

However, all of these changes are limited in that none of the trials were designed *a priori* to study the effects of fibrates on CHD in people with diabetes. In the following section, two trials which have been recently published will be discussed. Both the Diabetes Atherosclerosis Intervention Study (DAIS) and HPS trials have been specifically designed to address the important clinical question as to whether the use of lipid-regulating drugs (fenofibrate in DAIS and simvastatin in HPS) have beneficial effects on cardiovascular disease in patients with diabetes.

Effect of fenofibrate on progression of coronary-artery disease in type 2 diabetes: a randomized study.

Diabetes Atherosclerosis Intervention Study. *Lancet* 2001; **357**: 905–10.

BACKGROUND. Atherosclerosis is the most common complication of diabetes. Correction of hyperglycaemia helps to prevent microvascular complications but has little effect on macrovascular disease. *Post hoc* analyses of diabetic subpopulations in lipid intervention trials suggest that correction of lipoprotein abnormalities will lead to a decrease in coronary artery disease. DAIS was specifically designed to assess the effects of correcting lipoprotein abnormalities on coronary atherosclerosis in type 2 diabetes. A total of 731 men and women with type 2 diabetes were screened by metabolic and angiographic criteria. Of these, 418 were randomly assigned micronized fenofibrate (200 mg/day) or placebo for at least 3 years. They were in good glycaemic control (mean haemoglobin A1c 7.5%), had mild lipoprotein abnormalities, typical of type 2 diabetes, and at least one visible coronary lesion. Half had no previous clinical coronary disease. Initial and final angiograms followed a standard protocol and were analyzed by a computer-assisted quantitative approach. Missing data for the primary end-points (minimum lumen diameter, mean segment diameter, and mean percentage stenosis) were imputed. Analyses were by intention to treat. Total plasma cholesterol, HDL cholesterol, LDL cholesterol, and TG concentrations all changed significantly more from baseline in the fenofibrate group ($n = 207$) than in the placebo group ($n = 211$). The fenofibrate group showed a significantly smaller increase in percentage diameter stenosis than the placebo group (mean 2.11 [SEM 0.594] *vs* 3.65 [0.608]%, $P = 0.02$), a significantly smaller decrease in minimum lumen diameter (–0.06 [0.016] *vs* –0.10 (0.016) mm, $P = 0.029$), and a non-significantly smaller decrease in mean segment diameter (–0.06 [0.017] *vs* –0.08 [0.018] mm, $P = 0.171$). The trial was not powered to examine clinical end-points, but there were fewer in the fenofibrate group than the placebo group (38 *vs* 50).

INTERPRETATION. DAIS suggests that treatment with fenofibrate reduces the angiographic progression of coronary artery disease in type 2 diabetes. This effect is related, at least partly, to the correction of lipoprotein abnormalities, even those previously judged not to need treatment.

Comment

This was the first trial performed specifically in a cohort of type 2 diabetes and was performed in 11 centres in Canada and Europe; 418 (305 men, 113 women) patients with type 2 diabetes were screened by metabolic and angiographic criteria and randomized to micronized fenofibrate or placebo therapy for at least three years.

Previously *post hoc* subgroup analyses of the major secondary prevention trials using 3-hydroxy-3-methlgutamyl coenzyme A (HMG CoA) reductase inhibitors have demonstrated benefit to LDL-lowering over a wide range of cholesterol levels in the secondary prevention of CHD in diabetic patients. In the VA-HIT trial, the 627 diabetic patients (out of 2531 men with CHD) showed a significant reduction in the combined end-point of CHD death, non-fatal MI and confirmed stroke. The study population in the VA-HIT had relatively normal LDL cholesterol values (mean baseline 2.9 mmol/l) but low HDL cholesterol levels (mean baseline 0.8 mmol/l). Baseline TGs were 1.8 mmol/l. In the group treated with a fibrate (gemfibrozil), HDL cholesterol was 6% higher, total TGs 31% lower, total cholesterol 4% lower and no change in LDL cholesterol values.

DAIS was a double-blind, randomized placebo-controlled trial comparing fenofibrate 200 mg per day (micronized preparation) to placebo. Lipid entry criteria following withdrawal of any lipid-lowering drugs and maintenance of the National Cholesterol Education Program (NCEP) step 1 diet were that the total cholesterol/HDL cholesterol ratio should be $\geq$4.0, and either LDL cholesterol should be 3.5–4.5 mmol/l and TGs $\leq$5.2 mmol/l or LDL cholesterol $\leq$4.5 mmol/l. Glycaemic control was reasonable (mean HbA_{1c} [glycosylated haemoglobin] 7.5%). A total of 731 patients were screened for inclusion and 418 were included. Of these, 207 were randomized to fenofibrate and 211 to placebo. Compared to placebo, the main impact of fenofibrate therapy was a 29% decrease in TG and a 6% increase in HDL cholesterol.

DAIS was not designed as a clinical end-point study and the number of end-points was small. There were 38 subjects with events (death, MI, coronary revascularization and hospital admission for angina) in the fenofibrate group compared to 50 in the placebo group. This 23% reduction is not statistically significant due to the underpowering of the trial. However, the reduction is in line with the observed reduction in events in the larger end-point studies, Moreover, there was a consistently lower rate of each individual end-point with micronized fenofibrate compared to placebo except for hospital admission with angina. The DAIS study population was small with a relatively short follow-up period but the difference in the rate of end-points is very similar to that seen in the *post hoc* analyses of the larger and longer trials, VA-HIT, 4S and CARE. As the study population in DAIS was recruited on metabolic and angiographic criteria, and only 50% had clinically overt CHD, the clinical event rate was less than in the aforementioned trials.

Angiograms in the DAIS study were performed according to a standard protocol which allowed simultaneous computer assisted quantitative analysis of baseline and end-of-study films. On average, the number of coronary artery segments analyzed was 9.4 in the placebo group and 9.1 in the fenofibrate group. The decrease in the primary end-point which was the decrease in mean lumen diameter did not reach statistical significance in the DAIS study. However, there were significant reductions in progression of the minimum lumen diameter (40%, $P = 0.029$), and progression in percentage diameter stenosis (42% less, $P = 0.02$).

The angiographic changes observed in DAIS are of the same magnitude as several other angiographic trials in people without diabetes. In some of these trials, despite small populations and short duration, significant differences in clinical end-points were observed. However, pooling of the angiographic trials reveals a significant reduction in the clinical end-points. Whether this is methodologically justified or not, the positive findings in DAIS are encouraging and lend credence to the hypothesis that modifying the relatively mild lipid abnormalities found in patients with type 2 diabetes ameliorates the risk of CHD in those patients.

It is clear that the angiographic benefits seen in DAIS need to be confirmed in an end-point trial. However, the evidence from clinical trials of fibrates is limited, compared to the clear, compelling evidence available of benefit from statin therapy. The clinical indications for fibrate therapy are limited, even in patients with type 2 diabetes, whose typical atherogenic lipoprotein phenotype seems best suited to fibric acid derivate therapy. Fibrates are ligands for the perosisome proliferator activator receptor and lower plasma TG and raise HDL – a lipid-regulating effect that is particularly desirable in patients with type 2 diabetes.

Ongoing trials utilizing fibric acid derivatives include the Fenofibrate Intervention and Event Lowering in Diabetes Trial (FIELD). This is a randomized controlled trial comparing micronized fenofibrate (200 mg/day) *versus* placebo in 8000 men and women with type 2 diabetes. It is a combined primary and secondary trial, including 2000 patients with established CHD and 6000 without CHD. Trial duration will be 5 to 7 years and recruitment will be in Australia, New Zealand and Finland, and it should report in 2006.

An interesting trial, the Lipids in Diabetes Study (LDS), comparing fenofibrate, cerivastatin, drug combination and placebo has been discontinued due to the withdrawal of cerivastatin from clinical use in August 2001. The safety concerns raised by cerivastatin, particularly in the cases of rhabdomyolysis seen when the drug was used in combination with gemfibrozil, has resulted in the suspension of the trial and will make physicians much more circumspect in using a statin–fibrate combination in the future. Other trials in patients with diabetes which are currently in progress include the Collaborative Atorvastatin Diabetes Study (CARDS), expected to report in 2004, and the Atorvastatin as Prevention of Coronary Heart Diseases in Patients with Type 2 Diabetes Mellitus (ASPEN) which is due to be completed in 2003. Obviously, these are statin-based trials, using atorvastatin.

Need for large-scale randomized evidence about lowering LDL cholesterol in people with diabetes mellitus: MRC/BHF heart protection study and other major trials.

Armitage J, Collins R. *Heart* 2000; **84**: 357–60.

Background. There is good epidemiological evidence of continuous direct associations between cardiovascular disease risk and both blood pressure levels and LDL cholesterol concentrations throughout the ranges studied in different populations. Thus, lower levels of these risk factors might well, in the long term, be associated with lower levels of cardiovascular risk in people with diabetes.

Interpretation. This editorial outlines the rationale behind the MRC/BHF Heart Protection Study. The Heart Protection Study should provide the first large-scale prospective evidence as to whether reducing LDL cholesterol concentrations in a wide range of diabetic patients produces worthwhile benefits both in terms of morbidity and mortality.

Comment

This well-written editorial succinctly sets out the rationale for the inclusion of 5963 people aged 40–80 years with DM who were randomized into the Heart Protection Study (HPS; see next paper reviewed). According to pre-specified criteria, they were categorized into type 1 diabetes (10%) and type 2 diabetes (90%). The MRC/BHF Heart Protection Study provides the first large-scale prospective trial evidence of whether reducing LDL concentration in a wide range of diabetes patients produces worthwhile benefits. These data are in addition to the indirect evidence (from retrospective subgroup analyses) that suggest that the proportional effect on CHD risk in patients with pre-existing CHD and diabetes was similar to that observed among the other patients studied.

MRC/BHF Heart Protection Study of cholesterol lowering with simvastatin in 20 536 high-risk individuals: a randomized placebo-controlled trial.

Heart Protection Study Collaborative Group. *Lancet* 2002; **360**(9236): 7–22.

Background. Throughout the usual LDL cholesterol range in Western populations, lower blood concentrations are associated with lower cardiovascular disease risk. In such populations, therefore, reducing LDL cholesterol may reduce the development of vascular disease, largely irrespective of initial cholesterol concentrations. 20 536 UK adults (aged 40–80 years) with coronary disease, other occlusive arterial disease, or diabetes were randomly allocated to receive 40 mg simvastatin daily (average compliance: 85%) or matching placebo (average non-study statin use: 17%). Analyses are of the first occurrence of particular events, and compare all simvastatin-allocated

***versus* all placebo-allocated participants. These 'intention-to-treat' comparisons assess the effects of about two-thirds (85% minus 17%) of those taking a statin during the scheduled 5-year treatment period, which yielded an average difference in LDL cholesterol of 1.0 mmol/l (about two-thirds of the effect of actual use of 40 mg simvastatin daily). Primary outcomes were mortality (for overall analyses) and fatal or non-fatal vascular events (for subcategory analyses), with subsidiary assessments of cancer and of other major morbidity. All-cause mortality was significantly reduced (1328 [12.9%] deaths among 10 269 allocated simvastatin *versus* 1507 [14.7%] among 10 267 allocated placebo; *P* = 0.0003), due to a highly significant 18% (SEM = 5) proportional reduction in the coronary death rate (587 [5.7%] *vs* 707 [6.9%]; *P* = 0.0005), a marginally significant reduction in other vascular deaths (194 [1.9%] *vs* 230 [2.2%]; *P* = 0.07), and a non-significant reduction in non-vascular deaths (547 [5.3%] *vs* 570 [5.6%]; *P* = 0.4). There were highly significant reductions of about one-quarter in the first event rate for non-fatal myocardial infarction or coronary death (898 [8.7%] vs 1212 [11.8%]; *P* <0.0001), for non-fatal or fatal stroke (444 [4.3%] vs 585 [5.7%]; *P* <0.0001), and for coronary or non-coronary revascularization (939 [9.1%] vs 1205 [11.7%]; *P* <0.0001). For the first occurrence of any of these major vascular events, there was a definite 24% (SEM = 3; 95% CI 19–28) reduction in the event rate (2033 [19.8%] *vs* 2585 [25.2%] affected individuals; *P* <0.0001). During the first year the reduction in major vascular events was not significant, but subsequently it was highly significant during each separate year. The proportional reduction in the event rate was similar (and significant) in each subcategory of participant studied, including: those without diagnosed coronary disease who had cerebrovascular disease, or had peripheral artery disease, or had diabetes; men and, separately, women; those aged either under or over 70 years at entry; and – most notably – even those who presented with LDL cholesterol below 3.0 mmol/l (116 mg/dl), or total cholesterol below 5.0 mmol/l (193 mg/dl). The benefits of simvastatin were additional to those of other cardioprotective treatments. The annual excess risk of myopathy with this regimen was about 0.01%. There were no significant adverse effects on cancer incidence or on hospitalization for any other non-vascular cause.**

Table 6.1 Compliance with study simvastatin (≥80% taken) and/or use of non-study statin during follow-up

Follow-up (years)	Simvastatin-allocated	Placebo-allocated
1	8994/10 107 (89%)	389/10 088 (4%)
2	8457/9909 (85%)	889/9826 (9%)
3	8122/9664 (84%)	1608/9563 (17%)
4	7764/9388 (83%)	2262/9241 (24%)
5	6058/7370 (82%)	2345/7225 (32%)
Study average (SE)	85% (0.1)	17% (0.1)

For missing follow-up, non-compliance is assumed. Restriction of this analysis to those who had not yet suffered a major vascular event did not materially alter the estimated average use of statin therapy: 86% simvastatin-allocated versus 15% placebo-allocated (71% difference).

Source: Heart Protection Study Collaborative Group (2002).

Table 6.2 Differences in plasma concentrations of lipids during follow-up

Follow-up (years)	Mean (SE) difference in concentrations (simvastatin minus placebo)*					
	Total cholesterol	**LDL cholesterol**	**HDL cholesterol**	**Triglycerides**	**Apolipoprotein A_1**	**Apolipoprotein B**
1	−1.7 (0.08)	−1.3 (0.06)	0.02 (0.02)	−0.4 (0.08)	0.020 (0.015)	−0.36 (0.02)
3	−1.2 (0.09)	−0.9 (0.08)	0.02 (0.03)	−0.4 (0.11)	0.001 (0.016)	−0.29 (0.03)
5	−0.8 (0.03)	−0.7 (0.03)	0.02 (0.01)	−0.2 (0.04)	0.004 (0.026)	−0.18 (0.04)
Study average	−1.2 (0.02)	−1.0 (0.02)	0.03 (0.01)	−0.3 (0.03)	0.010 (0.007)	−0.28 (0.01)

*Intention-to-treat comparisons, with missing data imputed from initial pre-treatment screening values; mmol/l for total, LDL, HDL, and triglycerides, and g/l for apolipoproteins. Similar average differences with simvastatin versus placebo allocation were observed in the presence and absence of the study vitamins for total cholesterol (−1.20 [0.03] in those allocated vitamins and −1.18 [0.03] in those allocated matching placebo), for LDL cholesterol (−0.95 [0.02] and −0.97 [0.02]), for HDL cholesterol (0.02 [0.01] and 0.04 [0.01]), and for triglycerides (−0.31 [0.04] and −0.26 [0.04]).

Source: Heart Protection Study Collaborative Group (2002).

INTERPRETATION. Adding simvastatin to existing treatments safely produces substantial additional benefits for a wide range of high-risk patients, irrespective of their initial cholesterol concentrations. Allocation to 40 mg simvastatin daily reduced the rates of MI, of stroke, and of revascularization by about one-quarter. After making allowance for non-compliance, actual use of this regimen would probably reduce these rates by about one-third. Hence, among the many types of high-risk individual studied, 5 years of simvastatin would prevent about 70–100 people per 1000 from suffering at least one of these major vascular events (and longer treatment should produce further benefit). The size of the 5-year benefit depends chiefly on such individuals' overall risk of major vascular events, rather than on their blood lipid concentrations alone.

Comment

The HPS, with over 20 500 subjects, is the largest trial of statin therapy ever conducted. It provides important and definitive data on patients with diabetes, females, the elderly, and patients with low baseline total cholesterol and LDL cholesterol

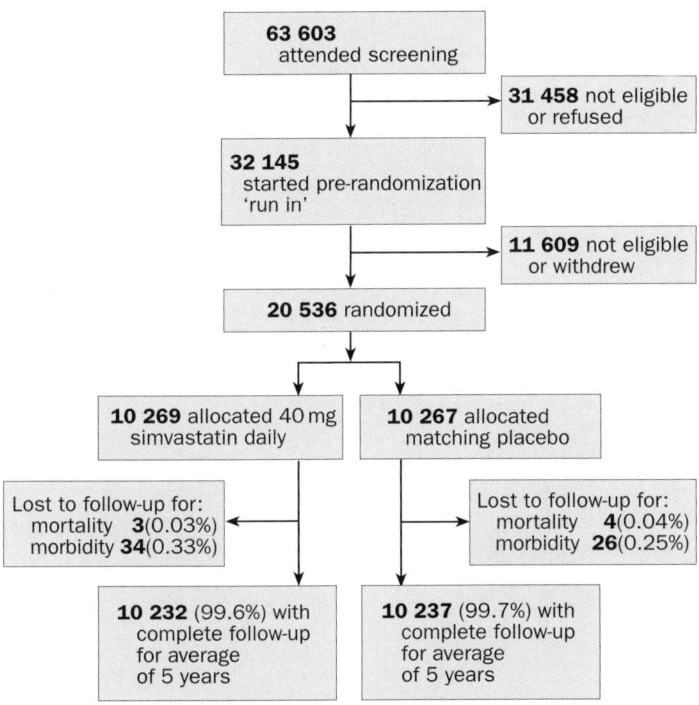

Fig. 6.1 Trial profile Numbers lost to follow-up relate to those without information to the end of the scheduled treatment period for mortality (as well as morbidity) and for morbidity alone. Source: Heart Protection Study Colloborative Group (2002).

Table 6.3 Average use of statin (study or non-study), and average plasma LDL cholesterol concentrations, during follow-up

Presenting feature	Use of study/non-study statin (%)			Plasma LDL cholesterol (mmol/l)		
	Simvastatin-allocated	Placebo-allocated	Absolute difference*	Simvastatin-allocated	Placebo-allocated	Absolute difference*
Prior disease						
Prior MI	87%	22%	65%	2.3	3.2	−0.9
Other CHD	85%	18%	67%	2.3	3.3	−1.0
No CHD	83%	11%	71%	2.3	3.3	−1.0
Sex						
Male	86%	18%	68%	2.2	3.2	−1.0
Female	82%	16%	65%	2.5	3.4	−0.9
Age (years)						
<65	85%	20%	64%	2.4	3.2	−0.9
≥65<70	87%	18%	69%	2.2	3.3	−1.0
≥70	84%	12%	72%	2.2	3.3	−1.1
Total cholesterol (mmol/l)						
<5.0	83%	5%	78%	1.8	2.6	−0.9
≥5.0<6.0	85%	15%	70%	2.1	3.1	−1.0
≥6.0	86%	26%	60%	2.7	3.7	−1.0
LDL cholesterol (mmol/l)						
<3.0	83%	8%	75%	1.8	2.7	−0.9
≥3.0<3.5	86%	16%	69%	2.2	3.2	−1.0
≥3.5	86%	26%	60%	2.7	3.7	−1.0
Pre-randomization LDL response						
Smaller (<38%)	83%	18%	64%	2.5	3.3	−0.8
Average	86%	18%	68%	2.3	3.3	−1.0
Larger (≥48%)	86%	16%	70%	2.1	3.2	−1.1
All patients	85%	17%	67%	2.3	3.3	−1.0

MI = myocardial infarction; CHD = coronary heart disease. *The absolute difference in LDL cholesterol that would be produced by full compliance with 40 mg simvastatin daily can be estimated as the ratio of these two columns. For example, −1.0/67% = −1.4 mmol/l for all patients; and, likewise, −1.2, −1.4, and −1.7 mmol/l, respectively, for those presenting with LDL cholesterol <3.0, ≥3.0<3.5, and ≥3.5 mmol/l.

Source: Heart Protection Study Collaborative Group (2002).

levels at randomization. Moreover, it included patients at high risk for CHD but with characteristics that excluded them from participation in previous trials (peripheral vascular disease, diabetes, hypertension, etc).

The trial was a prospective double-blind randomized controlled trial with a 2 × 2 factorial design investigating prolonged use (>5 years) of simvastatin 40 mg and a cocktail of antioxidant vitamins (vitamin E, vitamin C and β-carotene).

In essence, simvastatin 40 mg showed a benefit across all patient groups regardless of age, gender or baseline cholesterol values. It was safe and well tolerated. There appears to be no threshold cholesterol value below which statin therapy is not associated with benefit – even amongst patients with pre-treatment cholesterol levels below current recommended targets.

Over the study period, participants' own doctors were allowed to add in a non-study statin if they felt it was clinically indicated. This occurred in an increasingly progressive fashion over the 5½ year study period resulting in only two-thirds complying with the original intention-to-treat design. Nevertheless, the statin arm results were highly significant compared to the placebo (usual care comparator) arm. There appeared to be no benefit (nor harm) resulting from the antioxidant vitamin cocktail.

Clinical implications of the HPS data

Around 6000 patients with diabetes were included, and some 4000 had no prior clinical evidence of CHD. The data showed a significant reduction in cardiovascular and all-cause mortality in those taking simvastatin 40 mg compared to placebo. Vascular events in statin-treated and placebo-treated patients starting diverging at around 12 months and continued throughout the 6 years of observation without

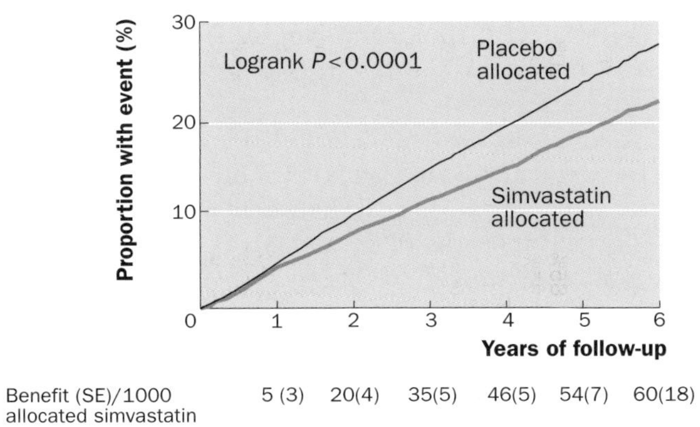

Fig. 6.2 Life-table plot of effects of simvastatin allocation on percentages having major vascular events. Source: Heart Protection Study Collaborative Group (2002).

a tendency to plateau. This study showed that 5 years' treatment with simvastatin 40 mg safely prevents major cardiovascular events in 100 of every 1000 patients with previous MI, in 80 out of every 1000 patients with other manifestations of CHD, 70 out of every 1000 patients with diabetes, 70 out of every 1000 patients with previous stroke and 70 out of every 1000 patients with other manifestations of peripheral vascular disease. The absolute risk reduction for all vascular mortality by simvastatin was 1.5% and this accounted for the reduction in total mortality. There were 152 fewer vascular deaths in the simvastatin group so the number needed to treat (NNT) to prevent one death was 68. An important secondary end-point was major vascular events (comprising of total CHD events, total stroke and revascularization procedures) which was reduced by 24%. The absolute risk reduction was 5.6%, or 564 less major vascular events as a result of simvastatin treatment. The calculated NNT for this secondary end-point was 18.

The data suggest that there is no threshold level for LDL cholesterol to achieve cardiovascular benefit. If there is, it must be substantially below 100 mg/dl (2.6 mmol/l). However, more definitive conclusions on this issue may be drawn when the results of trials such as Treat to New Targets (TNT) are available. Nevertheless, these data validate the general approach of the NCEP ATPIII (Adult Treatment Panel III) guidelines in identifying CHD risk equivalents such as diabetes, and non-coronary vascular disease and also supports the goal of therapy to be an LDL cholesterol <100 mg/dl (<2.6 mmol/l). The results from HPS fit into the concept of a log-linear relationship between total and LDL cholesterol levels with cardiovascular risk as implied from observational epidemiology, extending down to very low levels of cholesterol. However, the absolute benefit from statin therapy will diminish with decreasing absolute risk. The question of safety is also important. It appears that LDL cholesterol below 100 mg/dl (2.6 mmol/l) appears to be safe, but how are these low levels to be achieved? These results may encourage the use of high-dose statins, but caution should be exercised as adverse events are known to be dose related. Are low levels of LDL cholesterol best achieved by high dose monotherapy or drug combination of statin with other agents? This clearly is an unresolved question.

Finally, the results from this trial suggest that the British, European and American approaches to CHD event reduction with statins are more complicated than necessary and may deny care to many individuals who would benefit from intervention. Clinical guidelines benefit from simplicity and clarity. Dose titration of drugs can also be a relative disincentive to some clinicians.

These trial data, together with results from studies still to report, will require a revision and possibly a simplification of clinical guidelines in the near future.

Executive summary of the third report of The National Cholesterol Education Program (NCEP) expert panel on detection, evaluation, and treatment of high blood cholesterol in adults (Adult Treatment Panel III).

Expert Panel on Detection, Evaluation, and Treatment of High Blood Cholesterol in Adults. *JAMA* 2001; **285**: 2486–97.

BACKGROUND. The third report of the Expert Panel on Detection, Evaluation and Treatment of High Blood Cholesterol in Adults (Adult Treatment Panel III, or ATP III) constitutes the NCEP's updated clinical guidelines for cholesterol testing and management. The full ATP III document is an evidence-based and extensively referenced report that provides the scientific rationale for the recommendations contained in the executive summary. ATP III builds on previous ATP reports and expands the indications for intensive cholesterol-lowering therapy in clinical practice. It should be noted that these guidelines are intended to inform, not replace, the physician's clinical judgment, which must ultimately determine the appropriate treatment for each individual.

Comment

In 2001, the National Heart, Lung and Blood Institute's NCEP issued the first major update in years of the new clinical practice guidelines on the prevention and management of high cholesterol in adults.

Several key changes are included in the ATP III guidelines. They are summarized below:

1. There is a focus on primary intervention in persons with multiple risk factors. The Framingham function is used to estimate 10-year risk of developing CHD and this 10-year risk determines the intensity of the intervention employed.
2. The guidelines identified people with diabetes who have no clinically overt CHD as having a risk level of CHD equivalent to those patients without diabetes but with pre-existing coronary disease. Thus, a person having a diagnosis of diabetes is considered to have a CHD equivalent risk.
3. The guidelines identify individuals with the metabolic syndrome for intensified therapeutic lifestyle changes. Moreover, the metabolic syndrome is defined clinically in terms of abdominal obesity (waist circumference), impaired fasting glucose and categorical triglyceride and HDL levels.
4. ATPIII continues to identify elevated LDL cholesterol as the primary goal of therapy. The two major means of lowering LDL cholesterol are therapeutic lifestyle change (TLC) (nutrition, physical activity and weight control) and pharmacotherapy (usually a statin).

The identification of diabetes as a CHD equivalent is clearly going to have major implications for practising clinicians. The NCEP ATPIII expert panel was clearly

Table 6.4 New features of ATP III*

Focus on multiple risk factors

- Raises persons with diabetes without CHD, most of whom have multiple risk factors, to the risk level of CHD risk equivalent
- Uses Framingham projections of 10-year absolute CHD risk (i.e. the per cent probability of having a CHD event in 10 years) to identify certain patients with multiple (2+) risk factors for more intensive treatment
- Identifies persons with multiple metabolic risk factors (metabolic syndrome) as candidates for intensified therapeutic lifestyle changes

Modifications of lipid and lipoprotein classification

- Identifies LDL cholesterol $<$100 mg/dl as optimal
- Raises categorical low HDL cholesterol from $<$35 mg/dl to $<$40 mg/dl because the latter is a better measure of a depressed HDL
- Lowers the triglyceride classification cutpoints to give more attention to moderate elevations

Support for implementation

- Recommends a complete lipoprotein profile (total, LDL, and HDL cholesterol and triglycerides) as the preferred initial test, rather than screening for total cholesterol and HDL alone
- Encourages use of plant stanols/sterols and viscous (soluble) fibre as therapeutic dietary options to enhance lowering of LDL cholesterol
- Presents strategies for promoting adherence to therapeutic lifestyle changes and drug therapies
- Recommends treatment beyond LDL lowering for persons with triglycerides $\geq$200 mg/dl

*ATP indicates Adult Treatment Panel; CHD, coronary heart disease; LDL, low-density lipoprotein; and HDL, high-density lipoprotein.

Source: Expert Panel on Detection, Evaluation, and Treatment of High Blood Cholesterol in Adults (2001).

prescient in its review of the available evidence as no single prospective randomized control trial of CHD prevention in people with diabetes had reported at the time of publication of these guidelines. The results of the HPS clearly have corroborated their view.

Management of dyslipidaemia in adults with diabetes, American Diabetes Association Clinical Practice Recommendation 2002.

American Diabetes Association. *Diabetes Care* 2002; **25**: S74–S77.

BACKGROUND. The rationale for the treatment of diabetic dyslipidaemia is discussed in detail in this American Diabetes Association (ADA) technical review. Type 2 diabetes is associated with a two- to fourfold excess risk of coronary heart disease CHD. Although the degree of glycaemia in diabetic patients is strongly related to the

Table 6.5 Category of risk based on lipoprotein levels in adults with diabetes

Risk	LDL cholesterol	HDL cholesterol*	Triglyceride
High	≥130	<35	≥400
Borderline	100–129	35–45	200–399
Low	<100	>45	<200

Data are given in milligrams per decilitre. *For women, the HDL cholesterol values should be increased by 10 mg/dl.

Source: American Diabetes Association (2002).

risk of microvascular complications (retinopathy and renal disease), the relation of glycaemia to macrovascular disease in type 2 diabetes is more modest. The finding of increased cardiovascular risk factors before the onset of type 2 diabetes also suggests that aggressive screening for diabetes combined with improved glycaemic control alone will not be likely to completely eliminate excess risk of CHD in type 2 diabetic patients. Clearly, a multifactorial approach to prevention of CHD in type 2 diabetes will be necessary.

Comment

The position statement of the ADA (2002) defined categories of risk of CHD based on lipoprotein levels in adults with diabetes.

Optimal LDL cholesterol levels for adults with diabetes are <100 mg/dl (<2.60 mmol/l), optimal HDL cholesterol levels are >45 mg/dl (>1.15 mmol/l) and desirable TG levels are <150 mg/dl (<1.7 mmol/l).

The recommendations for treatment of elevated LDL cholesterol generally follow the guidelines set out by the NCEP ATPIII with a few caveats. The primary therapeutic goal is to lower LDL cholesterol levels to levels recommended for patients with pre-existing CHD (<2.6 mmol/l). The initial therapy for hypertriglyceridaemia is improved glycaemic control. Additional TG-lowering can be achieved by very high doses of statins or fibric acid derivatives.

Antioxidant supplements block the response of HDL to simvastatin–niacin therapy in patients with coronary artery disease and low HDL.

Cheung MC, Zhao XQ, Chait A, Albers JJ, Brown BG. *Arterioscler Throm Vasc Biol* 2001; **21**: 1320–6.

BACKGROUND. One strategy for treating coronary artery disease (CAD) patients with low HDL cholesterol (HDL-C) is to maximally increase the HDL-C:LDL-C ratio by combining lifestyle changes with niacin (N) plus a statin. Because HDL can prevent

Table 6.6 Baseline characteristics of patients in the four treatment groups

Variable	All placebo (n = 35)	Simvastatin-niacin (n = 38)	Antioxidants (n = 40)	Simvastatin-niacin + antioxidants (n = 40)
Age, y*	52.8 ± 7.5	52.5 ± 9.2	53.7 ± 8.6	54.8 ± 7.4
Sex, % male	94.2	86.8	87.5	85.0
BMI, kg/cm^2*	29.6 ± 3.9	29.1 ± 4.0	28.3 ± 3.2	29.2 ± 5.0
Smokers, %†	20.0	21.0	27.5	25.0
Cigarettes/day, No.†	17.4	18.8	16.6	14.4
Diabetes, %	17.1	5.3	15.0	25.0
High blood pressure, %	57.1	36.8	47.5	50.0
Baseline lipids, mg/dl*				
Plasma chol	199 ± 34	194 ± 37	189 ± 25	203 ± 37
Plasma TG	208 ± 109	195 ± 86	193 ± 116	231 ± 115
LDL-C	128 ± 27	129 ± 36	122 ± 28	131 ± 33
HDL-C	32 ± 5.5	31 ± 5.3	32 ± 5.3	31 ± 5.2

To convert cholesterol (chol, -C) and triglyceride (TG) levels to mmol/l, multiply the mg/dl values by 0.02589 and 0.0113, respectively.
*Values are mean ± SD.
†Current smokers only.

Source: Cheung *et al.* (2001).

LDL oxidation, the low-HDL state also may benefit clinically from supplemental antioxidants. Lipoprotein changes over 12 months were studied in 153 CAD subjects with low HDL-C randomized to take simvastatin and niacin (S–N), antioxidants (vitamins E and C, beta-carotene, and selenium), S–N plus antioxidants (S–N + A), or placebo. Mean baseline plasma cholesterol, TG, LDL-C, and HDL-C levels of the 153 subjects were 196, 207, 127, and 32 mg/dl, respectively. Without S–N, lipid changes were minor. The S–N and S–N + A groups had comparably significant reductions (*P* ≤0.001) in plasma cholesterol, TG, and LDL-C. However, increases in HDL-C, especially HDL(2)-C, were consistently higher in the S–N group than in the S–N + A group (25% *vs* 18% and 42% *vs* 0%, respectively). With S–N, but not with S–N + A, there was a selective increase in apolipoprotein (apo) A-I (64%) in HDL particles containing apoA-I but not A-II (Lp(A-I)) and their particle size. Thus, in CAD patients with low HDL-C, S–N substantially increased HDL(2)-C, Lp(A-I), and HDL particle size. These favourable responses were blunted by the antioxidants used owing to a striking selective effect on Lp(A-I).

Interpretation. This unexpected adverse interaction between antioxidants and lipid therapy may have important implications for the management of CAD.

Table 6.7 Plasma lipid and apolipoprotein levels at baseline and 12 months after simvastatin-niacin therapy with or without antioxidant supplements

Lipid and protein	Simvastatin-niacin (n = 38)			Simvastatin-niacin + antioxidants (n = 40)			
	Baseline	On therapy	% Median change	Baseline	On therapy	% Median change	P*
Plasma chol	194 ± 37	141 ± 25†	−25	203 ± 37	150 ± 36†	−26	0.780
Plasma TG	195 ± 86	113 ± 51†	−41	231 ± 115	170 ± 141†	−30	0.384
VLDL-C	34 ± 17	15 ± 7.9†	−57	41 ± 22	27 ± 30†	−42	0.315
IDL-C	14 ± 5.5	6.3 ± 3.9†	−55	15 ± 6.0	5.9 ± 5.8†	−67	0.204
LDL-C	129 ± 36	85 ± 21†	−34	131 ± 33	86 ± 21†	−31	0.715
HDL-C	31 ± 5.3	41 ± 11†	25	31 ± 5.2	37 ± 6.8†	18	0.057
HDL_2-C	3.7 ± 1.5	5.9 ± 3.8†	42	4.0 ± 1.7	4.3 ± 2.1	0	0.007
HDL_3-C	28 ± 4.1	35 ± 7.8†	27	27 ± 4.6	32 ± 5.4†	20	0.271
ApoA-I	107 ± 14	120 ± 25†	13	108 ± 15	119 ± 17†	10	0.672
ApoA-II	29 ± 3.8	28 ± 4.4	−3.4	31 ± 4.1	30 ± 5.1	−4.2	0.661
ApoB	116 ± 21	75 ± 17†	−35	120 ± 23	80 ± 21†	−34	0.799

Lipid and protein values are in mg/dl and represent mean ± SD.
*P values obtained by comparing the % changes between the 2 simvastatin-niacin-treated groups by the Mann-Whitney test.
†Different from baseline at $P \leq 0.001$ by the Wilcoxon matched-pairs signed-rank test. To convert the cholesterol (chol, -C) and triglyceride (TG) levels to mmol/l, multiply the mg/dl values by 0.02589 and 0.0113, respectively.

Source: Cheung *et al.* (2001).

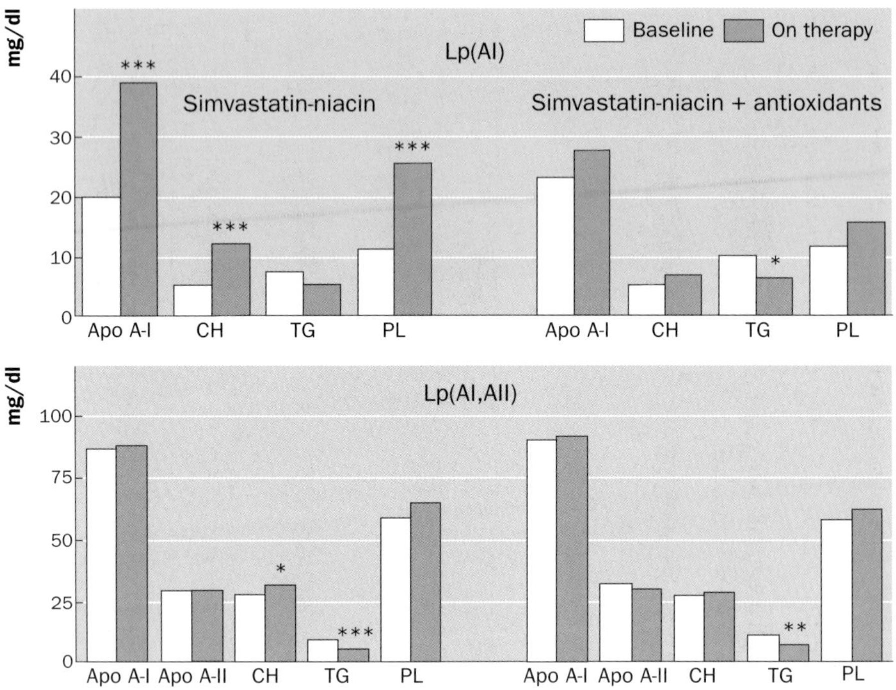

Fig. 6.3 Composition of Lp(A-I) (top) and Lp(A-I,A-II) (bottom) particles at baseline and 12 months after treatment with simvastatin and niacin (n =16) (left) or with combined simvastatin-niacin and antioxidant supplements (n =13) (right). *, **, and *** denote significant differences between baseline and treatment levels at $P \leq 0.01$, 0.005, and 0.001, respectively. CH indicates cholesterol; TG, triglyceride; PL, phospholipid. Source: Cheung *et al.* (2001).

Simvastatin and niacin, antioxidant vitamins, or the combination for the prevention of coronary disease.

Brown BG, Zhao XQ, Chait A, *et al. N Engl J Med* 2001; **345**: 1583–92.

BACKGROUND. Both lipid-modifying therapy and antioxidant vitamins are thought to have benefit in patients with coronary disease. We studied simvastatin–niacin and antioxidant-vitamin therapy, alone and together, for cardiovascular protection in patients with coronary disease and low plasma levels of HDL. In a three-year, double-blind trial, 160 patients with coronary disease, low HDL cholesterol levels and normal LDL cholesterol levels were randomly assigned to receive one of four regimens:

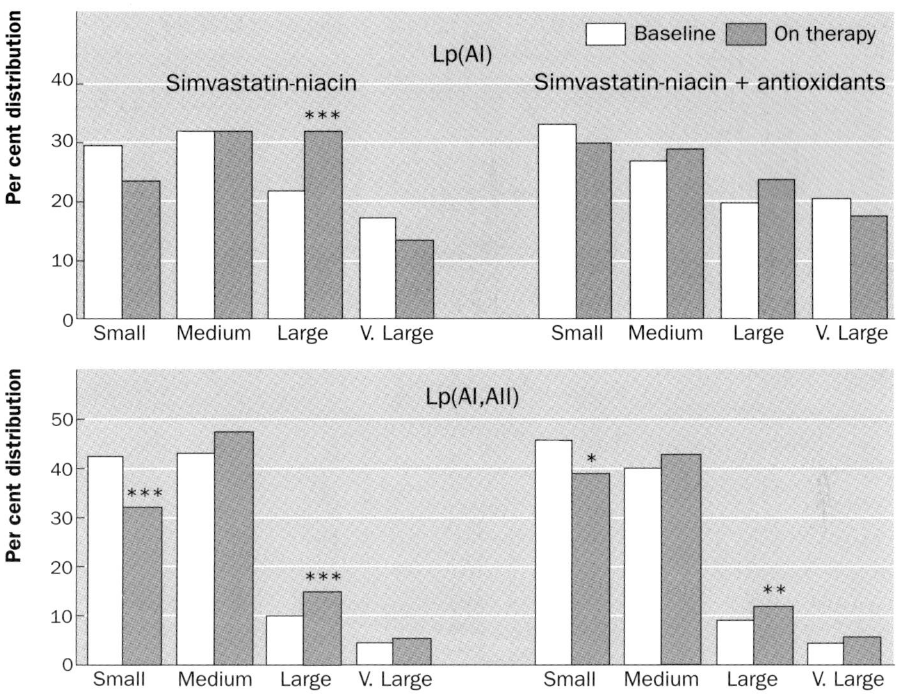

Fig. 6.4 Size distribution of Lp(A-I) (top) and Lp(A-I,A-II) (bottom) particles at baseline and 12 months after treatment with simvastatin and niacin ($n = 16$) (left) or with combined simvastatin-niacin and antioxidant supplements ($n = 13$) (right). *, **, and *** denote significant differences between baseline and treatment levels at $P \leq 0.01$, 0.005, and 0.001, respectively. Source: Cheung *et al.* (2001).

simvastatin plus niacin, vitamins, simvastatin–niacin plus antioxidants; or placebos. The end-points were arteriographic evidence of a change in coronary stenosis and the occurrence of a first cardiovascular event (death, MI, stroke, or revascularization). The mean levels of LDL and HDL cholesterol were unaltered in the antioxidant group and the placebo group; these levels changed substantially (by –42% and +26%, respectively) in the simvastatin–niacin group. The protective increase in HDL2 with simvastatin plus niacin was attenuated by concurrent therapy with antioxidants. The average stenosis progressed by 3.9% with placebos, 1.8% with antioxidants ($P = 0.16$ for the comparison with the placebo group), and 0.7% with simvastatin–niacin plus antioxidants ($P = 0.004$) and regressed by 0.4% with simvastatin–niacin alone ($P < 0.001$). The frequency of the clinical end-point was 24% with placebos; 3% with simvastatin–niacin alone; 21% in the antioxidant-therapy group; and 14% in the simvastatin–niacin-plus antioxidants group.

Interpretation. Simvastatin plus niacin provides marked clinical and angiographically measurable benefits in patients with coronary disease and low HDL levels. The use of antioxidant vitamins in this setting must be questioned.

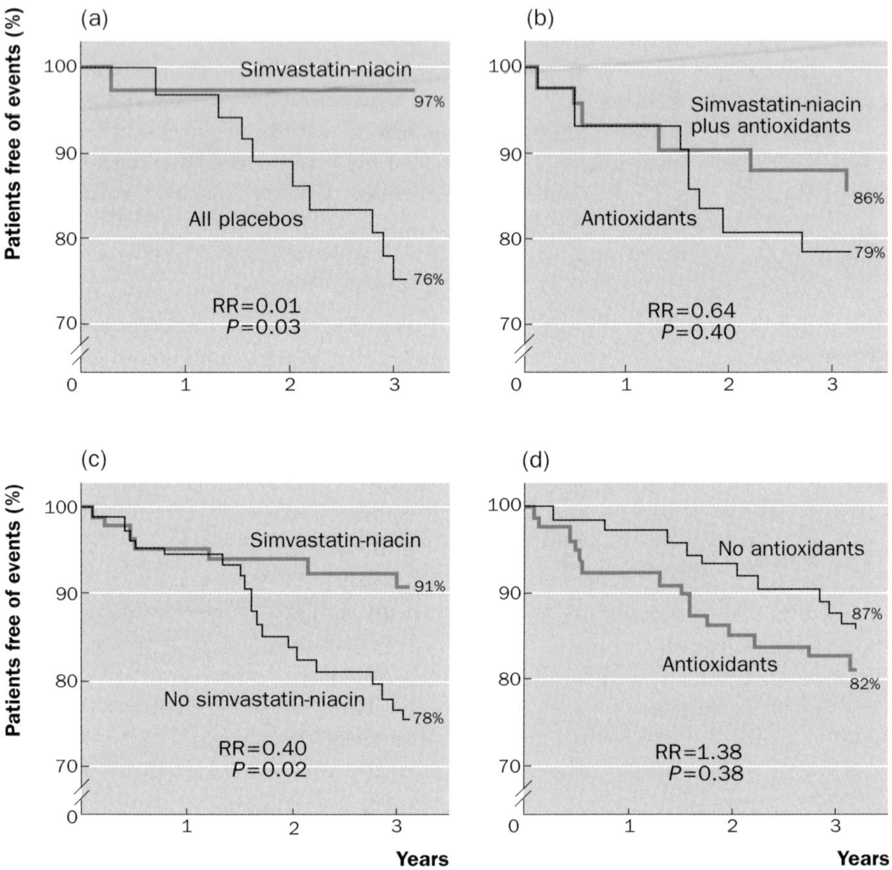

Fig. 6.5 Kaplan-Meier curves for the time to the first of the components of the composite primary clinical end-point (death from coronary causes, non-fatal myocardial infarction, confirmed stroke, or revascularization for worsening ischaemia). Panel (a) shows the curves for the 38 patients in the simvastatin–niacin group and for the 38 in the placebo group; the relative risk (RR) of an event was 0.10 (95% CI 0.01 to 0.81). Panel (b) shows the curves for the 42 patients assigned to receive simvastatin–niacin and antioxidants and for the 42 in the antioxidant group. Panel (c) shows the curves for all 80 patients who were assigned to receive simvastatin plus niacin and for the 80 who were not. Panel (d) shows the curves for the 84 patients who were assigned to receive antioxidants and for the 76 who were not. Source: Brown *et al.* (2001).

Comment (last two papers)

The first of the reports (Cheung *et al.*) revealed some surprising data, namely that supplemental antioxidants blunted the increase in apolipoprotein A-I, as well as the increase in particle size distribution of HDL, seen in response to simvastatin and niacin therapy. Specifically, the muted HDL2 cholesterol increase is of clinical concern as evidence in the literature relates a low level of HDL2 to CAD and the severity and progression of atherosclerotic lesions. The supplemental antioxidants were a cocktail of vitamins C, E, β-carotene and selenium and it is not possible from these data, to attribute this effect to any individual or combination antioxidant. However, there is a growing body of evidence that antioxidant supplements are, at best, ineffective in favourably influencing CAD risk and the growing suspicion that they may adversely affect risk can be entertained.

The second report (Brown *et al.*) provides an update with angiographic and outcome data. Brown *et al.* provide information on the effect of therapy with simvastatin and niacin with a combination of antioxidants alone and together, on angiographic changes and relative clinical end-points in patients with cardiovascular disease and low HDL cholesterol levels. The purpose of the combined therapy with a statin and niacin was to lower LDL cholesterol and raise HDL cholesterol simultaneously. The average age of the patients was 53 years and the majority were male; approximately 50% were hypertensive, former smokers, had a previous MI, had undergone angioplasty and 16% were diabetics.

Treatment with antioxidants alone lowered the HDL2 subfraction of total HDL and, moreover, blunted the increase of HDL2 seen with lipid-regulating therapy.

As compared to placebo, antioxidant therapy non-significantly attenuated the formation of plaque. When antioxidants were combined with simvastatin and niacin, arterial and clinical benefits tended to diminish, compared with those achieved using simvastatin and niacin alone. The adverse reaction between the two therapeutic strategies was significant in terms of angiographic end-points ($P = 0.02$) but not significant in terms of clinical end-points ($P = 0.13$). This interaction appears to result from a substantial and specific blunting by these vitamins of the expected increase in the level of protective HDL2 subfraction. The unexplained results in a clinical trial are often chance findings given the numerous possible analyses. The statistical test of significance is not meaningful in such an analysis because there was no *a priori* hypothesis before the data were reviewed. It is probably better to provide a point estimate of the effect and the confidence limits. These unusual findings however cannot be ignored just because they were not part of the planned analysis. It will require other investigators to either substantiate or refute these observations with different data sets.

In summary, in patients with low HDL cholesterol levels (commonly seen in patients with type 2 diabetes), treatment with a statin and niacin is associated with regression of plaque and a decrease in cardiac events. Antioxidant supplementation in such patients may abrogate or attenuate the benefits of lipid-lowering therapy. Extrapolation of this data to other clinical settings is fraught with difficulties

Table 6.8 Numbers of patients with at least one event among various categories of major clinical events during the 38-month follow-up period*

Cardiovascular event	Placebos (n = 38)	Simvastatin-niacin (n = 38)	Antioxidant vitamins (n = 42)	Simvastatin-niacin plus antioxidants (n = 42)
Death				
Non-cardiovascular causes	0	0	0	0
Cardiovascular causes	1	0	0	1
Non-fatal infarction				
Myocardial	4	1	1	3
Cerebral	2	0	2	0
Composite of death from cardiovascular causes or non-fatal infarction	7	1	3	3
Revascularization procedure				
Coronary bypass surgery	2	0	3	3
Coronary angioplasty or stenting	4	1	5	2
Carotid endarterectomy	0	0	0	0
Peripheral endarterectomy or grafting	0	0	2	0
Composite of death from cardiovascular causes, non-fatal infarction, or revascularization procedure†	9	1‡	9	6
Hospitalization for confirmed ischaemia without revascularization	3	0	4	1
Composite of death from cardiovascular causes, non-fatal infarction, revascularization procedure, or hospitalization for confirmed ischaemia	12	1§	11	6

*All enrolled patients were included in this intention-to-treat analysis.
†This composite was the primary clinical end-point.
‡P = 0.04 by Fisher's exact test for the comparison with the placebo group (adjusted for multiple comparisons).
§P = 0.003 by Fisher's exact test for the comparison with the placebo group (adjusted for multiple comparisons).

Source: Brown *et al.* (2001)

and clearly a larger trial is needed to address these intriguing issues more vigorously.

The initial enthusiasm for antioxidant therapy to prevent or treat cardiovascular disease has been substantially tempered by a series of negative clinical trials for both vitamin E and C and β-carotene. Clinical prudence would suggest against advocating pharmacological supplementation of antioxidants and vitamins in patients, particularly those patients who are taking lipid-regulating therapies that effectively raise HDL2 subfraction. However, these issues require further evaluation.

For further elucidation on the paper by Cheung *et al.* (see above), the reader is referred to the editorial by Lewis H Kuller in the same issue of the journal |**16**|.

Effects of pioglitazone and rosiglitazone on blood lipid levels and glycemic control in patients with type 2 diabetes mellitus: a retrospective review of randomly selected medical records.

Boyle PJ, King AB, Olansky L, *et al. Clin Ther* 2002; **24**: 378–96.

BACKGROUND. The antihyperglycemic effects of pioglitazone hydrochloride and rosiglitazone maleate are well documented. The results of clinical trials and observational studies have suggested, however, that there are individual differences in the effects of these drugs on blood lipid levels. This study evaluated the effects of pioglitazone and rosiglitazone on blood lipid levels and glycaemic control in patients with type 2 DM. This was a retrospective review of randomly selected medical records from 605 primary care practices in the United States in which adults with type 2 diabetes received pioglitazone or rosiglitazone between 1 August 1999, and 31 August 2000. The outcome measures were mean changes in serum concentrations of TGs, total cholesterol (TC), HDL-C, LDL-C, and glycosylated haemoglobin (HbA_{1c}) values. Treatment with pioglitazone was associated with a reduction in mean TG of 55.17 mg/dl, a reduction in TC of 8.45 mg/dl, an increase in HDL-C of 2.65 mg/dl, and a reduction in LDL-C of 5.05 mg/dl. Treatment with rosiglitazone was associated with a reduction in mean TG of 13.34 mg/dl, an increase in TC of 4.81 mg/dl, a reduction in HDL-C of 0.12 mg/dl, and an increase in LDL-C of 3.56 mg/dl. With the exception of HDL-C, the differences in mean changes in lipid parameters between treatment groups were statistically significant (*P* <0.001, pioglitazone *vs* rosiglitazone). Reductions in HbA1c were statistically equivalent between treatments (1.04% pioglitazone, 1.18% rosiglitazone).

INTERPRETATION. Treatment with pioglitazone was associated with greater beneficial effects on blood lipid levels than treatment with rosiglitazone, whereas glycaemic control was equivalent between the two treatments.

Comment

Troglitazone, the first of the thiazolidinediones (TZDs) available for clinical use, has been withdrawn because of drug-related idiosyncratic hepatic toxicity and hepatic failure that led to 63 reported deaths. The TZDs, a new class of antihyperglycaemic agents, exert their pharmacological actions via peroxisone proliferator-activator receptors (PPAR) that control genes involved in adipocyte differentiation, fatty acid metabolism and insulin regulation. Two TZDs are currently available in clinical practice – pioglitazone hydrochloride and rosiglitazone maleate.

The withdrawal of troglitazone provided an opportunity to compare and contrast the treatment effects of both pioglitazone and rosiglitazone in routine clinical settings, both in TZD naïve patients and in those who required a switch from troglitazone on its withdrawal.

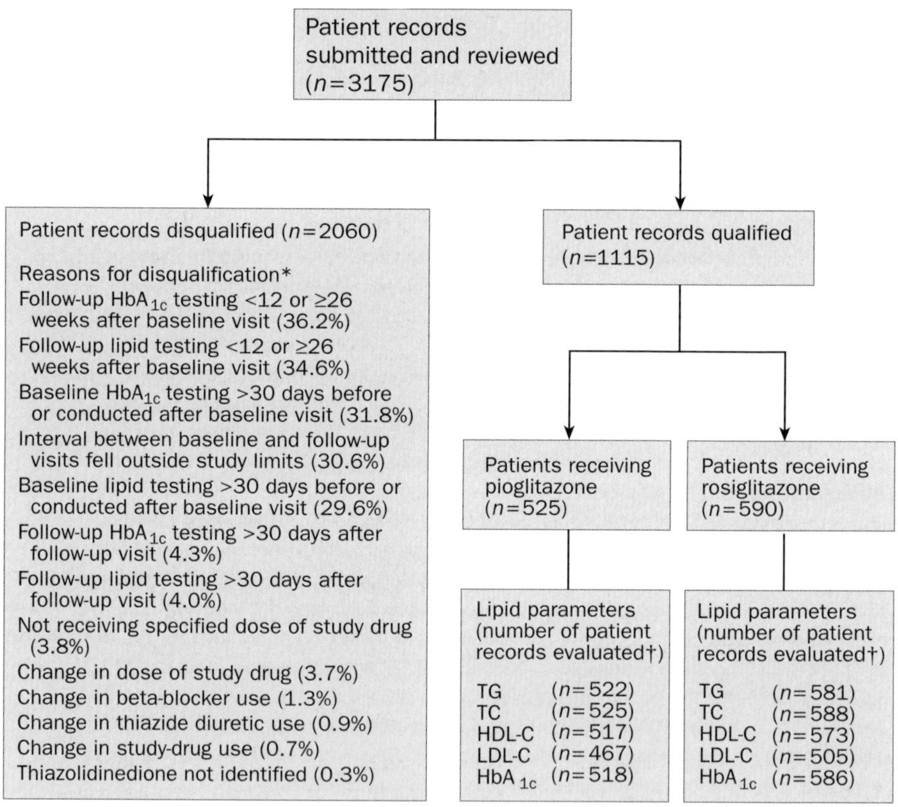

Fig. 6.6 Profile of study population. HbA1c =glycosylated haemoglobin; TG =triglycerides; TC =total cholesterol; HDL-C =high-density lipoprotein cholesterol; LDL=C =low-density lipoprotein cholesterol. *Patients may have been disqualified for >1 reason. Calculated percentages are based on total number of patient records submitted. †Assessment of each end-point included only qualified patients with complete baseline and follow-up laboratory values. Source: Boyle *et al.* (2002).

This is a retrospective review of randomly selected medical records from primary care practices in the US. The relative utility of retrospective research in evaluation of therapeutic interventions compared to prospective research has been a focus of heated debate. Critical assessments of the relative merits of both approaches show remarkable concordance in results between well-conducted observational studies and randomized control trials. However, there are some lingering doubts about the retrospective approach and concerns over confounding bias.

Nevertheless, this was a rigorously conducted study with practically identical demographic characteristics in the two treatment groups.

In this study, patients receiving pioglitazone showed significantly greater benefit in terms of changes in lipid profiles compared to those receiving rosiglitazone, but glycaemic control was similar in both groups. In essence, pioglitazone treatment was associated with a decrease in LDL cholesterol, whilst rosiglitazone treatment was associated with a slight rise in LDL cholesterol. Moreover, HDL levels increased in the pioglitazone group whilst HDL levels showed a minimal reduction in the rosiglitazone group. With respect to TG levels, pioglitazone treatment resulted in a 23% reduction and rosiglitazone treatment had a 6% reduction.

The limitations of this study have been discussed and the changes in lipid levels were over a 4-month period that began with initiation of TZD therapy. Clearly, longer-term prospective studies are required. Nevertheless, individual compounds in the TZD class of drugs cannot be assumed to cause equivalent changes in metabolic profiles – particularly as the mode of action of this class of drug is so complex. Each compound needs to be critically appraised, particularly with respect to macrovascular risk factors. At the end of the day, it is macrovascular disease which causes approximately 80% of mortality in type 2 diabetes.

A prospective, randomized comparison of the metabolic effects of pioglitazone or rosiglitazone in patients with type 2 diabetes who were previously treated with troglitazone.

Khan MA, St Peter JV, Xue JL. *Diabetes Care* 2002; **25**: 708–11.

BACKGROUND. The purpose of this study was to characterize potential differences in glycaemic control, plasma lipid level, and weight in a cohort of patients previously treated with troglitazone who were switched to either pioglitazone or rosiglitazone. After a 2-week washout from troglitazone, 186 patients were randomly assigned to receive either pioglitazone or rosiglitazone. Weight, HbA(1c), and fasting lipid profile were documented before discontinuing troglitazone and at 4 months after starting either pioglitazone or rosiglitazone. Secondarily, the effect of concurrent medications on study outcomes was assessed. A total of 127 patients completed follow-up: 67 individuals in the pioglitazone group (32 women, 35 men) and 60 individuals in the rosiglitazone group (33 women, 27 men). There were no significant differences in gender mix, age, weight, fasting lipid profile, or HbA(1c) between the rosiglitazone and

Table 6.9 Baseline characteristics of patients completing the protocol

	Rosiglitazone	**Pioglitazone**	***P* value**
Gender (male/female)	27/33	35/32	0.42
Age (years)	57.1 ± 12.1	57.8 ± 11.0	0.73
Weight (kg)	103.2 ± 24.8	101.4 ± 24.2	0.67
BMI (kg/m^2)	35.6 ± 7.4	35.2 ± 7.4	0.76
HbA_{1c}	7.9 ± 1.9	8.0 ± 1.7	0.58
Cholesterol (mg/dl)	190.7 ± 44.1	196.9 ± 44.5	0.43
HDL cholesterol (mg/dl)	45.3 ± 15.2	44.7 ± 15.6	0.83
LDL cholesterol (mg/dl)	105.9 ± 29.7	116.2 ± 38.0	0.10
Triglyceride (mg/dl)	236.0 ± 222.0	181.0 ± 110.1	0.19
HMG-CoA reductase inhibitor use	35 (58)	40 (60)	0.88
Metformin use	16 (27)	22 (33)	0.45
Insulin use	25 (42)	39 (58)	0.06
Sulfa use	40 (67)	41 (61)	0.52

Data are means ± SD or *n* (%).

Source: Khan *et al.* (2002)

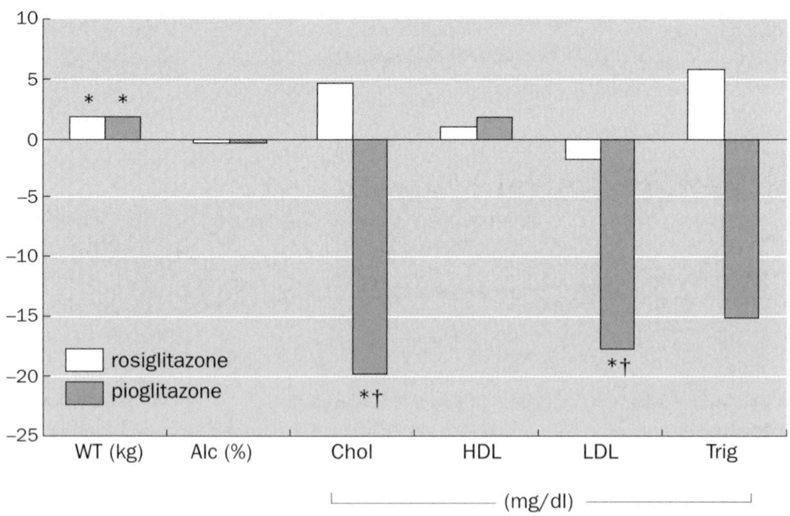

Fig. 6.7 Average changes from baseline through month 4 of randomized treatment after conversion from troglitazone to either pioglitazone or rosiglitazone. Units for each factor are listed. *$P<0.01$, within group comparison, baseline *vs* value at 4 months; †$P<0.01$; pioglitazone *vs* rosiglitazone. Source: Khan *et al.* (2002).

pioglitazone groups. After 4 months of randomized treatment, no change in HbA(1c) from baseline between or within groups was noted. Both groups experienced an equal and significant increase in weight from baseline of approximately 2.0 kg. TZD and HMG-CoA reductase inhibitor therapy had significant and independent effects on lipid profile (*P* <0.005). Significant improvements in lipid profile were noted in the pioglitazone group (*P* <0.01), whereas none were detected with conversion to rosiglitazone. Specifically, the pioglitazone group experienced an average decrease in total cholesterol of approximately 20 mg/dl.

Interpretation. Differing effects on lipid profile were apparent after random conversion from troglitazone to either pioglitazone or rosiglitazone, despite similar weight increase and glycaemic control. The clinical significance of these differences remains to be determined, and further comparative research is warranted.

Comment

This is a similar study to the previous paper by Boyle *et al.*, and subject to the same methodological reservations. After a 2-week washout from troglitazone therapy, 186 patients were randomly assigned to either pioglitazone or rosiglitazone therapy. Weight gain and glycaemic control were equivalent in both groups but again, a differential effect on the lipid profile was observed. The main differences were a significantly greater reduction in both LDL cholesterol and total cholesterol in the pioglitazone-treated group compared to the rosiglitazone-treated group.

The clinical significance of these differences remain to be evaluated and our previous comments (see Boyle *et al.*) that there is a need for longer-term prospective studies to evaluate the relative merits of these two TZDs are reinforced by these data.

Fibrate–statin-associated myotoxicity

Myotoxicity, in particular rhabdomyolysis, is a well-recognized complication of the combination of statins and fibrates, approximately 25% of the rhabdomyolysis-related mortality due to cerivastatin was associated with cerivastatin–gemfibrozil combination therapy. Initial evidence suggested that this interaction was purely pharmacodynamic. The fact that both statin and fibrate monotherapy each cause myopathy, could explain the observed higher risk of muscle toxicity when both drugs were taken together. The metabolism of fibrates is complex, and although it is attributed to the CYP 3A4 pathway, the precise mechanisms are controversial. Gemfibrozil and other fibrates are not however known to inhibit CYP 3A4-mediated statin elimination and thus should have no effect on systemic statin levels. Furthermore, statins are not known to significantly inhibit any CYP-450 pathways, thus statins should not significantly affect plasma fibrate levels and should not act as perpetrators in any pharmokinetic fibrate–statin interaction.

Recent evidence has, however, demonstrated that concomitant gemfibrozil and simvastatin therapy leads to increased plasma concentrations of simvastatin through non-CYP 3A4 mechanisms as *in vitro* evidence suggests that gemfibrozil has no effect on CYP 3A4 function. A further study demonstrates that gemfibrozil also leads to increased plasma levels of lovastatin, whilst bezafibrate has no effects on plasma lovastatin levels, an observation that supports the notion that there may be differences between individual fibrates. The mechanisms mediating these observations at present remain unclear. Glucuronidation is a metabolic pathway for simvastatin, atorvastatin and cerivastatin (but not pravastatin), and part of the effect of gemfibrozil on statin metabolism may be mediated by its ability to inhibit this process. Some preliminary *in vitro* data suggests that glucuronidation of cerivastatin may be more susceptible to inhibition by gemfibrozil than is the case with simvastatin and atorvastatin. Furthermore, there is accumulating *in vitro* data suggesting that CYP-2C8-mediated metabolism of cerivastatin, but not 3A4-mediated metabolism of atrovastatin or simvastatin, is inhibited by gemfibrozil and that these effects are more pronounced with gemfibrozil than with fenofibrate. Recently published studies also suggest that gemfibrozil may also inhibit the CYP-2C9 isoform, thus leading to potential interaction problems with fluvastatin.

Part of the hypolipidaemic effect of fibrates is mediated through activation of the peroxisome proliferator activated family of nuclear receptors (PPAR), in particular PPAR-α and γ. This aspect of fibrate action may contribute to the increased risk of interaction with statins, since the PPAR family of receptors have been shown to influence the regulation and function of CYP 2,3 and 4. The influence of the PPAR system on CYP function also raises important considerations relating to possible interactions between statins and the thiazolidinedione class of oral hypoglycaemic agents, which exert a therapeutic effect by activating the PPAR-γ receptor. Initial short-term studies involving relatively small numbers of subjects appear reassuring, with no evidence of an increase in adverse reactions related to combination statin and TZD therapy. These observations are however preliminary and a true evaluation of the safety of statin–TZD combination therapy can only be made following larger long-term studies combined with close post-marketing surveillance.

References

1. UK Prospective Diabetes Study 27. Plasma lipids and lipoproteins at diagnosis of NIDDM by age and sex. *Diabetes Care* 1997; **20**: 1683–7.
2. Evans M, Khan N, Rees A. Diabetic dyslipidaemia and coronary heart disease: new perspectives. *Curr Opin Lipidol* 1999; **10**: 387–91.
3. Rees A. Diabetic dyslipidaemia. In: Durrington P, Mackness M, Miller P, Rees A (eds). *The Year in Dyslipidaemia.* Oxford: Clinical Publishing Services Ltd, 2002: 187–228.
4. The Expert Panel. Executive summary of the Third Report of the NCEP Expert Panel on Detection, Evaluation and Treatment of High Blood Cholesterol in Adults (Adult Treatment Panel III). *JAMA* 2001; **285**: 2486–98.

5. Downs JR, Clearfield M, Weis S, *et al.* Primary prevention of acute coronary events with lovastatin in men, and women with average cholesterol levels: results of the AFCAPS/TexCAPS, Air Force/Texas Coronary Atherosclerosis Prevention Study. *JAMA* 1998; **279**: 1615–22.

6. Scandinavian Simvastatin Survival Study Group. Randomized trial of cholesterol lowering in 4444 patients with coronary heart disease. Scandinavian Simvastatin Survival study. *Lancet* 1994; **344**: 1383–9.

7. Sacks FM, Pfeffer MA, Moye LA, Rouleau JL, Rutherford JO, *et al.* for the Cholesterol and Recurrent Events Trial Investigators. The effect of Pravastatin on coronary events after myocardial infarction in patients with average cholesterol levels. Cholesterol and Recurrent Events Trials Investigators. *N Engl J Med* 1996; **335**: 1001–9.

8. Prevention of cardiovascular events and death with Pravastatin in patients with coronary heart disease and a broad range of initial cholesterol levels, The long term intervention with Pravastatin and Ischaemic Disease (LIPID) Study Group. *N Engl J Med* 1998; **339**: 1349–57.

9. Pyorala K, Pederson TR, Kjeksus J, Faergerman O, Olsson AG, Thorgeirsson G. Cholesterol lowering with Simvastatin improves prognosis of diabetic patients with coronary heart disease: a subgroup analysis of the Scandinavian Simvastatin Survival Study (4S). *Diabetes Care* 1997; **20**: 614–20.

10. Haffner SM, Alexander CMA, Cook TJ, *et al.* Reduced coronary events in simvastatin treated patients with coronary heart disease and diabetes or impaired fasting glucose levels. *Arch Intern Med* 1999; **159**: 2661–7.

11. Goldberg RB, Mellies MJ, Sacks FM, *et al.* Cardiovascular events and their reduction with Pravastatin in diabetic and glucose-intolerant myocardial infarction survivors with average cholesterol levels: subgroup analyses in the cholesterol and recurrent events (CARE) trial. The CARE Investigators. *Circulation* 1998; **98**: 2513–19.

12. Sacks F, Tonkin A, Shepherd J, *et al.* Effects of Pravastatin on coronary disease events in subgroups defined by coronary risk factors: The Prospective Pravastatin Pooling Project. *Circulation* 2000; **102**: 1893–1900.

13. Koskinen P, Manttari M, Manninen V, Huttunen JK, Heinonen OP, Frick MH. Coronary heart disease incidence in NIDDM patients with the Helsinki Heart Study. *Diabetes Care* 1992; **15**: 820–5.

14. Rubins HB, Robins SJ, Collins D, *et al.* Gemfibrozil for the secondary prevention of coronary heart disease in men with low levels of high-density lipoprotein cholesterol. Veterans Affairs High-Density Lipoprotein Cholesterol Intervention Trial Study Group. *N Engl J Med* 1999; **341**: 410–18.

15. Elkeles RS, Diamond JR, Poulter C, *et al.* Cardiovascular outcomes in type 2 diabetes: a double-blind placebo-controlled study of Bezafibrate. The St. Mary's, Ealing, Northwick Park Diabetes Cardiovascular Disease Prevention (SENDCAP) Study. *Diabetes Care* 1998; **21**: 641–8.

16. Kuller LH. A time to stop prescribing antioxidant vitamins to prevent and treat heart disease? *Arterioscler Thromb Vasc Biol* 2001; **21**: 1253.

Part III

Lipids and atherosclerosis

Lipids and atherosclerosis

Introduction

Lipids and atherosclerosis is an extremely wide field encompassing many scientific disciplines. An attempt has been made in this section to concentrate on a few major themes of interest to the author. It is therefore not 'all inclusive'. However, in future issues of *The Year in Dyslipidaemia* other themes will emerge and will be included. The themes for this issue are the pleitrophic effects of statins, the antiatherogenic effects of high-density lipoproteins and antioxidants, and the cardiovascular effects of oxidized low-density lipoprotein. Also included for the first time is a section on clinical chemistry, outlining some important methodological advances.

7

Non-lipid lowering effects of statins

Introduction

The so-called pleitrophic effects of statins have again been on centre stage in this review period. Doubts have been expressed that these effects are truly pleitrophic i.e. independent of the inhibition of 3-hydroxy-3-methylglutaryl coenzyme A (HMG-CoA) reductase or that, indeed, the concentrations of statins used in *in vitro* experiments are truly achieveable in the peripheral tissues *in vivo* at currently used pharmacological statin doses. It remains fairly obvious that statins have effects that are independent of their reduction of low-density lipoprotein (LDL) cholesterol. However, the biochemical pathways that mediate these pleitrophic effects are probably the same as those inhibiting cholesterol biosynthesis with some interesting extra effects. This chapter, therefore, includes manuscripts which investigate the mechanisms behind the non-lipid lowering effects of statins.

Statin-induced inhibition of the Rho-signaling pathway activates PPARα and induces HDL apo A-I.

Martin G, Duez H, Blanquart C, *et al. J Clin Invest* 2001; **107**: 1423–32.

BACKGROUND. Statins are inhibitors of the rate-limiting enzyme in cholesterol synthesis, HMG-CoA reductase. In addition to reducing LDL cholesterol, statin treatment increases the levels of antiatherogenic high-density lipoprotein (HDL) and its major apolipoprotein apoA-I. Here, the molecular mechanisms of apoA-I regulation by statins was investigated. Treatment with statins increased apoA-I mRNA (messenger ribonucleic acid) levels in human HepG2 hepatoma cells and this effect was reversed by the addition of mevalonate, implicating HMG-CoA reductase as the relevant target of these drugs. Pre-treatment with actinomycin D abolished the increase of apoA-I mRNA, indicating that statins act at the transcriptional level. Indeed, statins increased the human apoA-I promoter activity in transfected cells, and a statin response element was identified that coincides with a peroxisome proliferator-activated receptor-α (PPARα) response element known to confer fibrate responsiveness to this gene. The statin effect could be abolished not only by mevalonate, but also by geranylgeranyl pyrophosphate, whereas inhibition of geranylgeranyl transferase activity or treatment with an inhibitor of the

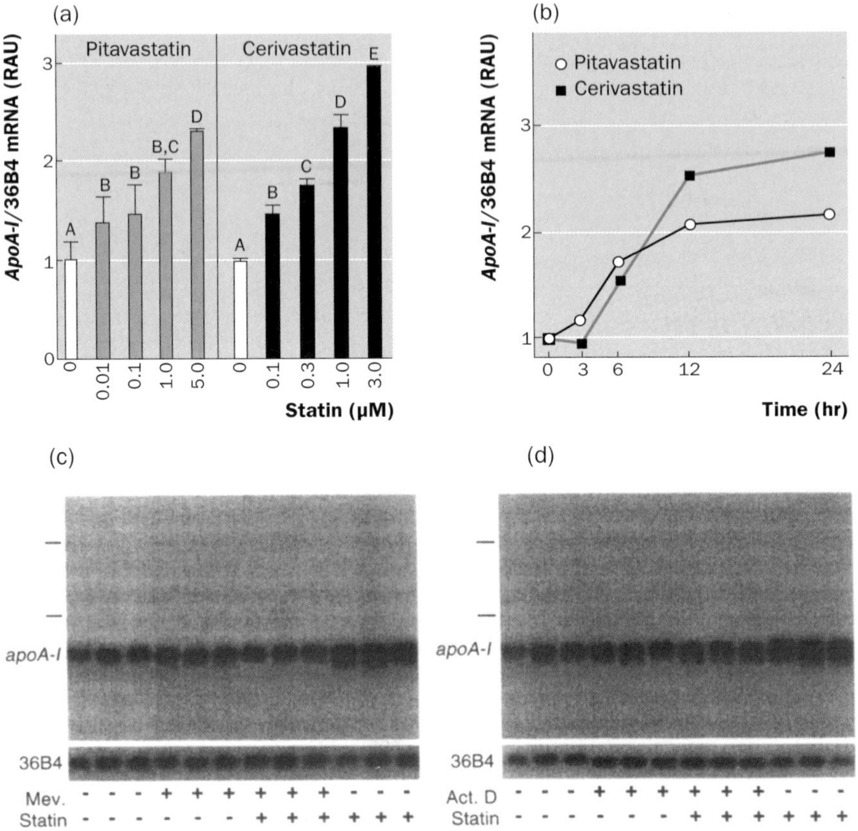

Fig. 7.1 Statins induce apoA-I gene expression at the transcriptional level in HepG2 cells by inhibition of HMG-CoA reductase. (a) Statins induce apoA-I mRNA in HepG2 cells in a dose-dependent manner. HepG2 cells were treated for 24 h with the indicated doses of cerivastatin or pitavastatin or appropriate solvent in serum-free medium. RNA levels were quantified and expressed (means $\pm$ SD, $n = 3$/point) relative to the untreated control set at 1. Values without a common superscript are significantly different, $P < 0.05$. RAU, relative arbitrary units. (b) Statins induce apoA-I mRNA in HepG2 cells in a time-dependent manner. HepG2 cells were treated for 3, 6, 12 or 24 h with 5 μM of cerivastatin, or pitavastatin, or appropriate solvent in serum-free medium. RNA levels were quantified and expressed relative to the untreated controls at each time point set at 1. (c) Mevalonate reverses the induction of apoA-I mRNA by statins. HepG2 cells were treated for 24 h with cerivastatin (5 μM) and/or mevalonate (Mev; 3mM) or appropriate solvent in serum-free medium. (d) Statins regulate apoA-I gene expression in HepG2 cells at the transcriptional level. HepG2 cells were treated for 24 h with cerivastatin (5 μM) and/or mevalonate (Mev; 3μM) or appropriate solvent in serum-free medium. Actinomycin D (Act D; 5 μg/ml) was added to the medium 90 min before treatment. For all, total RNA (10 μg) was subjected to Northern blot analysis using human apoA-I and 36B4 cDNA probes. Source: Martin *et al.* (2001).

Rho GTP-binding protein family increased PPARα activity. Using dominant negative forms of these proteins, it was found that Rho A itself mediates this response.

INTERPRETATION. Because cotreatment with statins and fibrates activated PPARα in a synergistic manner, these observations provide a molecular basis for combination treatment with statins and fibrates in coronary heart disease (CHD).

Comment

This is the first evidence of cross-talk between the statin and PPARα signalling systems and indicates PPARα as a target for both fibrates and statins which could therefore be combined for more effective treatment of CHD. Statins induced apoA-I transcription in a cultured human hepatoma cell line (Hep G2) in a dose-dependent manner (Fig. 7.1)

This is not, however, true in people treated with statins, where increases in HDL do not appear to be strictly dose-dependent due probably to large individual variations in response. New generations of statins could be engineered to have larger HDL-increasing ability and dose dependency.

HMG-CoA reductase inhibitors prevent migration of human coronary smooth muscle cells through suppression of increase in oxidative stress.

Yasunari K, Maeda K, Minami M, Yoshikawa J. *Arterioscler Thromb Vasc Biol* 2001; **21**: 937–42.

BACKGROUND. ***In vitro*** **and** ***in vivo*** **evidence of a decrease in vascular smooth muscle cell (SMC) migration induced by CoA reductase inhibitors has been reported. When added to SMC cultures for six hours, the HMG-CoA reductase inhibitors fluvastatin, simvastatin, and pravastatin at 1 micromol/l resulted in a 48%, 50%, and 16% suppression, respectively, of human coronary SMC migration; these reductions mirrored the suppression in oxidative stress induced by 1 micromol/l lysophosphatidylcholine (lyso-PC) of 50%, 53% and 19%, respectively. The hydroxylated metabolites of fluvastatin, M(2) and M(3), at 1 micromol/l also suppressed the enhancement of SMC migration by 58% and 45% and the increase in oxidative stress induced by lyso-PC of 58% and 49%, respectively. Lyso-PC activated phospholipase D and protein kinase C (PKC), and this activation was also suppressed by HMG-CoA reductase inhibitors. The inhibition of phospholipase D and PKC was reversed by 100 micromol/l mevalonate, its isoprenoid derivative, farnesol, and geranylgeraniol but not by 10 micromol/l squalene. Antisense oligodeoxynucleotides at 5 micromol/l to PKC-α, but not those to the PKC-α isoform, suppressed the lyso-PC-mediated increases in SMC migration and oxidative stress.**

INTERPRETATION. These findings suggest that HMG-CoA reductase inhibitors have direct antimigratory effects on the vascular wall beyond their effects on plasma lipids and that they might exert such antimigratory effects via suppression of the phospholipase D-

and PKC (possibly PKC-α)-induced increase in oxidative stress, which might in turn prevent significant coronary artery disease.

Comment

This study indicates that statins can work by inhibiting oxidized LDL-mediated intra-cellular signalling, probably by inhibiting prenylation of signalling proteins (Fig. 7.2).

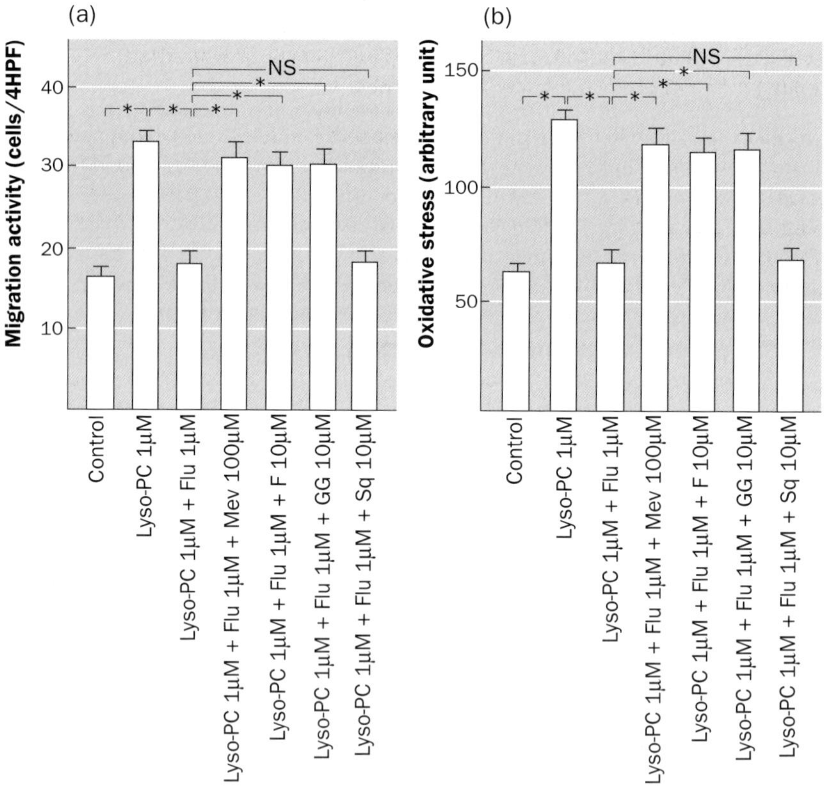

Fig. 7.2 Effects of Mevalonate (Mev), its isoprenoid derivatives, farnesol (F), geranylgeraniol (GG), and squalene (Sq) on the suppression of coronary SMC migration (a), and oxidative stress (b) by fluvastatin (Flu). Values are mean ± SD of 8 determinations. *$P < 0.05$, significant difference between the bracketed groups. NS indicates not significant.. Source: Yasunari *et al.* (2001).

In vivo anti-inflammatory effect of statins is mediated by nonsterol mevalonate products.

Diomede L, Albani D, Sottocorno M, *et al. Arterioscler Thromb Vasc Biol* 2001; **21**: 1256–8.

BACKGROUND. This study set out to clarify whether the inhibition of sterol or non-sterol derivatives arising from mevalonate biotransformation plays a major role in the *in vivo* anti-inflammatory action of statins. Hepatic synthesis of all these derivatives was inhibited in mice by administered statins, whereas squalestatin inhibited only sterol derivatives. Using a short-term treatment schedule, it was found that statins reduced the hepatic activity of HMG CoA reductase without affecting blood cholesterol. This treatment inhibited lipopolysaccharide- and carrageenan-induced pouch leukocyte recruitment and the exudate production of interleukin-6, monocyte chemotactic protein-1, and RANTES. Coadministration of mevalonate reversed the effect of statin on leukocyte recruitment.

INTERPRETATION. The inhibition of sterol synthesis by squalestatin did not have any anti-inflammatory effect, indicating that the biosynthesis of non-sterol compounds arising from mevalonate is crucial for the *in vivo* regulation of cytokine and chemokine production by statins. Their inhibition by statins may account for the reported anti-inflammatory effects of these drugs and may provide a biochemical basis for the recently reported effects of statins in the prevention of cardiovascular disease and mortality.

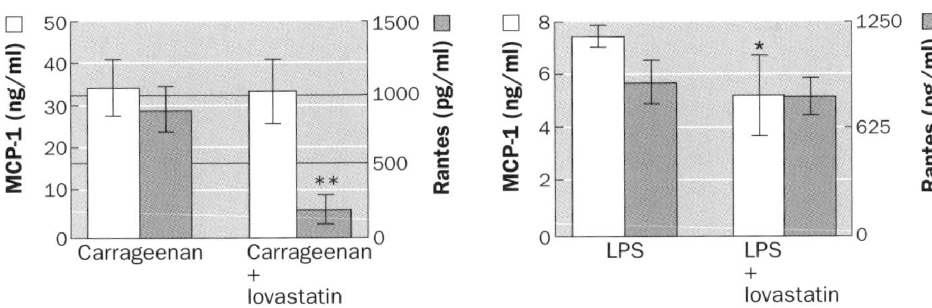

Fig. 7.3 Effect of lovastatin on chemokine production. Subcutaneous dorsal pouches were created in CD1 mice. Inflammation was induced by injection of 1 mL 1% carrageenan in saline per mouse or 0.2 μg LPS in 1 ml 0.5% CMC per mouse into the pouch. Controls received the corresponding vehicle alone. Lovastatin, 10 mg/kg, was given orally to mice 0.5 h before and 8 and 20 h after carrageenan or 20, 12, and 0.5 h before LPS. Histograms show the mean ± SD of MCP-1 (white bars) and RANTES (grey bars) exudate levels of at least 6 animals.*$P<0.05$, **$P<0.01$ *vs* LPS or carrageenan alone, Student's *t* test.Source: Diomede *et al.* (2001).

Comment

This paper gives a further indication that non-lipid lowering effects of statins are due to the inhibition of isoprenoid production via the inhibition of HMG-CoA reductase (Fig. 7.3).

Simvastatin exerts both anti-inflammatory and cardioprotective effects in apolipoprotein E-deficient mice.

Scalia R, Gooszen ME, Jones SP, *et al. Circulation* 2001; **103**: 2598–603.

BACKGROUND. Simvastatin attenuates ischaemia and reperfusion in normocholesterolaemic animals by stabilizing endothelial nitric oxide synthase activity and inhibiting neutrophil-mediated injury. Because endothelial dysfunction is a detrimental effect of hypercholesterolaemia, these investigators examined whether short-term treatment with simvastatin could inhibit leukocyte–endothelium interaction and attenuate myocardial ischaemia-reperfusion injury in apoE-deficient (apoE(-/-))

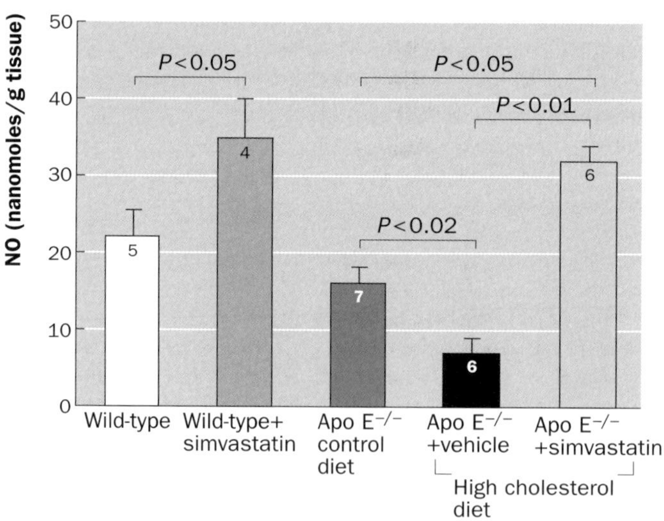

Fig. 7.4 Basal release of NO expressed as nmol/g tissue. NO release was measured in isolated mouse aortic rings obtained from wild-type mice, wild-type mice receiving simvastatin, apoE-/- mice fed a control diet, and apoE-/- mice fed a high-cholesterol diet and given either vehicle or 1 mg/kg simvastatin 18 h before study. Short-term simvastatin treatment significantly increased NO release in wild-type and hypercholesterolaemic apoE-/- mice. Source: Scalia *et al.* (2001).

mice fed a high-cholesterol diet. ApoE(-/-) mice fed a high-cholesterol diet exhibited higher blood cholesterol levels, which were not affected by short-term simvastatin treatment. However, the increased leukocyte rolling and adherence that occurred in cholesterol-fed apoE(-/-) mice (*P* <0.001 *versus* control diet) were significantly attenuated by simvastatin treatment (*P* <0.01 *versus* vehicle). Cholesterol-fed apoE(-/-) mice subjected to myocardial ischaemia-reperfusion also experienced increased myocardial necrosis (*P* <0.01 *versus* control diet), which was significantly attenuated by simvastatin (*P* <0.01 *versus* vehicle). Simvastatin therapy also significantly increased vascular nitric oxide production in apoE(-/-) mice.

INTERPRETATION. Simvastatin attenuates leukocyte-endothelial cell interactions and ameliorates ischaemic injury in hypercholesterolaemic mice independently of lipid-lowering actions.

Comment

Statins may exert protective effects by increasing nitric oxide production, which is inhibited by hypercholesterolaemia. Nitric oxide is required for normal function of the cardiovascular system and has cardioprotective effects (Fig. 7.4).

Pravastatin treatment increases collagen content and decreases lipid content, inflammation, metalloproteinases, and cell death in human carotid plaques: implications for plaque stabilization.

Crisby M, Nordin-Fredriksson G, Shah PK, Yano J, Zhu J, Nilsson J.
Circulation 2001; **103**: 926–33.

BACKGROUND. The clinical benefits of lipid lowering with statins are attributed to changes in plaque composition leading to lesion stability, but supporting clinical data from human studies are lacking. Therefore, this study investigated the effect of three months of pravastatin treatment on composition of human carotid plaques removed during carotid endarterectomy. Consecutive patients with symptomatic carotid artery stenosis received 40 mg/d pravastatin ($n = 11$) or no lipid-lowering therapy ($n = 13$; control subjects) for three months before scheduled carotid endarterectomy. Carotid plaque composition was assessed with special stains and immunocytochemistry with quantitative image analysis. Plaques from the pravastatin group had less lipid by oil red O staining (8.2 ± 8.4% *vs* 23.9 ± 21.1% of the plaque area, *P* <0.05), less oxidized LDL immunoreactivity (13.3 ± 3.6% *vs* 22.0 ± 6.5%, *P* <0.001), fewer macrophages (15.0 ± 10.2% *vs* 25.3 ± 12.5%, *P* <0.05), fewer T cells (11.2 ± 9.3% *vs* 24.3 ± 13.4%, *P* <0.05), less matrix metalloproteinase 2 (MMP-2) immunoreactivity (3.6 ± 3.9% *vs* 8.4 ± 5.3%, *P* <0.05), greater tissue inhibitor of metalloproteinase 1 (TIMP-1) immunoreactivity (9.0 ± 6.2% *vs* 3.1 ± 3.9%, *P* <0.05), and a higher collagen content by Sirius red staining (12.4 ± 3.1% *vs* 7.5 ± 3.5%, *P* <0.005). Cell death by TUNEL staining was reduced in the pravastatin group (17.7 ± 7.8% *vs* 32.0 ± 12.6%, *P* <0.05).

Interpretation. Pravastatin decreased lipids, lipid oxidation, inflammation, MMP-2, and cell death and increased TIMP-1 and collagen content in human carotid plaques, confirming its plaque-stabilizing effect in humans.

Comment

Plaque stabilization by statins is a potentially hugely important property as it would lead to a reduction of acute events and could potentially lead on to plaque regression, one of the holy grails of cardiovascular research.

Discussion

The recent large primary and secondary prevention trials of statins (see Chapter 12) have shown profound reduction of LDL cholesterol accompanied by a reduction in cardiovascular events of approximately 30%. However, statins also have non-lipid lowering effects as described in this chapter. These appear to add to the overall effect of statins in reducing both cardiovascular and non-cardiovascular mortality. As the use of statins increases, as it undoubtedly will as physicians take on board the evidence base, these effects should enhance the effectiveness of statin therapy. It will be interesting to see in the future whether statins will be of benefit in the treatment of non-cardiovascular diseases.

8

High-density lipoprotein and atherosclerosis

Introduction

The mechanism(s) by which high-density lipoprotein (HDL) protects against the development of atherosclerosis continues to be the centre of research. Whether this is by reverse-cholesterol transport or the anti-oxidative and anti-inflammatory properties of HDL or a combination of all is the subject of much debate. It is probable that all mechanisms contribute to the anti-atherosclerotic properties of HDL. This chapter, therefore, contains a number of papers representing this broad field, and includes further epidemiological and experimental evidence for a role of HDL in a much wider context than the prevention of coronary heart disease (CHD). This includes protection against stroke and lipid-peroxidation of cell membranes. In this regard, the similarity in structure between cell membranes and lipoproteins may be more important in the anti-atherosclerotic mechanism of HDL than previously believed.

High-density lipoprotein cholesterol and ischemic stroke in the elderly: the Northern Manhattan Stroke Study.

Sacco RL, Benson RT, Kargman DE, *et al. JAMA* 2001; **285**: 2729–35.

BACKGROUND. Elevated high-density lipoprotein cholesterol (HDL-C) levels have been shown to be protective against cardiovascular disease. However, the association of specific lipoprotein classes and ischaemic stroke has not been well defined, particularly in higher-risk minority populations. The objective of this study was to evaluate the association between HDL-C and ischaemic stroke in an elderly, racially or ethnically diverse population. Population-based, incident case–control study was conducted from July 1993 to June 1997. A multiethnic community in northern Manhattan, New York, NY was studied. Cases (*n* = 539) of first ischaemic stroke (67% aged ≥65 years; 55% women; 53% Hispanic, 28% black, and 19% white) were enrolled and matched by age, gender and race or ethnicity to stroke-free community residents (controls; *n* = 905).

After risk factor adjustment, a protective effect was observed for HDL-C levels of at least 35 mg/dl (0.91 mmol/l) (odds ratio [OR], 0.53; 95% confidence interval [CI],

0.39–0.72). A dose–response relationship was observed (OR, 0.65; 95% CI, 0.47–0.90 and OR, 0.31; 95% CI, 0.21–0.46) for HDL-C levels of 35 to 49 mg/dl (0.91–1.28 mmol/l) and at least 50 mg/dl (1.29 mmol/l), respectively. The protective effect of a higher HDL-C level was significant among participants aged 75 years or older (OR, 0.51; 95% CI, 0.27–0.94), was more potent for the atherosclerotic stroke subtype (OR, 0.20; 95% CI, 0.08–0.50), and was present in all three racial or ethnic groups studied.

INTERPRETATION. Increased HDL-C levels are associated with reduced risk of ischaemic stroke in the elderly and among different racial or ethnic groups. These data add to the evidence relating lipids to stroke and support HDL-C as an important modifiable stroke risk factor.

Comment

This manuscript provides further strong epidemiological evidence that HDL is protective against cardiovascular disease. Interestingly, the data (Fig. 8.1) shows that HDL protection against stroke is independent of ethnic origin; higher HDL levels provide stronger protection.

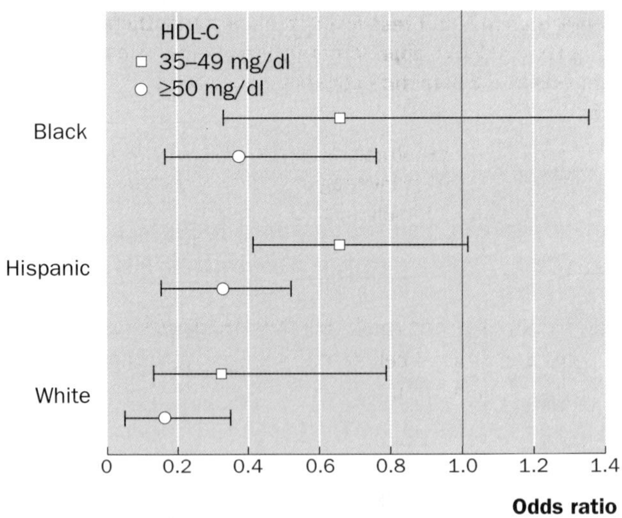

Fig. 8.1 Protective-dose response relationship of high-density lipoprotein cholesterol (HDL-C) and stroke risk stratified by race and ethnicity. Matched for age and sex and adjusted for hypertension, diabetes mellitus, body mass index, CHD, physical activity, and educational level. To convert HDL-C from mg/dl to mmol/l, multiply by 0.0259. Error bars indicate 95% CI. Source: Sacco *et al.* (2001).

Elevation of plasma high-density lipoprotein concentration reduces interleukin-1-induced expression of E-selectin in an *in vivo* model of acute inflammation.

Cockerill GW, Huehns TY, Weerasinghe A, *et al. Circulation* 2001; **103**: 108–12.

BACKGROUND. Although there is strong evidence that plasma HDL levels correlate inversely with the incidence of coronary artery disease, the precise mechanism(s) for the protective effect of HDLs remains unclear. The authors recently showed that HDLs inhibit endothelial cell expression of cytokine-induced leukocyte adhesion molecules *in vitro*. The current study therefore sought to test the hypothesis that elevating the level of circulating HDLs would inhibit endothelial cell activation *in vivo*. A porcine model of inflammation previously established in the same laboratory was used, in which the level of vascular endothelial cell expression of E-selectin in interleukin (IL)-1alpha-induced skin lesions was measured by the uptake of a radiolabelled anti-E-selectin antibody (1.2B6). Porcine plasma HDL levels were elevated by use of a bolus injection of reconstituted discoidal HDL (recHDL). These particles resemble nascent HDL particles in shape and contain apolipoprotein (apo)A-I as the sole protein and soybean phosphatidylcholine as the sole phospholipid. It was found that recHDLs inhibited the expression of IL-1alpha-induced E-selectin by porcine aortic endothelial cells *in vitro*, confirming that the inhibitory effect is conserved with synthetic HDLs and demonstrating that the phenomenon is not restricted to human endothelial cells. *In vivo*, elevating the circulating level of HDLs approximately 2-fold led to significant inhibition of basal and IL-1alpha-induced E-selectin expression by porcine microvascular endothelial cells.

INTERPRETATION. These observations demonstrate the potential anti-inflammatory action of HDLs and provide support for the further investigation of the mechanisms underlying the inhibitory effects of HDLs on endothelial cell activation.

Comment

This is one of the few reports to date to confirm *in vivo* that HDL acts as an anti-inflammatory agent and reduces endothelial cell activation by inhibiting the pro-inflammatory action of interleukin-1 (Fig. 8.2).

Time sequence of the inhibition of endothelial adhesion molecule expression by reconstituted high density lipoproteins.

Clay MA, Pyle DH, Rye K-A, Vadas MA, Gamble JR, Barter PJ. *Atherosclerosis* 2001; **157**: 23–29.

BACKGROUND. This investigation used discoidal reconstituted high-density lipoproteins (rHDL) containing apoA-I and dimyristoyl phosphatidylcholine (DMPC) as a tool to investigate the time sequence of the HDL-mediated inhibition of vascular cell

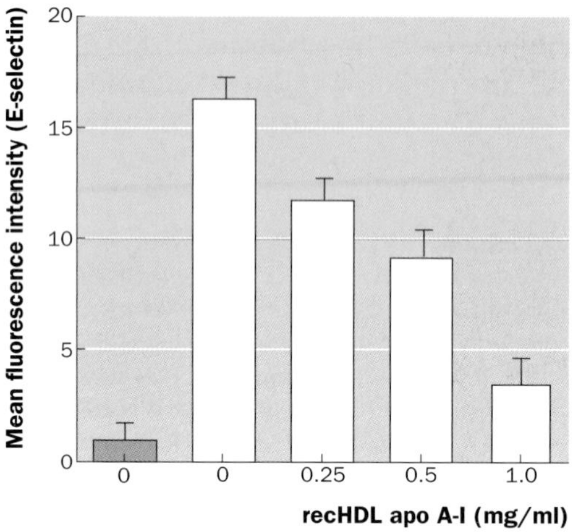

Fig. 8.2 RecHDLs inhibit IL-1α–induced expression of E-selectin by PAECs. Concentration-dependent effect of recHDLs on IL-1–induced expression of E-selectin by PAEs. Confluent monolayers of endothelial cells were incubated in presence of a range of recHDLs for 3 h. IL-1 (10 ng/ml) was then added to culture medium. Cell-surface expression of E-selectin was measured by flow cytometry 4 h later. Values represent mean fluorescence intensities of E-selectin in response to IL (white bars) relative to unstimulated basal expression (grey bar). Values are mean ± SD, $n = 6$. Statistical analysis using 1-factor ANOVA, followed by Dunnett's *t* test, $P < 0.01$. Data are representative of 3 similar experiments. Source: Cockerill *et al.* (2001).

adhesion molecule (VCAM)-1 and E-selectin expression in cytokine-activated human umbilical vein endothelial cells (HUVECs). Specifically, a few questions were asked: (i) how long do the cells need to be exposed to the rHDL before adhesion molecule expression is inhibited and (ii) how long does the inhibition persist after removing the rHDL from the cells? When the cells were not pre-incubated with rHDL, there was no inhibition. The magnitude of the inhibition increased progressively with increasing duration of pre-incubation up to 16 h. Inhibition did not require the rHDL to be physically present during the activation of adhesion molecule expression by tumour necrosis factor (TNF)-α, excluding the possibility that the rHDL was merely interfering with the interaction between TNF-α and the cells. When HUVECs were pre-incubated for 16 h with rHDL, the inhibition remained substantial even if the rHDL was removed from the medium up to 8 h prior to addition of TNF-α. The HDL-mediated inhibition of VCAM-1 in HUVECs was unaffected by the presence of puromycin, an inhibitor of protein synthesis, excluding the possibility that HDL may have acted by stimulating the synthesis of a cell protein that itself inhibits adhesion molecule expression.

Interpretation. These results have important implications in terms of understanding the mechanism(s) of the HDL-mediated inhibition of endothelial adhesion molecule expression.

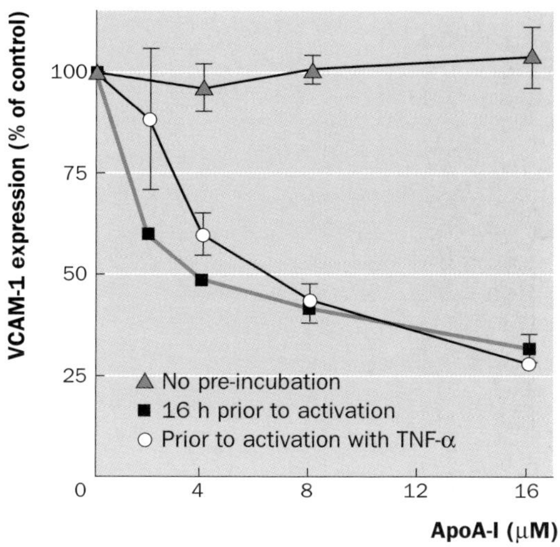

Fig 8.3 Requirement for pre-incubation on the inhibition of VCAM-1 by A-I/DMPC rHDL. Increasing concentrations of A-I/DMPC rHDL were added to HUVECs immediately prior to activation of the cells with TNF-α or 16 h prior to activation of the cells. An additional set of cells was pre-incubated with A-I/DMPC rHDL for 16 h, but the rHDL were removed from the cells prior to activation with TNF-α. VCAM-1 expression was determined 5 h after addition of TNF-α. Results are expressed as mean ± SEM (n = 6). Source: Clay *et al.* (2001).

Comment

The ability of HDL to prevent endothelial cell VCAM-1 expression is an important anti-inflammatory property. The requirement for endothelial cells to be pre-incubated with HDL to inhibit VCAM-1 expression (Fig. 8.3) indicates that HDL is donating some protective factor. The nature of this factor is unknown but its discovery could have important implications for cardiovascular and other inflammatory diseases.

On the ability of high density lipoproteins to remove phospholipid peroxidation products from erythrocyte membranes.

Klimov AN, Kozhevnikova KA, Kuzmin AA, Kuznetsov AS, Belova EV. *Biochemistry (Moscow)* 2001; **66**: 371–7.

BACKGROUND. The aim of this investigation was to study the transfer of oxidized phospholipids from cell membranes to HDL. Human Cu^{2+}-oxidized erythrocyte

Table 8.1 Content of LPP in HDL_3 subfraction precipitated from the supernatant after incubation of HDL_3 with Cu^{2+}-oxidized erythrocyte membranes for 17h at 37 °C (average data of 6 experiments, M ± SD)

Incubated material	**LPP**					
	CD		**LHP**		**MDA**	
	nmole/ml sample	**nmole/mg protein**	**nmole/ml sample**	**nmole/mg protein**	**nmole/ml sample**	**nmole/mg protein**
HDL_3 (control)	44.25 ± 9.76	7.83 ± 1.58	47.22 ± 10.49	8.36 ± 1.34	12.85 ± 1.74	2.30 ± 0.44
HDL_3 + Cu^{2+}-oxidized membranes	40.59 ± 6.09 (ns)	7.66 ± 1.38 (ns)	62.12 ± 4.84*	11.69 ± 1.19**	16.31 ± 1.77***	3.09 ± 0.54*

Note: Incubation medium contained sodium salt of benzylpenicillin (0.1 IU/ml) and streptomycin sulfate (1 μg/ml); ns, non-significant differences.
* $P < 0.02$.
** $P < 0.002$.
*** $P < 0.01$.

Source: Klimov *et al.* (2001).

membranes were incubated with HDL_3 subfraction for 17 h at 37°C followed by isolation of the supernatant, precipitation from it of HDL_3, and determination of lipid peroxide products (LPP) in them. The incubation increased the content of lipid hydroperoxides in HDL_3 significantly by (32 and 40% calculated per ml of sample or mg of protein) and of malondialdehyde (by 27 and 34%, respectively) compared to control (incubation of HDL_3 alone). The content of conjugated dienes did not change significantly. Fluorescence analyses of isolated HDL_3 particles showed that the content of fluorescent products (exitation wavelength $\lambda_{ex} = 365$nm, emission wavelength $\lambda_{em} = 430$nm) in them was 2.5 times higher than in control, and the number of binding sites for the 1-anilinonaphthalene-8-sulfonic acid probe decreased by 22%. This also confirms accumulation of LPP in the lipoprotein subfraction. It seems likely that an increase in LPP (at least hydroperoxides) in HDL_3 after their incubation with oxidized membranes occurs via transport of phospholipids containing LPP from erythrocyte membranes to lipoproteins.

Interpretation. The data on the ability of HDL_3 to accept LPP from erythrocyte membranes *in vitro* suggest that HDL_3 may have a protective action on cell membranes undergoing oxidation *in vivo* as well.

Comment

HDL is ubiquitously distributed in tissue fluid. This important manuscript provides evidence that HDL can protect cell membranes from oxidative damage (Table 8.1). It is highly likely that the ability of HDL to protect LDL from oxidative damage, is due to the resemblance of LDL to a cell membrane.

Apolipoprotein A-I promotes the formation of phosphatidylcholine core aldehydes that are hydrolysed by paraoxonase (PON1) during high density lipoprotein oxidation with a peroxynitrite donor.

Ahmed Z, Ravandi A, Maguire GF, *et al. J Biol Chem* 2001; **276**: 24473–81.

Background. HDL is rich in polyunsaturated phospholipids that are sensitive to oxidation. However, the effect of apoA-I and paraoxonase-1 (PON1) on phosphatidylcholine oxidation products has not been identified. These investigators subjected native HDL, trypsinized HDL, and HDL lipid suspensions to oxidation by the peroxynitrite donor, 3-morpholinosydnonimine. HDL had a basal level of phosphatidylcholine mono- and di-hydroperoxides that increased to a greater extent in HDL, compared with either trypsinized HDL or HDL lipid alone. Phosphatidylcholine core aldehydes, which were present in small amounts, increased ten-fold during oxidation of native HDL, compared with trypsinized HDL ($P = 0.004$), and four-fold compared with HDL lipid suspension ($P = 0.0021$). In addition, the content of lysophosphatidylcholine increased 300% during oxidation of native HDL, but only 80 and 25%, respectively, during oxidation of trypsinized HDL and HDL lipid suspensions. Phosphatidylcholine isoprostanes accumulated in comparable amounts during the oxidation of all three preparations. Incubation of apolipoprotein AI with 1-palmitoyl-2-linoleoyl

glycerophosphocholine proteoliposomes in the presence of 3-morphollinosydnonimine or apoA-I with phosphatidylcholine hydroperoxides resulted in a significant increase in phosphatidylcholine core aldehydes with no formation of lysophosphatidylcholine. It was proposed that apoA-I catalyzes a one-electron oxidation of alkoxyl radicals. Purified PON1 hydrolyzed phosphatidylcholine core aldehydes to lysophosphatidylcholine.

Interpretation. It was concluded that, upon HDL oxidation with peroxynitrite, apoA-I increases the formation of phosphatidylcholine core aldehydes that are subsequently hydrolyzed by PON1.

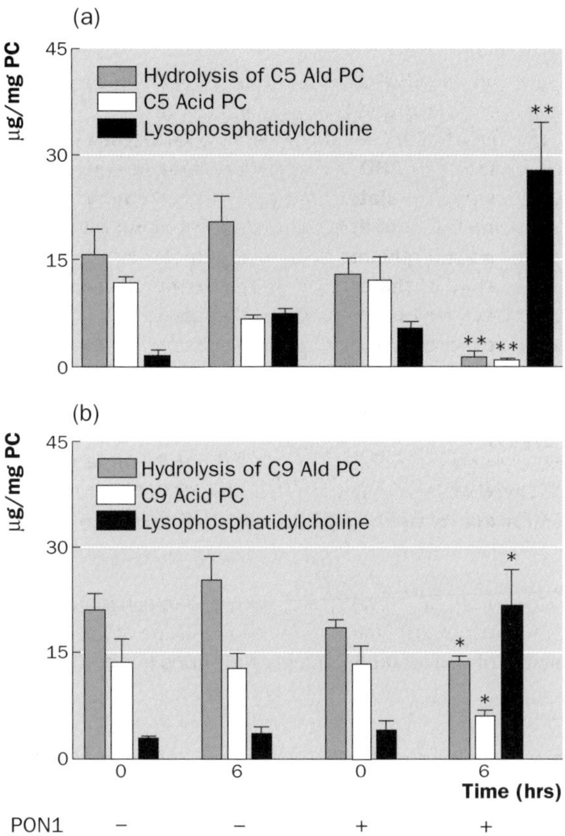

Fig. 8.4 The effect of PON1 on the core aldehydes and carboxylate proteoliposomes. (a) Hydrolysis of C5 Ald PC (grey bar) and C5 Acid PC (white bar) by PON1 to produce lysophosphatidylcholine (black bar). (b) Hydrolysis of C9 Ald PC (grey bar) and C9 Acid PC (white bar) by PON1 to produce lyso PC (black bar). Values represent the mean ± SEM of three separate experiments. Source: Ahmed *et al.* (2001).

Comment

The manuscript provides some insight into the mechanism by which PON1 protects against oxidative damage by hydrolyzing oxidatively derived 9 carbon aldehydes of phosphatidylcholine (Fig. 8.4).

Paraoxonase status in coronary heart disease: are activity and concentration more important than genotype?

Mackness B, Davies GK, Turkie W, *et al. Arterioscler Thromb Vasc Biol* 2001; **21**: 1451–7.

BACKGROUND. Human serum paraoxonase (PON1) hydrolyzes oxidized lipids in low-density lipoprotein (LDL) and could therefore retard the development of atherosclerosis. In keeping with this hypothesis, several case–control studies have shown a relationship between the presence of CHD and polymorphisms at amino acid positions 55 and 192 of PON1, which was associated with a decreased capacity of PON1 to protect LDL against the accumulation of lipid peroxides, but some other studies have not. However, the PON1 polymorphisms are only one factor in determining the activity and concentration of the enzyme. Only 3 of the previous 18 studies directly determined PON1 activity and concentration. Therefore, this investigation studied PON1 activity, concentration, and gene distribution in 417 subjects with angiographically proven CHD and in 282 control subjects. It was found that PON1 activity and concentration were significantly lower in subjects with CHD than in control subjects (activity to paraoxon 122.8 [3.3–802.8] *versus* 214.6 [26.3–620.8] nmol/min/ml, P <0.01; concentration 71.6 [11.4–489.3] *versus* 89.1 [16.8–527.4] μg/ml, P <0.001). There were no differences in the PON1-55 and -192 polymorphisms or clusterin concentration between patients with CHD and control subjects.

INTERPRETATION. These results indicate that lower PON1 activity and concentration and, therefore, the reduced ability to prevent LDL lipid peroxidation may be more important in determining the presence of CHD than paraoxonase genetic polymorphisms.

Comment

There have been many case–control studies investigating PON1 as a risk factor for CHD, however, very few have actually measured PON1 activity. This manuscript shows low PON1 (and not PON1 gene polymorphisms) to be a risk factor for CHD (Table 8.2) and indicates the importance of measuring activity in all future case–control and prospective studies.

Table 8.2 PON1 parameters and clusterin in the study populations

	Control	**CHD**
PON1		
Concentration, μg/ml	89.1 (16.8–527.4)	71.6* (11.4–489.3)
Paraoxon hydrolysis, nmol · min^{-1} · ml serum^{-1}	214.6 (26.3–620.8)	122.8* (3.3–802.8)
Specific activity, nmol · min^{-1} · μg^{-1}	2.10 (0.22–20.2)	1.70* (0.1–26.4)
Clusterin, μg/ml	115.3 ± 32	111.0 ± 62
Clusterin/PON1 activity	0.54 (0.03–4.87)	0.85 (0.01–15.22)
ApoAI/PON1 activity	0.62 (0.13–3.79)	0.67 (0.02–6.47)
HDL-C/PON1 activity	$7.2 \times 10^{-3} \pm 4.1 \times 10^{-3}$	$12.1 \times 10^{-3} \pm 5.3 \times 10^{-3}$
Clusterin/PON1 concentration	1.48 ± 0.06	1.85 ± 0.08
ApoAI/PON1 concentration	1.49 ± 0.07	1.36 ± 0.05
HDL-C/PON1 concentration	$18.0 \times 10^{-3} \pm 0.07 \times 10^{-3}$	$24.2 \times 10^{-3} \pm 0.09 \times 10^{-3}$

Values are mean ± SD or median (range).
**P* <0.001 *vs* control.

Source: Mackness *et al.* (2001).

Atherogenic role of elevated CE transfer from HDL to VLDL(1) and dense LDL in type 2 diabetes : impact of the degree of triglyceridemia.

Guerin M, Le Goff W, Lassel TS, Van Tol A, Steiner G, Chapman MJ.
Arterioscler Thromb Vasc Biol 2001; **21**: 282–8.

Background. Plasma cholesteryl ester transfer protein (CETP) facilitates intravascular lipoprotein remodelling by promoting the heteroexchange of neutral lipids. To determine whether the degree of triglyceridaemia may influence the CETP-mediated redistribution of HDL cholesteryl ester (CE) between atherogenic plasma lipoprotein particles in type 2 diabetes, CE mass transfer from HDL to apoB-containing lipoprotein acceptors was evaluated in the plasma of type 2 diabetes subjects (n = 38). In parallel, the potential relationship between CE transfer and the appearance of an atherogenic dense LDL profile was investigated. The diabetic population was divided into three subgroups according to fasting plasma triglyceride (TG) levels: group 1 (G1), TG <100 mg/dl; group 2 (G2), TG <200 mg/dl; and group 3 (G3), TG >200 mg/dl. Type 2 diabetes patients displayed an asymmetrical LDL profile in which the dense LDL subfractions predominated. Plasma levels of dense LDL subfractions were strongly positively correlated with those of plasma TG (r = 0.471; P = 0.0003). The rate of CE mass transfer from HDL to apoB-containing lipoproteins was significantly enhanced in G3 compared with G2 or G1 (46.28.1, 33.65.3, and 28.22.7 μg CE transferred/h/ml in G3, G2, and G1, respectively; P <0.0001 G3 *versus* G1, P = 0.0001 G2 *versus* G1, and P = 0.02 G2 *versus* G3). The relative

capacities of very low-density lipoprotein (VLDL) and LDL to act as acceptors of CE from HDL were distinct between type 2 diabetes subgroups. LDL particles represented the preferential CE acceptor in G1 and accounted for 74% of total CE transferred from HDL. By contrast, in G2 and G3, TG-rich lipoprotein subfractions accounted for 47% and 72% of total CE transferred from HDL, respectively. Moreover, the relative proportion of CE transferred from HDL to VLDL(1) in type 2 diabetes patients increased progressively with increase in plasma TG levels. The VLDL(1) subfraction accounted for 34%, 43%, and 52% of total CE transferred from HDL to TG-rich lipoproteins in patients from G1, G2, and G3, respectively. Finally, dense LDL acquired an average of 45% of total CE transferred from HDL to LDL in type 2 diabetes patients.

INTERPRETATION. CETP contributes significantly to the formation of small dense LDL particles in type 2 diabetes by a preferential CE transfer from HDL to small dense LDL, as well as through an indirect mechanism involving an enhanced CE transfer from HDL to VLDL(1), the specific precursors of small dense LDL particles in plasma.

Comment

CETP catalyzed transfer from HDL to apoB containing lipoproteins is elevated in type 2 diabetes (Table 8.3) and results in a more atherogenic lipoprotein profile, providing further evidence for a pro-atherogenic role of CETP.

Table 8.3 Rates of cholesteryl ester transfer from HDL to apoB-containing lipoproteins in type 2 diabetic patients and control subjects

	μg CE transferred · h^{-1} · ml plasma^{-1}			
Acceptor lipoprotein fraction	**G0**	**G1**	**G2**	**G3**
Total apoB-containing lipoproteins	49.1 ± 6.4	28.2 ± 2.7‡	33.6 ± 5.3‡‖	46.2 ± 8.1¶#
Total TRLs	13.9 ± 2.6	7.4 ± 2.4‡	15.8 ± 2.4†¶	33.5 ± 9.1‡¶††
VLDL$_1$ (Sf 60–400)	9.0 ± 1.6	2.5 ± 1.2‡	6.8 ± 2.2*¶	17.4 ± 8.0†¶††
VLDL$_2$ (Sf 20–60)	2.2 ± 0.8	1.9 ± 1.0	3.6 ± 1.1†‖	7.5 ± 1.4‡¶††
IDL (Sf 12–20)	2.7 ± 0.8	3.0 ± 1.3	5.4 ± 1.8‡‖	8.6 ± 2.4‡¶**
Total LDL (*d* 1.109–1.063 g/ml)	35.2 ± 4.9	20.8 ± 2.5‡	17.8 ± 3.1‡§	12.7 ± 2.1‡¶††
LDL$_1$ (*d* 1.019–1.023 g/ml)	3.6 ± 0.8	1.8 ± 0.6‡	1.7 ± 0.5‡	1.6 ± 0.4‡
LDL$_2$ (*d* 1.023–1.029 g/ml)	9.9 ± 1.7	2.8 ± 0.7‡	2.5 ± 0.7‡	1.7 ± 0.4‡¶**
LDL$_3$ (*d* 1.029–1.039 g/ml)	12.3 ± 2.0	7.3 ± 1.3‡	5.9 ± 1.5‡§	3.3 ± 1.0‡¶††
LDL$_4$ (*d* 1.039–1.050 g/ml)	6.0 ± 0.9	5.7 ± 1.3	5.3 ± 1.0	3.9 ± 0.9‡‖**
LDL$_5$ (*d* 1.050–1.063 g/ml)	3.4 ± 0.7	3.2 ± 0.6	2.4 ± 0.5‡‖	2.2 ± 0.7‡‖

Rates of CE mass transfer determined in normolipidaemic subjects (G0, $n = 14$) and in 3 subgroups of type 2 diabetes patients: G1, G2, and G3. ApoB-containing lipoproteins are defined as total d <1.063 g/ml lipoproteins and include VLDL$_1$, VLDL$_2$, IDL, and LDL subfractions. Values are mean ± SD.
*P <0.05; †P <0.005, and ‡P <0.0005 *vs* G0; §P <0.05, ‖P <0.005, and ¶P <0.0005 *vs* G1;
#P <0.05, **P <0.005, and ††P <0.0005 *vs* G2.

Source: Guerin *et al.* (2001).

Variation at the cholesteryl ester transfer protein gene in relation to plasma high density lipoproteins cholesterol levels and carotid intima-media thickness.

Kakko S, Tamminen M, Paivansalo M, *et al.Eur J Clin Invest* 2001; **31**: 593–602.

BACKGROUND. CETP plays a major role in lipoprotein metabolism. This investigation screened the CETP gene for mutations and polymorphisms regulating HDL-C levels and the development of atherosclerosis, and found some polymorphisms (I405V and R451Q) to have minor effects. The purpose of this study was to investigate the combined effect of the several polymorphisms of the CETP gene so far found on HDL-C levels and carotid intima-media thickness (IMT), and, in addition, to study whether the recently found functional polymorphism in the promoter region of the CETP gene (C to A, -629 relative to the first transcribed nucleotide) explains the previous associations due to linkage disequilibrium. The genotypes were determined in a population sample of 481 men and women. There were no significant differences in plasma CETP activity or carotid IMT between the genotypes of the promoter polymorphism. The women with the CC genotype of the promoter polymorphism had the lowest HDL-C levels (*P*<0.001), but no such difference was seen in men. Detected polymorphisms of the CETP gene explained about 8% of the variation in HDL-C in women and about 7 and 10% of the variation in carotid IMT in women and men, respectively. The associations of the promoter, I405V and R451Q-A373P polymorphisms with HDL-C and carotid IMT seemed to be independent of each other. The associations with IMT were independent of total HDL-C levels, suggesting that HDL subfractions may have more effect on IMT.

INTERPRETATION. The CETP gene locus was found to be polymorphic and its polymorphisms explained a reasonable proportion of the variation in the degree of carotid atherosclerosis.

Comment

Polymorphisms of the CETP gene were related to carotid atherosclerosis, as measured by intima-media thickness (Fig. 8.5), however, the mechanism of this effect is uncertain as the polymorphisms had only a small effect on CETP activity. Further research into CETP and atherosclerosis is warranted.

Discussion

HDL appears to protect against atherosclerosis by several mechanisms. These include reverse-cholesterol transport and mechanisms which prevent inflammation and lipid-peroxidation amongst others |**1**|. Central to these functions of HDL appear to be the lipid transfer proteins and enzymes associated with HDL. Chief amongst these is the enzyme paraoxonase (PON1) which appears to be the primary

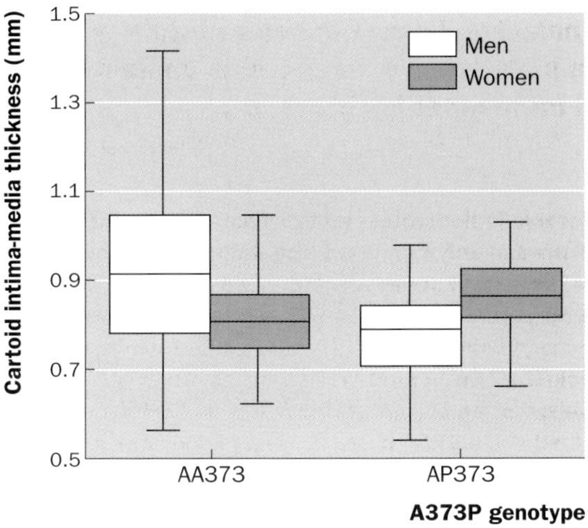

Fig. 8.5 Box plots showing the association of the A373P genotype with carotid IMT in men (white boxes) and in women (grey boxes). The box represents the interquartile range which contains 50% of the values. The lines extend from the box to the highest and lowest values, excluding outliers. A line across the box indicates the median. Source: Kakko *et al.* (2001).

anti-oxidative/anti-inflammatory enzyme of HDL |**2**|. However, HDL now appears to have protective functions not only against LDL oxidation but also that of cell membranes and its anti-inflammatory properties have been proven *in vivo*. HDL is ubiquitous throughout the tissue fluid and therefore is in a position to protect against damaging membrane lipid peroxidation in a large number of cell types. This will be an extremely interesting and potentially profitable field of exploration as also will be efforts to raise HDL in humans; the next few years will provide an extremely stimulating research environment.

References

1. Mackness MI, Durrington PN, Mackness B. How HDL protects against the effects of lipid peroxidation. *Curr Opin Lipidol* 2000; **11**: 383–8.

2. Durrington PN, Mackness B, Mackness MI. Paraoxonase and atherosclerosis. *Arterioscler Thromb Vasc Biol* 2001; **21**: 473–80.

9

Antioxidants and atherosclerosis

Introduction

The recently announced Heart Protection Study (HPS) (see Chapter 12) was a body blow to those who advocate the use of pharmacological doses of antioxidants for the prevention of coronary disease. It follows several other recent large-scale trials of antioxidant vitamins which were unable to reduce coronary risk [1,2]. *In vitro* studies continue to supply a mountain of evidence for the oxidation theory of atherosclerosis (see Chapter 10) and that antioxidants prevent many of the pro-atherogenic effects of oxidation *in vitro*. All of this leaves health-care professionals in a quandary and many more years of research will be required to reconcile the apparently contradictory findings. This is discussed in more detail in the concluding remarks to Chapter 11.

However, inhibition of lipid oxidation by physiological proteins, as described for paraoxonase in Chapter 8 or by apolipoprotein (apo)E and apoA-IV in this section may be more important than that of dietary vitamins or polyphenolic antioxidants. Recent evidence from the HPS (see concluding remarks, Chapter 11) has certainly thrown doubt on the use of pharmacological doses of antioxidant vitamins for the prevention of cardiovascular disease, however, physicians and the general public can at least take heart from the fact that they are not harmful.

Is the emperor wearing clothes? Clinical trials of vitamin E and the LDL oxidation hypothesis.

Heinecke JW. *Arterioscler Thromb Vasc Biol* 2001; **21**: 1261–4.

BACKGROUND. A wealth of evidence indicates that oxidized low-density lipoprotein (LDL) may be of central importance in animal models of atherogenesis. In recent clinical trials, however, dietary vitamin E supplements have not consistently prevented cardiac events in humans with established coronary artery disease. Such mixed results have led many to question the role of LDL oxidation in human atherosclerosis, although this interpretation assumes that the doses of vitamin E used in the studies inhibited lipid oxidation *in vivo*. In fact, there is remarkably little evidence indicating that those particular regimens effectively inhibit lipid peroxidation in healthy humans.

Moreover, evidence of increased oxidative stress was not a criterion for inclusion in the trials; therefore, vitamin E may have benefited only a subset of the participants.

Interpretation. These uncertainties raise doubts about the ability of vitamin E to augment antioxidant defence mechanisms *in vivo* and leave many questions about LDL oxidation and atherosclerosis unanswered.

Comment

This paper provides a review of clinical trials using vitamin E and provides some possible explanations for their failure.

Consumption of flavonoids in onions and black tea: lack of effect on F2-isoprostanes and autoantibodies to oxidized LDL in healthy humans.

O'Reilly JD, Mallet AI, McAnlis GT, *et al. Am J Clin Nutr* 2001; **73**: 1040–4.

Background. Oxidative damage to lipids *in vivo* may be involved in the development of atherosclerosis and cancer. Onions and black tea are foods rich in flavonoids, predominantly the flavonoid quercetin, which is a potent *in vitro* inhibitor of membrane lipid peroxidation and LDL oxidation. Thirty-two healthy humans were studied in a randomized cross-over design. Indexes of oxidative damage used were plasma F2-isoprostanes (a biomarker of lipid peroxidation *in vivo*) and the titre of antibodies to malondialdehyde (MDA)-modified LDL. There were no significant differences in the intake of macronutrients or assessed micronutrients, plasma F2-isoprostane concentrations, and MDA-LDL autoantibody titre between the high flavonoid (HF) and low flavonoid (LF) dietary treatments. In the men, however, plasma concentrations of the F2-isoprostane 8-epi-prostaglandin F2alpha were slightly higher after the HF treatment phase than after the LF treatment (0.31–0.029 nmol/l [111–10.4 ng/l] compared with 0.26–0.022 nmol/l [92–7.8 ng/l]; *P* = 0.041). In all subjects, plasma quercetin concentrations were significantly higher after the HF treatment phase than after the LF treatment: 221.6–37.4 nmol/l compared with less than the limit of detection of 66.2 nmol/l.

Interpretation. Flavonoid consumption in onions and tea had no significant effect on plasma F2-isoprostane concentrations and MDA-LDL autoantibody titre in this study and thus does not seem to inhibit lipid peroxidation in humans.

Comment

This is one of several publications to show no effect of increased consumption of dietary antioxidants on lipid peroxidation in humans (Table 9.1). Other studies using lutein, lycophene and β-carotene also showed no effect |**3**|. Interest in the potential health benefits of moderate alcohol consumption, in particular, red wine, because of its high content of antioxidant flavonols has also been dented by a study

Table 9.1 Plasma 8-*epi*-prostaglandin F_{2a} (8-*epi*-PGF_{2a}), malondialdehyde-modified (MDA)-LDL autoantibody titer, and quercetin concentrations after low-flavonoid and high-flavonoid dietary treatment periods*

	Low-flavonoid diet	**High-flavonoid diet**
8-*epi*-PGF_{2a}		
(ng/l)	107 ± 8.4	117 ± 9.7
(nmol/l)	0.30 ± 0.024	0.33 ± 0.027
MDA-LDL autoantibody titer	1.34 ± 0.02	1.33 ± 0.02
Quercetin (nmol/l)	$<$LOD ± —	$221.6^{\dagger} \pm 37.4$

* $\bar{x} \pm$ SEM. $n = 32$ except for 8-*epi*-PGF_{2a}. $n = 20$. LOD, limit of detection.
† Significantly different from low-flavonoid diet, $P < 0.01$ (paired *t* test).
Source: O'Reilly *et al.* (2001).

indicating that red wine is a poor source of bioavailable flavonols in humans compared to onions or even tea |**4**|.

Antioxidative and antiatherosclerotic effects of human apolipoprotein A-IV in apolipoprotein E-deficient mice.

Ostos MA, Conconi M, Vergnes L, *et al. Arterioscler Thromb Vasc Biol* 2001; **21**: 1023–8.

BACKGROUND. Mice expressing human apolipoprotein A-IV (apoA-IV) mainly in the intestine were obtained in an apolipoprotein E-deficient (apoE(0)) background (apoA-IV/E(0) mice). Quantification of aortic lesions and plasma lipid determination showed that compared with their control apoE(0) counterparts, the apoA-IV/E(0) mice are protected against atherosclerosis without an increase in high-density lipoprotein cholesterol. Because oxidized lipoproteins play an important role in atherogenesis, it was tested whether the protection observed in these animals is accompanied by an *in vivo* reduction of the oxidation parameters. The lag time in the formation of conjugated dienes during copper-mediated oxidation, the aggregation state of LDL, and the presence of anti-oxidized LDL antibodies were measured. The presence of oxidized proteins in tissues and the presence of oxidation-specific epitopes in heart sections of atherosclerotic lesions were also analyzed. Except for lag time, the results showed that the oxidation parameters were reduced in the apoA-IV/E(0) mice compared with the apoE(0) mice.

INTERPRETATION. This suggests that human apoA-IV acts *in vivo* as an antioxidant. In addition, human apoA-IV accumulation was detected in the atherosclerotic lesions of apoA-IV/E(0) mice, suggesting that apoA-IV may inhibit oxidative damage to local tissues, thus decreasing the progression of atherosclerosis.

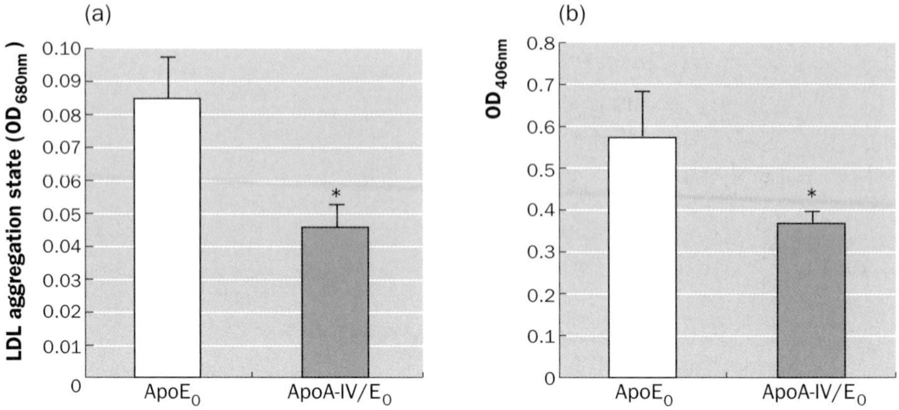

Fig. 9.1 Effect of human apoA-IV over-expression on LDL aggregation state and on plasma antioxidized LDL antibody levels. (a) LDL aggregation state in apoE$_0$ and apoA-IV/E$_0$ mice. LDL was obtained from the pooled plasma of 5 male mice. The turbidity generated by LDL aggregates was measured spectrophotometrically at 680 nm. Results are mean ± SD of 3 different experiments. $*P < 0.05$ (ANOVA). (b) Quantification of antioxidized LDL antibodies in plasma from 6-month-old apoE$_0$ ($n = 10$) and apoA-IV/E$_0$ ($n = 10$) male mice. OD indicates optical density. Data are expressed as mean ± SD. $*P < 0.02$ (ANOVA). . Source: Ostos *et al.* (2001).

Comment

The body contains numerous enzymes that are capable of acting as antioxidants. ApoA-IV has however been little studied compared to other apolipoproteins. In this interesting publication, human apoA-IV was over-expressed in the atherosclerosis susceptible apoE null mouse. ApoA-IV expression reduced atherosclerosis development due to it being able to reduce LDL aggregation and oxidation (Fig. 9.1) to pro-atherosclerotic states of LDL.

Reduction of isoprostanes and regression of advanced atherosclerosis by apolipoprotein E.

Tangirala RK, Pratico D, FitzGerald GA, *et al.* *J Biol Chem* 2001; **276**: 261–6.

BACKGROUND. ApoE is a multifunctional protein synthesized by hepatocytes and macrophages. Plasma apoE is largely liver-derived and known to regulate lipoprotein metabolism. Macrophage-derived apoE has been shown to reduce the progression of atherosclerosis in mice. The hypothesis that liver-derived apoE could directly induce regression of pre-existing advanced atherosclerotic lesions without reducing plasma cholesterol levels was tested. Aged LDL receptor-deficient (LDLR(-/-)) mice were fed a

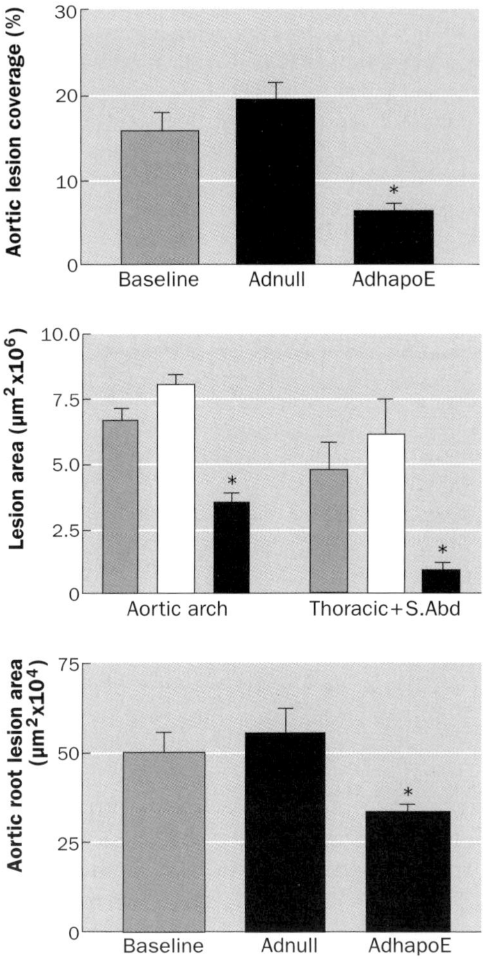

Fig. 9.2 Extent of atherosclerosis in LDLR $^{-/-}$ mice fed high fat diet for 14 weeks (baseline) and an additional 6 weeks after injection with a control (Adnull) and human apoE3 expressing virus (AdhapoE). (a) The atherosclerotic lesion area covering the aortic surface was quantified in *en* face aortic preparations and expressed as percentage of total aortic area. (b) Evaluation of apoE-induced regression in aortic arch and thoracic + supra-abdominal (S.Abd) segments of the aorta. The lesion areas in each of these segments are expressed as μm^2. The reduction in lesion areas in AdhapoE group (black bars) relative to baseline group (grey bars) in arch and thoracic + supra-adbominal portions of aorta were 48 and 81%, respectively (white bars represent Adnull group). (c) Aortic root lesion area measured in serial cross-sections through the aortic origin. The mean lesion areas per each group ($n = 9$) were shown. All data are mean ± SE with $n = 9$ in each group. Asterisks indicate statistically significant difference from baseline and Adnull groups ($P < 0.001$). Source: Tangirala *et al.* (2001).

western-type diet for 14 weeks to induce advanced atherosclerotic lesions. One group of mice was sacrificed for evaluation of atherosclerosis at baseline, and two other groups were injected with a second generation adenoviruses encoding human apoE3 or a control empty virus. Hepatic apoE gene transfer increased plasma apoE levels four-fold at one week, and apoE levels remained at least two-fold higher than controls at six weeks. There were no significant changes in plasma total cholesterol levels or lipoprotein composition induced by expression of apoE. The liver-derived human apoE gained access to and was retained in the arterial wall. Compared with baseline mice, the control group demonstrated progression of atherosclerosis; in contrast, hepatic apoE expression induced highly significant regression of advanced atherosclerotic lesions. Regression of lesions was accompanied by the loss of macrophage-derived foam cells and a trend toward increase in extracellular matrix of lesions. As an index of *in vivo* oxidant stress, we quantitated the isoprostane iPF(2 alpha)-VI and found that expression of apoE markedly reduced urinary, LDL-associated, and arterial wall iPF(2 alpha)-VI levels.

INTERPRETATION. These results demonstrate that liver-derived apoE directly induced regression of advanced atherosclerosis and has anti-oxidant properties *in vivo* that may contribute to its anti-atherogenic effects.

Comment

ApoE has been shown to be involved in several physiological processes such as lipoprotein transport and metabolism, promoting cellular cholesterol efflux, cell membrane repair (particularly in neuronal cells) and as a risk factor for Alzheimer's disease. This manuscript illustrates that apoE acts also as an antioxidant reducing the formation of isoprostanes (non-enzymatically formed oxidation products of arachidonic acid) and decreasing atherosclerosis (Fig. 9.2).

Discussion

The papers included in this section indicate an important aspect of physiology. This is one of redundancy i.e. several different systems doing the same job but in a different manner. It may well be that dietary antioxidants i.e. vitamins and (largely vegetable-derived) polyphenolics are less important in preventing, overall, cardiovascular disease than antioxidant proteins, produced by the body and which have a longer life-span. Chain-breaking antioxidants may provide the first line of defence against peroxidative damage, however, once these are exhausted the second line of defence in PON1, apoE etc. are there. There is an alternative explanation to this which is discussed in the concluding remarks to Chapter 11.

References

1. Dietary supplementation with n-3 polyunsaturated fatty acids and vitamin E after myocardial infarction: results of the GISSI-Prevenzione trial. Gruppo Italiano per lo Studio della Sopravvivenza nell'Infarto miocardico. *Lancet* 1999; **354**: 447–55.
2. Yusuf S, Dagenais G, Pogue J, Bosch J, Sleight P. Vitamin E supplementation and cardiovascular events in high risk patients. The Heart Outcomes Prevention Evaluation Study Investigators. *N Engl J Med* 2000; **342**: 154–60.
3. Hininger IA, Meyer-Wenger A, Moser U, Wright A, Southon S, Thurnham D, Chopra M, Van Den Berg H, Olmedilla B, Favier AE, Roussel AM. No significant effects of lutein, lycopene or β-carotene supplementation on biological markers of oxidative stress and LDL oxidizability in healthy adult subjects. *J Am Coll Nutr* 2001; **20**: 232–8.
4. de Vris JHM, Hollman PCH, Van Amersfoort I, Olthof MR, Katan MB. Red wine is a poor source of bioavailable flavonols in men. *J Nurtr* 2001; **131**: 745–8.

10

Oxidized low-density lipoprotein and atherosclerosis

Introduction

This chapter illustrates some of the more interesting papers on this subject, which is currently the favoured theory for the development and progression of atherosclerosis. It appears to be more favoured by *in vitro* data rather than *in vivo*, which will be dealt with in further detail in the concluding remarks to Chapter 11.

Mitogen-activated protein kinase phosphatase 1 activity is necessary for oxidized phospholipids to induce monocyte chemotactic activity in human aortic endothelial cells.

Reddy S, Hama S, Grijalva V, *et al. J Biol Chem* 2001; **276**: 17030–5.

Background. Entrapment and oxidation of low-density lipoproteins (LDL) in the sub-endothelial space is a key process in the initiation of atherosclerotic lesion development. Functional changes induced by oxidized lipids in endothelial cells are early events in the pathogenesis of atherosclerosis. Oxidized-l-alpha-1-palmitoyl-2-arachidonoyl-sn-glycero-3-phosphocholine (ox-PAPC), a major component of minimally modified/oxidized LDL (MM-LDL) mimics the biological activities assigned to MM-LDL both *in vitro* in a co-culture model as well as *in vivo* in mice. It was hypothesized that ox-PAPC initiates gene expression changes in endothelial cells that result in enhanced endothelial/monocyte interactions. To analyze the gene expression changes that oxidized lipids induce in endothelial cells, a suppression subtractive hybridization procedure was used to compare mRNA from PAPC-treated human aortic endothelial cells (HAEC) with that of ox-PAPC-treated cells. The identification of a gene was reported, mitogen-activated protein kinase phosphatase 1 (MKP-1), that is rapidly and transiently induced in ox-PAPC-treated HAEC. Inhibition of MKP-1 using either the phosphatase inhibitor sodium orthovanadate or antisense oligonucleotides prevents the accumulation of monocyte chemotactic activity in ox-PAPC-treated HAEC supernatants. Furthermore, it was shown that decreased monocyte chemotactic activity in HAEC treated with sodium orthovanadate or MKP-1 antisense oligonucleotides is due to decreased MCP-1 protein.

INTERPRETATION. These results implicate a direct role for MKP-1 in ox-PAPC-induced signalling pathways that result in the production of MCP-1 protein by ox-PAPC-treated HAEC.

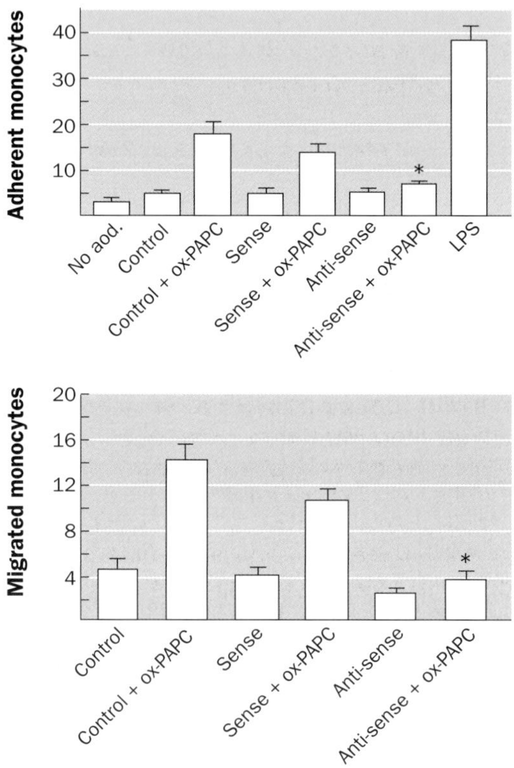

Fig. 10.1 Inhibition of MKP-1 prevents ox-PAPC-induced endothelial-monocyte interactions in HAEC. Supernatants from ox-PAPC-treated HAEC pre-treated with antisense oligonucleotides to MKP-1 do not promote monocyte adhesion and monocyte chemotaxis. HAEC were transfected with either antisense or sense phosphorothioate oligonucleotides (100 nM) to human MKP-1. Control and transfected cells were either left untreated or treated with ox-PAPC (50 μg/ml) for an additional 6 h. After the treatments, cell supernatants were collected and analyzed for monocyte adhesion (upper panel) and monocyte chemotaxis (lower panel). The results were similar in three independent experiments. The asterisk indicates a *P* value <0.05. LPS, lipopolysaccharide. Source: Reddy *et al.* (2001).

Comment

The discovery of the cellular pathways which lead to the upregulation of MCP-1, one of the first processes in atherogenesis, (Fig. 10.1) has important implications for the development of anti-atherosclerotic drugs for the future treatment of cardiovascular disease.

Oxidized alkyl phospholipids are specific, high affinity peroxisome proliferator-activated receptor gamma ligands and agonists.

Davies SS, Pontsler AV, Marathe GK, *et al.* *J Biol Chem* 2001; **276**: 16015–23.

BACKGROUND. Synthetic high affinity peroxisome proliferator-activated receptor (PPAR) agonists are known, but biological ligands are of low affinity. Oxidized low-density lipoprotein (oxLDL) is inflammatory and signals through PPARs. It was shown, by phospholipase A(1) digestion, that PPARγ agonists in oxLDL arise from the small pool of alkyl phosphatidylcholines in LDL. An abundant oxidatively fragmented alkyl phospholipid in oxLDL, hexadecyl azelaoyl phosphatidylcholine (azPC), was identified as a high affinity ligand and agonist for PPARγ. [(3)H]azPC-bound recombinant PPARγ with an affinity (K(d)((app)) approximately 40 nm) that was equivalent to rosiglitazone (BRL49653), and competition with rosiglitazone showed that binding occurred in the ligand-binding pocket. azPC induced PPAR response element (PPRE) reporter gene expression, as did rosiglitazone, with a half-maximal effect at 100 nm. Overexpression of PPARγ or PPARγ revealed that azPC was a specific PPARγ agonist. The scavenger receptor CD36 is encoded by a PPRE-responsive gene, and azPC enhanced expression of CD36 in primary human monocytes. It was found that anti-CD36 inhibited azPC uptake, and it inhibited PPRE reporter induction. Results with a small molecule phospholipid flippase mimetic suggest azPC acts intracellularly and that cellular azPC accumulation was efficient.

INTERPRETATION. Certain alkyl phospholipid oxidation products in oxLDL are specific, high affinity extracellular ligands and agonists for PPARγ that induce PPAR-responsive genes.

Comment

This manuscript establishes a new relationship between LDL-oxidation and the induction of PPARγ-regulated genes (Fig. 10.2) which could maintain or increase the inflammatory response in atherosclerosis and be a potential target for anti-atherosclerotic drug treatment.

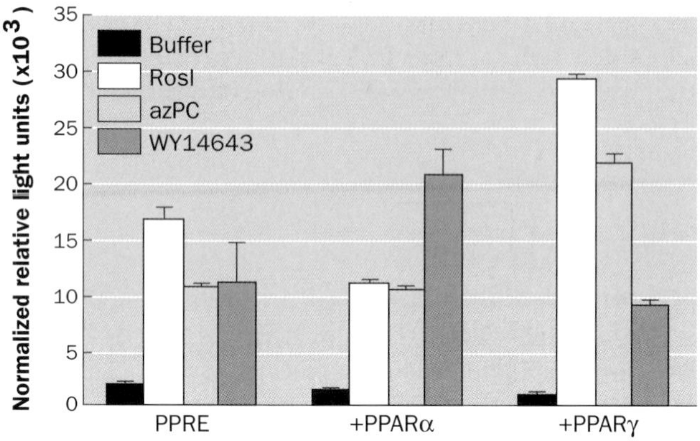

Fig. 10.2 azPC is a selective PPARγ agonist. CV-1 cells were transfected with acyl-CoA oxidase-PPRE-luciferase and SV40-β-galactosidase without or with cotransfection with PPARα or PPARγ expression plasmids. These cells were then exposed to buffer, 1 μM rosiglitazone (Rosi), or azPC, or 5 μm WY14643 for 24 h before the ratio of luciferase to β-galactosidase was determined as before. These results are representative of two separate experiments. Source: Davies *et al.* (2001).

Expression of interleukin-15 in mouse and human atherosclerotic lesions.

Wuttge DM, Eriksson P, Sirsjo A, Hansson GK, Stemme S. *Am J Pathol* 2001; **159**: 417–23.

Background. Atherosclerotic lesions are characterized by prominent macrophage and T-cell infiltration and atherosclerosis is widely recognized as an inflammatory disease. The cytokine interleukin-15 (IL-15) has T-cell chemotactic and pro-inflammatory properties and promotes the recruitment of T cells to sites of inflammation. Therefore IL-15 expression was examined in the atherosclerotic ApoE-deficient mouse model as well as in human atherosclerotic lesions. In gene expression arrays, a transcript corresponding to IL-15 mRNA was elevated in atherosclerotic aortas of ApoE-deficient mice fed a western diet for 10 and 20 weeks, corresponding to lesions of the fatty streak and fibrofatty plaque stage, respectively. Immunostaining for IL-15 localized to aortic smooth muscle cells in non-atherosclerotic C57BL/6 mice, whereas both macrophages and smooth muscle cells stained positive for IL-15 in atherosclerotic lesions of ApoE-deficient mice. Finally, advanced atherosclerotic lesions of human carotid arteries were immunostained to determine whether IL-15 is involved in human disease. IL-15 protein was present also in the human lesions with a distribution primarily overlapping that of macrophages.

Interpretation. IL-15 is upregulated in both human and animal atherosclerotic lesions and may contribute to the recruitment of T cells and their activation during atherogenesis.

Comment

Interleukin-15 expression is upregulated in atherosclerotic lesions (Fig. 10.3) where it may mediate inflammation during the progression of atherosclerosis.

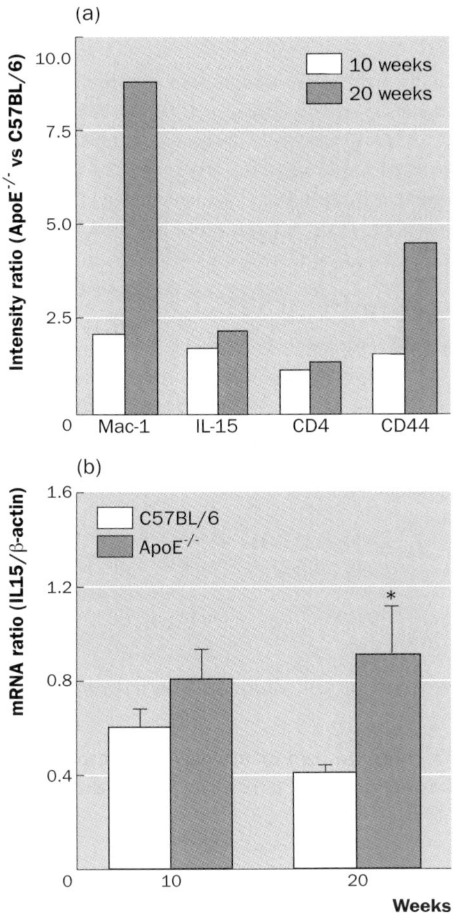

Fig. 10.3 (a) Gene expression analysis of aortas of ApoE-/- mice with the Clontech mouse gene array detected increased expression levels of Mac-1, IL-15, CD4, and CD44 mRNAs compared with aortas of C57BL/6 mice. The bar graph shows intensity ratios of gene expression in ApoE-/- mice divided by the gene expression in C57BL/6 mice at 10 (white bars) and 20 weeks (grey bars) of diet. (b) Quantitative real-time PCR analysis of IL-15 mRNA of the aortas of both ApoE-/- (grey bars) and C57BL/6 (white bars) mice after 10 and 20 weeks of diet showed a similar pattern to the gene array. The *y* axis indicates the mRNA ratio of IL-15 divided by the β-actin. Mean ± SEM of three pooled samples each containing five mice for each time point. At 20 weeks, the IL-15 mRNA levels were significantly increased in the ApoE-/- mice compared to non-atherogenic C57BL/6 mice. Source: Wuttge *et al.* (2001).

Circulating oxidized LDL is a useful marker for identifying patients with coronary artery disease.

Holvoet P, Mertens A, Verhamme P, *et al. Arterioscler Thromb Vasc Biol* 2001; **21**: 844–8.

BACKGROUND. The aim of this study was to determine the usefulness of circulating oxLDL in the identification of patients with coronary artery disease (CAD). A total of 304 subjects were studied: 178 patients with angiographically proven CAD and 126 age-matched subjects without clinical evidence of cardiovascular disease. The Global Risk Assessment Score (GRAS) was calculated on the basis of age, total and high-density lipoprotein cholesterol, blood pressure, diabetes mellitus, and smoking. Levels of circulating oxLDL were measured in a monoclonal antibody 4E6-based competition enzyme linked immunosorbent assay (ELISA). Compared with control subjects, CAD patients had higher levels of circulating oxLDL (*P* <0.001) and a higher GRAS (*P* <0.001). The sensitivity for CAD was 76% for circulating oxLDL (55% for men and 81% for women) compared with 20% (24% for men and 12% for women) for GRAS, with a specificity of 90%. Logistic regression analysis revealed that the predictive value of oxLDL was additive to that of GRAS (*P* <0.001). Ninety-four per cent of the subjects with high (exceeding the 90th percentile of distribution in control subjects) circulating oxLDL and high GRAS had CAD (94% of the men and 100% of the women).

INTERPRETATION. circulating oxidized LDL is a sensitive marker of CAD. Addition of oxidized LDL to the established risk factors may improve cardiovascular risk prediction.

Table 10.1 Logistic regression analysis of relationship between CAD and potential cardiovascular risk factors

Covariate	χ^2	*P*	OR (95% CI)
Multivariate model 1			
Male sex	15	<0.001	3.1 (1.8–5.6)
Dyslipidaemia	8.2	0.0042	2.6 (1.4–5.0)
Age	6.1	0.013	1.1 (1.0–1.1)
Hypercholesterolaemia	4.3	0.037	1.9 (1.0–3.4)
Multivariate model 2			
Oxidized LDL	59	<0.001	7.0 (4.3–11)
Dyslipidaemia	6.0	0.014	3.0 (1.3–7.4)
Male sex	6.2	0.013	3.1 (1.3–7.7)
Age	4.3	0.038	1.1 (1.0–1.1)

The multivariate model 1 contained age, sex, hypertension, diabetes type 2, hypercholesterolaemia, dyslipidaemia, smoking, and BMI as covariates. The R^2 value of this model was 0.22. Overall, 40% of the patients were predicted correctly at a classification cut-off at 0.9. The multivariate model 2 contained levels of circulating oxidized LDL and all other covariants included in the first model. The R^2 value of this model was 0.67. Overall, 72% of patients were predicted correctly at a classification cut-off of 0.9.

Source: Holvoet *et al.* (2001).

Comment

Plasma oxLDL is a sensitive marker for CAD that correlates well with more established CAD risk factors (Table 10.1). The ELISA described, which is commercially available, should allow the determination of oxLDL concentration in large prospective studies where it was not feasible previously.

Human-derived anti-oxidized LDL autoantibody blocks uptake of oxidized LDL by macrophages and localizes to atherosclerotic lesions *in vivo.*

Shaw PX, Horkko S, Tsimikas S, *et al. Arterioscler Thromb Vasc Biol* 2001; **21**: 1333–9.

BACKGROUND. Autoantibodies to oxidation-specific epitopes of LDL, such as malondialdehyde-modified LDL (MDA-LDL), occur in plasma and atherosclerotic lesions

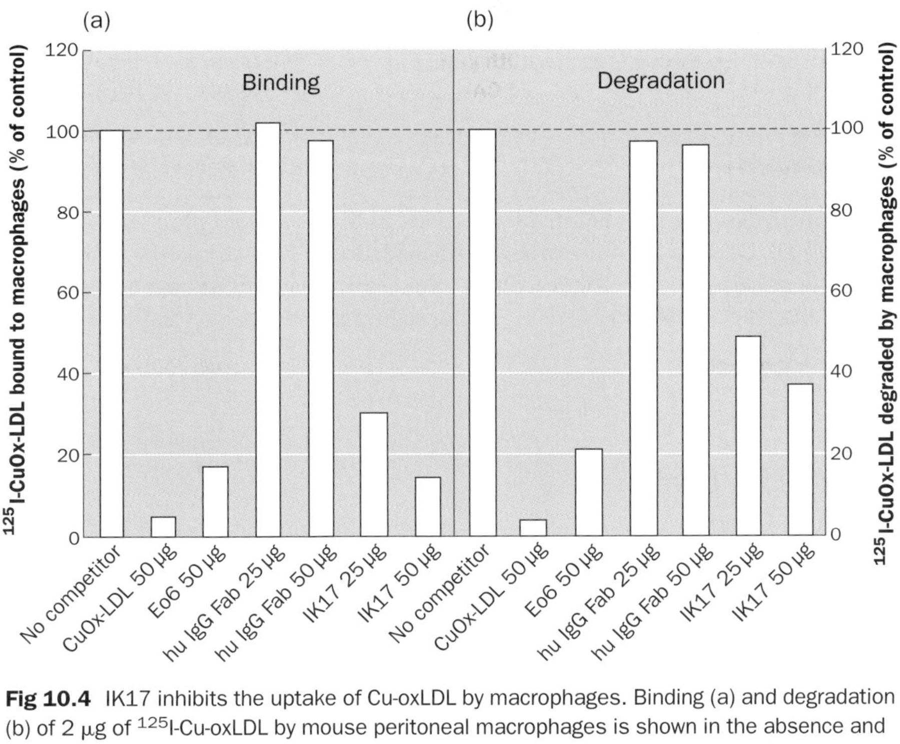

Fig 10.4 IK17 inhibits the uptake of Cu-oxLDL by macrophages. Binding (a) and degradation (b) of 2 μg of ^{125}I-Cu-oxLDL by mouse peritoneal macrophages is shown in the absence and presence of indicated competitors. hu indicates human. The data are expressed as percentage of ^{125}I-Cu-oxLDL associated with cells and degraded in the absence of competitor. Each bar is the mean of triplicate determinations. Source: Shaw *et al.* (2001).

of humans and animals. Plasma titres of such antibodies are correlated with atherosclerosis in murine models, and several such autoantibodies have been cloned. However, human-derived monoclonal antibodies to epitopes of oxLDL have not yet been reported. These investigators constructed a phage display antibody library from a patient with high plasma anti-MDA-LDL titres and isolated three monoclonal IgG Fab antibodies, which specifically bound to MDA-LDL. One of these, IK17, also bound to intact oxLDL as well as to its lipid and protein moieties but not to those of native LDL. IK17 inhibited the uptake of oxLDL by macrophages and also bound to apoptotic cells and inhibited their phagocytosis by macrophages. IK17 strongly immunostained necrotic cores of human and rabbit atherosclerotic lesions. When (125)I-IK17 was injected intravenously into LDL receptor-deficient mice, its specific uptake was greatly enriched in atherosclerotic plaques *versus* normal aortic tissue.

Interpretation. Human autoantibodies to oxLDL have important biological properties that could influence the natural course of atherogenesis.

Comment

This antibody could have therapeutic potential either to block the uptake of oxLDL by macrophages (Fig. 10.4) or to target anti-atherosclerotic drugs to where they are needed most, i.e. the atherosclerotic lesion.

Discussion

The papers described in this chapter are part of the mountain of *in vitro* evidence that oxLDL causes atherosclerosis, *in vivo* evidence in humans has been harder to find and remains elusive. The reasons for this are unclear, but we may be looking in the wrong place. This is discussed in the Concluding Remarks to Chapter 11.

11

Clinical chemistry

Introduction

This chapter includes some papers of interest in methodology/interpretation of results which may improve subsequent research studies and/or improve diagnosis of coronary heart disease (CHD) risk factors in the clinic.

Lipoprotein(a) further increases the risk of coronary events in men with high global cardiovascular risk.

von Eckardstein A, Schulte H, Cullen P, Assmann G. *JACC* 2001; **37**: 434–9.

BACKGROUND. The role of elevated lipoprotein(a) (Lp(a)) as a risk factor for CHD is controversial. In addition, little attention has been paid to the interaction of Lp(a) with other risk factors. A total of 788 male participants of the Prospective Cardiovascular Munster (PROCAM) study aged 35 to 65 years were followed for 10 years. Both Lp(a) and traditional cardiovascular risk factors (e.g. age, low-density lipoprotein (LDL) cholesterol, high-density lipoprotein (HDL) cholesterol, triglycerides, systolic blood pressure, cigarette smoking, diabetes mellitus, angina pectoris, and family history of myocardial infarction) were evaluated in 44 men who suffered from myocardial infarction, and in 744 men who survived without major coronary events or stroke. A multiple logistic function algorithm was used to estimate global cardiovascular risk by the combined effects of traditional risk factors. Overall, the risk of a coronary event in men with an Lp(a) ≥0.2 g/l was 2.7 times that of men with lower levels (95% confidence interval (CI): 1.4–5.2). This increase in risk was most prominent in men with LDL cholesterol level ≥4.1 mmol/l (relative risk [RR]: 2.6; 95% CI: 1.2–5.7), with HDL cholesterol ≤0.9 mmol/l (RR 8.3; 95% CI: 2.0–35.5), with hypertension (RR 3.2; 95% CI: 1.4–7.2), or within the two highest global risk quintiles (RR: 2.7; 95% CI: 1.3–5.7).

INTERPRETATION. Lp(a) increases the coronary risk, especially in men with high LDL cholesterol, low HDL cholesterol, hypertension and/or high global cardiovascular risk.

Comment

Recently has been a relatively quiet period in Lp(a) research which may be re-awoken by this report (Fig. 11.1). There have been several problems with Lp(a) studies, not least of which are problems with analysis and stability (see next paper). These need

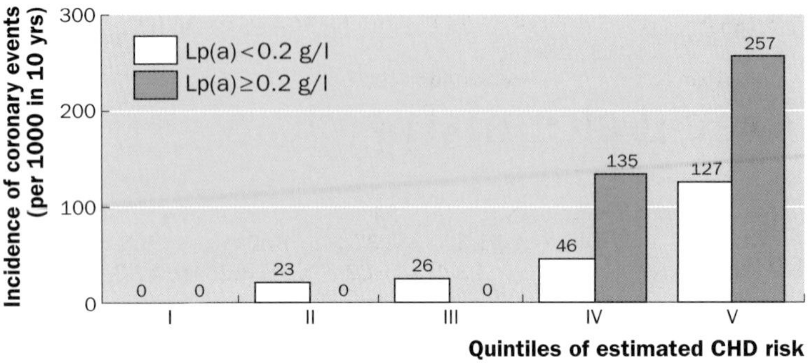

Fig. 11.1 Incidence of major coronary events in male Prospective Cardiovascular Munster study (PROCAM) participants aged 35 to 65 years over a 10-year period according to risk estimated by multiple logistic function and Lp(a): 41 events among 761 men. Lp(a) = lipoprotein(a); 95% CI = 95% confidence interval. Source: von Eckardstein *et al.* (2001).

to be overcome before (i) measurement of Lp(a) can be consistent between studies and (ii) Lp(a) can be included in diagnostic testing for CHD.

Instability of lipoprotein(a) in plasma stored at –70 degrees C: effects of concentration, apolipoprotein(a) genotype, and donor cardiovascular disease.

Simo JM, Camps J, Vilella E, Gomez F, Paul A, Joven J. *Clin Chem* 2001; **47**: 1673–8.

Background. There is considerable evidence to suggest that plasma Lp(a) concentration is a cardiovascular risk factor. Confusing results in epidemiological studies, however, suggest that the effects of storage should be further investigated. The influence of the assay method, the initial plasma Lp(a) concentration, and the apolipoprotein A (apoA) genotype are all factors that should be considered. Blood was obtained from 65 survivors of premature myocardial infarction and 95 age-matched controls. The plasma samples were stored in sterile conditions at –70°C for 5 years in the presence of antioxidant and antiproteolytic substances. Plasma Lp(a) was measured by immunoturbidimetry, and apoA alleles were determined by pulsed-field gel electrophoresis and Southern blotting. Plasma Lp(a) was significantly higher in patients. The mean kringle number for the smallest isoform was also lower in patients than in controls, but no differences were found in the distribution of the largest isoform. All patients and controls were heterozygotes. During storage, mean Lp(a) decreased significantly in samples from patients (–23%; *P* <0.001) but not in

Table 11.1 Percentage of change in plasma Lp(a) concentration between first and second measurements in each quartile of initial plasma Lp(a) concentration

	Initial plasma Lp(a) concentration, mg/l			
	≤40	**41–163**	**164–345**	**≥346**
Patients				
n	21	6	14	24
Mean ± SD, %	1.4 ± 6.5	−40.1 ± 38.5	−35.7 ± 40.2	−28.1 ± 30.2
95% confidence interval, %	(−1.3 to 4.0)	(−120.5 to 23.4)	(−89.5 to −18.2)	(−42.3 to −13.6)
Controls				
n	33	20	26	16
Mean ± SD, %	2.7 ± 11.0	−13.5 ± 23.4	−5.1 ± 22.2	−22.9 ± 60.7
95% confidence interval, %	(−1.4 to 6.6)	(−104.2 to 41.2)	(−13.9 to 3.8)	(−66.1 to 12.7)
P	0.485	0.043	0.0038	0.532

Source: Simo *et al.* (2001).

samples from controls (–9%; *P*, not significant). This was not related to the kringle number and was limited to samples with initial plasma Lp(a) concentrations between 41 and 345 mg/l.

Interpretation. Plasma Lp(a) from patients is less stable than Lp(a) from controls, and the difference is not related to distribution of apoA genotypes but may be concentration-dependent. Differential sample stability may complicate the interpretation of several studies.

Comment

These findings will add to the difficulty in interpreting the role of Lp(a) in cardiovascular disease; many results published may simply be due to methodological problems inherent in determining Lp(a) concentration (Table 11.1).

High apoliprotein B, low apolipoprotein A-I, and improvement in the prediction of fatal myocardial infarction (AMORIS study): a prospective study.

Walldius G, Jungner I, Holme I, Aastveit AH, Kolar W, Steiner E. *Lancet* 2001; **358**: 2026–33.

Background. Apolipoprotein B (apoB) and apolipoprotein A-I (apoA-I) are thought to be better predictors of acute myocardial infarction than total cholesterol and LDL cholesterol. This study investigated whether apoB and apoA-I are predictors of risk of

fatal myocardial infarction. It also aimed to establish whether apoB and apoA-I add further information about risk of fatal myocardial infarction to that obtained with total cholesterol, triglycerides, and LDL cholesterol. The investigators recruited 175 553 individuals mainly from screening programmes. Concentrations of apoB, apoA-I, total cholesterol, and triglycerides were measured, and apoB/apoA-I ratio and concentrations of LDL cholesterol and HDL cholesterol were calculated. The relationship between death from acute myocardial infarction and initial values for apoB, apoA-I, and other lipids was examined. Mean follow-up was 66.8 months (standard deviation (SD) 41.3) for 98 722 men and 64.4 months (41.4) for 76 831

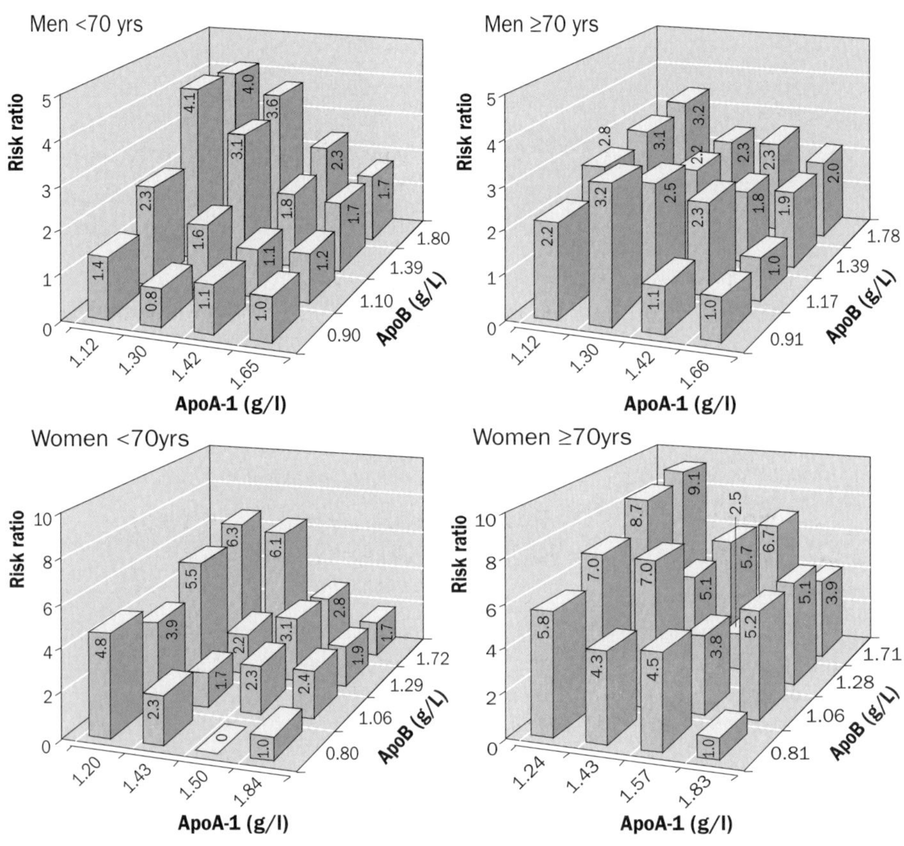

Fig. 11.2 Relative risk ratios of fatal myocardial infarction by quartiles of apolipoprotein B (ApoB) and apolipoprotein A-I (Apo-I) in men and women aged less than 70 years and 70 years or older. Adjusted for age, total cholesterol, and triglycerides by the Cox model. The lowest reference cell was used for those with the lowest ApoB and the highest ApoA-I quartile. Source: Walldius *et al.* (2001).

women. A total of 864 men and 359 women had fatal myocardial infarction. In univariate analyses adjusted for age and in multivariate analyses adjusted for age, total cholesterol, and triglycerides the values of apoB and apoB/apoA-I ratio were strongly and positively related to increased risk of fatal myocardial infarction in men and women. ApoA-I was noted to be protective. In multivariate analysis, apoB was a stronger predictor of risk than LDL cholesterol in both sexes.

INTERPRETATION. Although LDL cholesterol and HDL cholesterol are known risk factors, it was suggested that apoB, apo B/apoA-I, and apoA-I should also be regarded as highly predictive in evaluation of cardiac risk. Although increased throughout the range of values of LDL cholesterol, apoB and apoA-I might be of greatest value in diagnosis and treatment in men and women who have common lipid abnormalities, but have normal or low concentrations of LDL cholesterol.

Comment

The ratio of apoA-I and apoB more accurately predicts fatal myocardial infarction than do the more usually measured risk factors (total cholesterol, triglycerides, HDL cholesterol and LDL cholesterol) in both men and women regardless of age (Fig. 11.2).

Unfortunately, apolipoprotein measurements are quite expensive which limits their routine clinical utility. However, if apolipoprotein measurements become more popular, the price would almost certainly drop, making them more available in risk factor analysis.

Non-high-density lipoprotein cholesterol level as a predictor of cardiovascular disease mortality.

Cui Y, Blumenthal RS, Flaws JA, *et al. Arch Intern Med* 2001; **161**: 1413–9.

BACKGROUND. Non-HDL cholesterol contains all known and potential atherogenic lipid particles. Therefore, non-HDL cholesterol level may be as good a potential predictor of risk for cardiovascular disease (CVD) as LDL cholesterol. Data are from the Lipid Research Clinics Program Follow-up Study, a mortality study with baseline data gathered from 1972 to 1976, and mortality ascertained through 1995. A total of 2406 men and 2056 women aged 40 to 64 years at entry were observed for an average of 19 years, with CVD death as the main outcome measure. A total of 234 CVD deaths in men and 113 CVD deaths in women occurred during follow-up. Levels of HDL cholesterol and non-HDL cholesterol at baseline were significant and strong predictors of CVD death in both sexes. In contrast, LDL cholesterol level was a somewhat weaker predictor of CVD death in both. Differences of 0.78 mmol/l (30 mg/dl) in non-HDL cholesterol and LDL cholesterol levels corresponded to increases in CVD risk of 19% and 15%, respectively, in men. In women, differences of 0.78 mmol/l (30 mg/dl) in non-HDL cholesterol and LDL cholesterol levels corresponded to increases in CVD risk of 11% and 8%, respectively.

Table 11.2 Comparison of lipid levels in predicting CVD mortality in men and women

	Coefficient (SE)	RR (95% CI)†	χ^2 for addition to model‡
Men			
Non-HDL-C§	0.17 (0.03)	1.19 (1.13–1.26)	24.3
LDL-C§	0.11 (0.05)	1.11 (1.02–1.22)	5.0
TC§	0.15 (0.03)	1.16 (1.08–1.23)	14.4
HDL-C‖	−0.26 (0.06)	0.77 (0.69–0.86)	23.2
Women			
Non-HDL-C§	0.14 (0.04)	1.15 (1.06–1.25)	8.3
LDL-C§	0.08 (0.06)	1.08 (0.96–1.22)	1.8
TC§	0.09 (0.05)	1.10 (0.99–1.22)	2.8
HDL-C‖	−0.26 (0.06)	0.77 (0.69–0.88)	18.5

†Relative risks are adjusted for age at baseline as a continuous variable using Cox proportional hazards models.
‡χ^2 for difference in −2 logarithm likelihood between models with age only and age plus lipid level.
§Measured by an increase of 0.78 mmol/l (30 mg/dl).
‖Measured by an increase of 0.26 mmol/l (10 mg/dl).

Source: Cui *et al.* (2001).

Interpretation. Non-HDL cholesterol level is a somewhat better predictor of CVD mortality than LDL cholesterol level. Screening for non-HDL cholesterol level may be useful for CVD risk assessment.

Comment

This method appears to be more accurate in determining CHD risk than LDL cholesterol (Table 11.2). It is more convenient methodologically for large-scale clinical studies, and could in future provide researchers with an extra option in determining cardiovascular risk.

Discussion

This small selection of papers illustrates some of the problems of providing an adequate analytical service to the general lipid clinic. Firstly, Lp(a) is still a research tool and needs more development before it is useful to the physician. Secondly, new methods for the analysis of risk factors are being developed which may be more accurate than those in current use, but which are either too expensive or technologically difficult to provide to the routine lipid clinic. If we are to be more accurate in diagnosing CHD risk, these issues need to be addressed at a national level, to determine the cost-to-benefit ratio.

Concluding remarks

The Heart Protection Study (HPS) has proved to be a great vindication for the use of statins in the prevention of CHD but was a body-blow for those advocating the use of pharmacological doses of antioxidants for the prevention of CHD |**1–3**|. It has also cast doubt on what has become to be known as the oxidation theory of atherosclerosis |**4,5**|. However, although there is much convincing *in vitro* evidence for the theory (as described in chapter 10), *in vivo* evidence has been hard to obtain. Several researchers have turned to supplementing the diets of experimental animals with various antioxidants. These studies have invariably shown that dietary supplementation with antioxidants inhibits the development/progression of atherosclerosis |**6**|. Unfortunately, human studies, supplementing the diet with pharmacological doses of antioxidant vitamins have at best been disappointing and mostly not effective |**7,8,9,10**|. The latest study to report is The Medical Research Council/ British Heart Foundation HPS |**1–3**|. In this study 20 536 people were assigned in a 2×2 factorial design to receive either placebo, 40 mg simvastatin, antioxidants (650 mg vitamin E, 250 mg vitamin C and 20 mg β-carotene) or simvastatin and antioxidants. Antioxidant vitamin supplementation had no effect on incidence of CVD, and did not add to the effect of simvastatin, although vitamins had no harmful effects. The question is, therefore, is there any future for antioxidant vitamin supplementation for the prevention of atherosclerosis?

The answer may be slightly different than expected. Life-long high intake of antioxidant vitamins and other naturally occurring dietary antioxidants, as in the 'Mediterranean diet', is very probably beneficial in delaying the onset of such diseases as cancer and atherosclerosis and thus increasing longevity. However, the use of pharmacological doses of antioxidant vitamins, given, probably too late in life, is insufficient to prevent atherosclerosis which is in the process of developing or has already been developed.

An alternative explanation is that several lipoprotein-associated enzymes such as paraoxonase (PON1), platelet-activating factor acetylhydrolase (PAFAH) and lecithin:cholesterol acyltransferase (LCAT) have been shown to inhibit the oxidation of LDL |**1,11,12,13**|. The protection these enzymes provide lasts for a considerably longer time than that of physical (i.e. chain-breaking, non-enzymatic) antioxidants. However, with time, these enzymes are inhibited by their substrates, oxidized phospholipids and cholesteryl esters. Aviram and colleagues have shown that lipid-soluble antioxidant vitamins and dietary polyphenolics protect PON1 from inhibition by oxidized lipids |**14**|. This, therefore, may be the primary function of dietary antioxidants, to protect antioxidant enzymes and increase their effective lifespan.

Although it does appear that the HPS is the final nail in the coffin for pharmacological antioxidant vitamin supplementation for the prevention of atherosclerosis, this may not be the ultimate test of the oxidative theory of atherosclerosis. This may lie in the role of antioxidant enzymes associated with lipoproteins and their interrelationship with dietary antioxidants.

This, once again, brings the antioxidant role of HDL to the fore. Statins tend to increase HDL by approximately 10% |**2,3**|. Undoubtedly pharmaceutical companies are searching for agents which increase HDL to a significantly greater extent and should therefore be anti-atherogenic via the protective mechanisms of HDL. Several groups are also investigating expressing individual HDL proteins through adenoviral systems with the aim of boosting short-term protection against restenosis |**15**|. To this author's mind, this area is where future research on HDL will concentrate, i.e. the mechanistic investigation of HDL function coupled with attempts to increase its concentration as an anti-atherogenic strategy.

Atherosclerosis is an inflammatory disease |**16**| and another up-and-coming field of study is the measurement of inflammatory markers such as C-reactive protein (CRP) and their relationship to atherosclerosis. It has been suggested that raised CRP is a diagnostic marker for the presence of atherosclerosis |**17**| and a great deal of research is being aimed at determining the mechanisms behind the raised CRP. Interestingly, one of the non-lipid-lowering effects of statins is to decrease the raised CRP found in patients with atherosclerosis |**18**|. This will be a fruitful field of research in the future.

In conclusion, although this is very much the year of the HPS there have been many interesting and exciting developments in the broad field of lipids and atherosclerosis and more to come in the future.

References

1. Collins R, Peto R, Armitage J. The MRC/BHF Heart Protection Study: preliminary results. *Int J Clin Pract* 2002; **56**: 53–6.

2. Heart Protection Study Collaborative Group. MRC/BHF Heart Protection Study of cholesterol lowering with simvastatin in 20536 high-risk individuals: a randomized placebo-controlled trial. *Lancet* 2002; **360**: 7–22.

3. Heart Protection Study Collaborative Group. MRC/BHF Heart Protection Study of antioxidant vitamin supplementation in 20536 high-risk individuals: a randomized placebo-controlled trial. *Lancet* 2002; **360**: 23–33.

4. Steinberg D. Is there a potential therapeutic role for vitamin E or other antioxidants in atherosclerosis. *Curr Opin Lipidol* 2000; **11**: 603–7.

5. Steinberg D, Parthasarathy S, Carew TE, Khoo JC, Witztum JL. Beyond cholesterol modifications of low-density lipoprotein that increase its atherogenicity. *New Engl J Med* 1989; **320**: 915–24.

6. Stocker R. Dietary and pharmacological antioxidants in atherosclerosis. *Curr Opin Lipidol* 1999; **10**: 589–97.

7. Gruppo Italiano per lo Studio della Sopravvivenza nell'Infarto miocardico. Dietary supplementation with n-3 polyunsaturated fatty acids and vitamin E after myocardial infarction: results of the GISSI-Prevenzione trial. *Lancet* 1999; **354**: 447–55.

8. Yusuf S, Dagenais G, Pogue J, Bosch J, Sleight P. Vitamin E supplementation and cardiovascular events in high risk patients. The Heart Outcomes Prevention Evaluation Study Investigators. *N Engl J Med* 2000; **342**: 154–60.
9. The α-Tocopherol, β-Carotene Cancer Prevention Study Group. The effect of vitamin E and β-carotene on the incidence of lung cancer and other cancers in male smokers. *N Engl J Med* 1994; **330**: 1029–35.
10. Boaz M, Smetana S, Weinstein T, Matas Z, Gafter U, Iaina A, Knecht A, Weissgarten Y, Brunner D, Fainaru M. Secondary prevention with antioxidants of cardiovascular disease in end-stage renal disease (SPACE): randomized placebo-controlled trial. *Lancet* 2000; **356**: 1213–18.
11. Durrington PN, Mackness B, Mackness MI. Paraoxonase and atherosclerosis. *Arterioscler Thromb Vasc Biol* 2001; **21**: 473–80.
12. Mackness MI, Durrington PN. High density lipoprotein, its enzymes and its potential to influence lipid peroxidation. *Atherosclerosis* 1995; **115**: 243–53.
13. Navab M, Hama SY, Hough GP, *et al.* High density lipoprotein associated enzymes: their role in vascular biology. *Curr Opin Lipidol* 1998; **9**: 449–56.
14. Aviram M, Rosenblat M, Scott B, Drogul J, Sorenson R, Bisgaier CL, Newton RS, La Du B. Human serum paraoxonase (PON1) is inactivated by oxidised low density lipoprotein and preserved by antioxidants. *Free Rad Biol Med* 1999; **26**: 892–904.
15. Ishiguro H, Yoshida H, Major AS, Zhu T, Babaev VR, Linton MF, Fazio S. Retrovirus-mediated expression of apolipoprotein-AI in the macrophage protects against atherosclerosis *in vivo*. *J Biol Chem* 2001; **276**: 36742–8.
16. Ross R. The pathogenesis of atherosclerosis: a perspective for the 1990s. *Nature* 1993; **362**: 801–9.
17. Rifai N, Ridker PM. Inflammatory markers and coronary heart disease. *Curr Opinon Lipidol* 2002; **13**: 383–9.
18. Jialal I, Stein D, Balis D, Grundy SM, Adams-Huet B, Devaraj S. Effect of hydroxymethyl glutaryl, co-enzyme a reductase inhibitor therapy on high sensitive C-reactive protein levels. *Circulation* 2001; **103**: 1933–35.

Part IV

Lipid-lowering trials

12

Trials with clinical end-points

Introduction

This year has seen the publication of a remarkable number of reports from existing and new clinical trials of lipid-modifying therapy which has greatly extended their clinical indications, particularly those of statins. The Long-term Intervention with Pravastatin in Ischaemic Disease (LIPID) study group have published their eight-year results, the GREek Atorvastatin and Coronary-heart-disease Evaluation (GREACE) trial achieved the biggest average decrease in serum cholesterol in any statin trial and the Heart Protection Study (HPS) has reported results of simvastatin treatment in more than 20 000 high-risk patients. Taken together these trials emphasize that statins not only prevent coronary heart disease (CHD), but also stroke, that patients with peripheral arterial disease or at high CHD risk from any combination of risk factors can benefit, that women as well as men benefit, that older patients benefit and that this all applies regardless of initial low-density lipoprotein (LDL) cholesterol levels. The HPS also provided confirmation that antioxidant vitamins do not prevent cardiovascular disease. The trials do not agree that the greatest reduction in LDL cholesterol achieves the greatest decrease in cardiovascular risk, but really only GREACE was designed to test this by following the National Cholesterol Education Program (NCEP) recommendations to achieve LDL cholesterol levels of <2.6 mmol/l (100 mg/dl) by dose titration: the outcome of this trial which was a halving of CHD event rates in only three years would appear to strongly support this target.

When to start statins after an acute coronary syndrome (ACS) is an issue which was not addressed by the randomized, placebo-controlled statin trials which were designed to examine longer-term effects. MIRACL (Myocardial Ischaemia Reduction with Aggressive Cholesterol Lowering) which examined the clinical effects of intervention with atorvastatin 80 mg daily is the first randomized placebo-controlled trial in this area |**1**|. Olson and Schwartz this year reviewed the results of MIRACL together with case–control studies in which statins were commenced early and concluded that there was overall benefit in the prevention of early CHD recurrence. Whilst Newby and colleagues were not able to find in their case–control study that there was evidence of benefit, they like Olsen and Schwartz found no adverse effect. In clinical practice therefore, in order to ensure that patients are not later overlooked and thus do not benefit from the longer-term use of statins, initiation of such therapy whilst in hospital with an ACS seems justified. The case–control study of Chen *et al.* also led to similar conclusions about the introduction of statin therapy before coronary angioplasty and stenting. Lordahl and coworkers found that the higher mortality associated with acute myocardial infarction (MI) in diabetes was prevented if diabetic patients were already established on statin therapy before the event.

Epidemiological studies show a weaker relationship between cholesterol and coronary risk with advancing age. Whilst this would predict a diminishing benefit for statin therapy with advancing age, statin trials have not shown that the relative risk reduction achieved does get less within the age ranges so far randomized. Furthermore because the absolute CHD event rate increases with age, more events are prevented as age increases. Lemaitre and colleagues provide data to support this and the design of the Prospective Study of Pravastatin in the Elderly at Risk (PROSPER) will further allow this to be tested in a randomized, placebo-controlled trial specifically in the elderly.

The risk of new CHD events both before and after the development of CHD is related not only to LDL cholesterol, but also to HDL cholesterol and triglycerides. Ballantyne and colleagues this year confirmed this was the case in the Simvastatin Scandinavian Survival Study (4S) study and showed patients with the triad of raised LDL cholesterol, low HDL cholesterol and raised triglycerides derived the greatest benefit from simvastatin. Four articles have been included which address the importance of lipid-lowering effects on HDL cholesterol in the context of CHD prevention which appeared this year. Sacks and Rubins both reviewed evidence from statin and fibrate trials. Robbins addressed more specifically the findings of the Veterans Affairs Cooperative HDL Cholesterol Intervention Trial (VA-HIT) |**2**|. The Expert Group on HDL attempted to provide clinical guidance on the management of patients with low HDL. Most patients at high CHD risk should receive statin therapy so the tenor of these articles is very much towards whether some patients might benefit from a combination of statin and fibrate or nicotinic acid treatment.

Great attention this year has been paid to the findings of randomized, placebo-controlled trials of hormone replacement therapy (HRT). Grady and colleagues reported that combined oestrogen and progestogen treatment did not prevent CHD events in the Heart and Estrogen/progestin Replacement Study (HERS) secondary prevention trial even in the long term. On the other hand, women receiving statin treatment in the same trial were reported by Herrington to have a reduced risk both of coronary events and venous thromboembolism. Worryingly, combined HRT not only failed to prevent CHD but increased its risk and that of stroke in a primary prevention study reported by the Women's Health Initiative Investigators.

Omega–3 fatty acids employed in the Gruppo Italiano per lo Studio della Sopravvivenza nell'Infarto Miocardico (GISSI) study of MI survivors prevented sudden death probably due to a decrease in cardiac dysrrhythmias. An analysis of the time-course of this effect was published by Marchioli *et al.* (see p. 248).

Long-term effectiveness and safety of pravastatin in 9014 patients with coronary heart disease and average cholesterol concentrations: the LIPID trial follow-up.

The Lipid Study Group, *Lancet* 2002; **359**: 1379–87.

BACKGROUND. The LIPID study showed that pravastatin therapy over six years reduced mortality and cardiovascular events in patient with previous ACSs and average cholesterol concentrations. The longer-term effects of initial treatment with pravastatin on further cardiovascular events and mortality were assessed over a total follow-up period of eight years. In the main trial, 9014 patients with previous MI or unstable angina and a baseline plasma cholesterol concentration of 4.0–7.0 mmol/l were randomly assigned pravastatin 40 mg daily or placebo and followed up for six years. Subsequently, all patients were offered open-label pravastatin for two more years. Major cardiovascular events and adverse events were compared according to initial treatment assignment. A total of 7680 (97% of those still alive) had two years of extended follow-up; 3766 (86%) of those assigned placebo and 3914 (88%) assigned pravastatin agreed to take open-label pravastatin. During this period, patients originally assigned pravastatin had almost identical cholesterol concentration to those assigned placebo, but a lower risk of death from all causes (219 [5.6%] *vs* 255 [6.8%], *P* = 0.029), CHD death (108 [2.8%] *vs* 137 [3.6%], *P* = 0.026), and CHD death or non-fatal MI (176 [4.5%] *vs* 196 [5.2%], *P* = 0.08). Over the total eight-year period, all-cause mortality was 888 (19.7%) in the group originally assigned placebo and 717 (15.9%) in the group originally assigned pravastatin, CHD mortality was 510 (11.3%) *vs* 395 (8.8%), MI was 570 (12.7%) *vs* 435 (9.6%; each *P* <0.001), and stroke was 272 (6.0%) *vs* 224 (5.0%; *P* = 0.015). Stronger evidence of separate treatment benefits than in the main trial was seen in important pre-specified subgroups (women, patients aged ≥70 years, and those with total cholesterol <5.5 mmol/l). Pravastatin had no significant adverse effects.

INTERPRETATION. The evidence of sustained treatment benefits and safety of long-term pravastatin treatment reinforces the importance of long-term cholesterol-lowering treatment for almost all patients with previous CHD events.

Comment

The LIPID study was a six-year placebo-controlled randomized trial of pravastatin 40 mg daily in the secondary prevention of CHD |**3**|.

This study extends the period of observation of patients on pravastatin 40 mg daily to eight years during which time no clinically significant side-effects were encountered. The placebo group as well as the active intervention group received pravastatin for the last two years of the eight years. Despite this, the decrease in total mortality, combined fatal and non-fatal MI and stroke was even more evident than that observed in the first six years when patients in the original active treatment group were compared with those in the former placebo group.

The benefit of treating women, older patients and those with low cholesterol levels was also clearer than in the original trial. A similar message with regard to women and older patients was produced in a similar extension of the 4S |**4**| except that pre-treatment cholesterol levels were not as low as those in LIPID. The conclusions of these extension studies clearly foreshadow GREACE and HPS (see p. 226) raising issues, many of which these later trials answer.

Treatment with atorvastatin to the national cholesterol educational program goal *versus* 'usual' care in secondary coronary heart disease prevention. The GREek Atorvastatin and Coronary-heart-disease Evaluation (GREACE) Study.

Athyros VG, Papageorgiou AA, Mercouris BR, *et al. Curr Med Res Opin* 2002; **19**: 220–8.

Background. Atorvastatin is very effective in reducing plasma LDL cholesterol levels. However, there is no long-term survival study that has evaluated this statin. To assess the effect of atorvastatin on morbidity and mortality (total and coronary) of patients with established CHD, 1600 consecutive patients were randomized either to atorvastatin or to 'usual' medical care. The dose of atorvastatin was titrated from 10 to 80mg/day, in order to reach the National Cholesterol Education Program (NCEP) goal of LDL cholesterol <100 mg/dl (2.5 mmol/l). All patients were followed up for a mean period of three years. Primary end-points of the study were defined as death, non-fatal MI, unstable angina, congestive heart failure, revascularization (coronary morbidity) and stroke. Secondary end-points were the safety and efficacy of the hypolipidaemic drugs as well as the cost-effectiveness of atorvastatin. The mean dosage of atorvastatin was 24 mg/day. This statin reduced total cholesterol by 36%, LDL cholesterol by 46%, triglycerides by 31% and non-high-density lipoprotein cholesterol (non-HDL-C) by 44%, while it increased HDL cholesterol by 7%; all these changes were significant. The NCEP LDL cholesterol and non-HDL-C treatment goals were reached by 95% (n = 759) and 97% (n = 776), respectively, of patients on atorvastatin. Only 14% of the 'usual' care patients received any hypolipidaemic drugs throughout the study and 3% of them reached the NCEP LDL cholesterol treatment goal. The cost per quality-adjusted life-year gained with atorvastatin was estimated at $US 8350. During this study 196 (24.5%) CHD patients on 'usual' care had a CHD recurrent event or died *vs* 96 (12%) CHD patients on atorvastatin; risk ratio (RR) 0.49, confidence interval (CI) 0.27–0.73, $P<0.0001$. In detail, atorvastatin reduced, in comparison to 'usual' care, total mortality (RR 0.57, CI 0.39–0.78, $P = 0.002$), coronary mortality (RR 0.53, CI 0.29–0.74, $P = 0.0016$) coronary morbidity (RR 0.46, CI 0.25–0.71, $P<0.0001$), and stroke RR 0.53, CI 0.30–0.82, $P = 0.0018$). All subgroups of patients (women, those with diabetes mellitus, arterial hypertension, age 60 to 75 years, congestive heart failure, recent unstable angina or prior revascularization) benefited from treatment with atorvastatin. Withdrawal of patients because of side-effects from the atorvastatin group was low (0.75%) and similar to that of the 'usual' care group (0.4%).

Interpretation. Long-term treatment of CHD patients with atorvastatin to achieve NCEP lipid targets significantly reduces total and coronary mortality, coronary morbidity and stroke, in comparison to patients receiving 'usual' medical care. Treatment with atorvastatin is well tolerated and cost-effective.

Comment

The GREACE trial, unheralded by any previous publication of its protocol, provided answers to many outstanding issues about statin therapy which had previously required extrapolation, not to say speculation, by those bodies making national and international recommendations for CHD prevention. In the trial, patients with established CHD were randomized to receive 'usual' care or atorvastatin titrated in doses from 10–80 mg daily with the aim of achieving the NCEP LDL cholesterol target of ≤2.6 mmol/l (100 mg/dl). Throughout the trial 14% of the 'usual' care group received lipid-lowering therapy. However, whereas 95% of

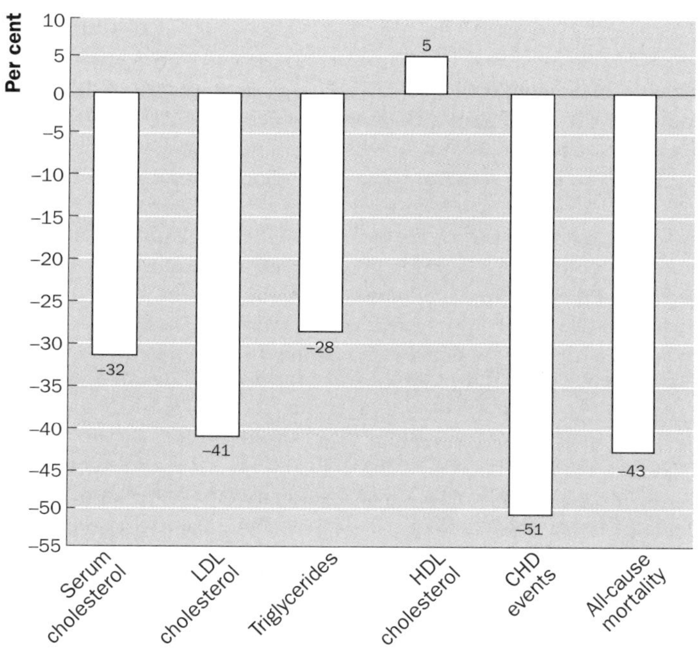

Fig. 12.1 Results of the GREek Atorvastatin and Coronary-heart-disease Evaluation (GREACE) Study. Differences are between the active intervention group in which 95% of patients achieved LDL cholesterol <2.6 mmol/l (100 mg/dl) and 'usual' care in which only 3% did. Source: Athyros *et al.* (2002).

patients randomized to atorvastatin achieved their LDL target, only 3% of the 'usual' care patients did so. This resulted in average differences of –32%, –41%, –28% and +5% in serum cholesterol, LDL cholesterol, triglyceride and HDL cholesterol concentrations respectively between the atorvastatin group and the 'usual' care group. There was a 51% decrease in fatal and non-fatal CHD events on atorvastatin (Fig. 12.1). This benefit was shared by women, older patients and diabetic patients. This is the closest trial to the NCEP recommendations for statin therapy and those of many other countries. It appears to be a strong vindication not only of treating patients with established CHD with statin therapy almost regardless of their initial cholesterol levels but also of aiming for a low LDL cholesterol target. The result represents the largest decrease in CHD risk in any trial to date and it was achieved in only three years.

MRC/BHF Heart Protection Study of cholesterol lowering with simvastatin in 20 536 high-risk individuals: a randomised placebo-controlled trial.

Heart Protection Study Collaborative Group. *Lancet* 2002; **360**: 7–22.

BACKGROUND. Throughout the usual LDL cholesterol range in Western populations, lower blood concentrations are associated with lower cardiovascular disease risk. In such populations, therefore, reducing LDL cholesterol may reduce the development of vascular disease, largely irrespective of initial cholesterol concentrations. A total of 20 536 UK adults (aged 40–80 years) with coronary disease, other occlusive arterial disease, or diabetes were randomly allocated to receive 40 mg simvastatin daily (average compliance: 85%) or matching placebo (average non-study statin use: 17%). Analyses were made of the first occurrence of particular events, and all simvastatin-allocated *versus* all placebo-allocated participants were compared. These 'intention-to-treat' comparisons assess the effects of about two-thirds (85% minus 17%) taking a statin during the scheduled five-year treatment period, which yielded an average difference in LDL cholesterol of 1.0 mmol/l (about two-thirds of the effect of actual use of 40 mg simvastatin daily). Primary outcomes were mortality (for overall analyses) and fatal or non-fatal vascular events (for subcategory analyses), with subsidiary assessments of cancer and of other major morbidity. All-cause mortality was significantly reduced (1328 [12.9%] deaths among 10 269 allocated simvastatin *versus* 1507 [14.7%] among 10 267 allocated placebo; *P* = 0.0003), due to a highly significant 18% (SEM 5) proportional reduction in the coronary death rate (587 [5.7%] *vs* 707 [6.9%]; *P* = 0.0005), a marginally significant reduction in other vascular deaths (194 [1.9%] *vs* 230 [2.2%]; *P* = 0.07), and a non-significant reduction in non-vascular deaths (547 [5.3%] *vs* 570 [5.6%]; *P* = 0.4). There were highly significant reductions of about one-quarter in the first event rate for non-fatal MI or coronary death (898 [8.7%] vs 1212 [11.8%]; *P* <0.0001), for non-fatal or fatal stroke (444 [4.3%] vs 585 [5.7%]; *P* <0.0001), and for coronary or non-coronary revascularization (939 [9.1%] *vs* 1205 [11.7%]; *P* <0.0001). For the first occurrence of any of these major vascular events, there was a definite 24% (SEM 3; 95% CI 19–28) reduction in

the event rate (2033 [19.8%] *vs* 2585 [25.2%] affected individuals; *P* <0.0001). During the first year the reduction in major vascular events was not significant, but subsequently it was highly significant during each separate year. The proportional reduction in the event rate was similar (and significant) in each subcategory of participant studied, including: those without diagnosed coronary disease who had cerebrovascular disease, or had peripheral artery disease, or had diabetes; men and, separately, women; those aged either under or over 70 years at entry; and – most notably – even those who presented with LDL cholesterol below 3.0 mmol/l (116 mg/dl), or total cholesterol below 5.0 mmol/l (193 mg/dl). The benefits of simvastatin were additional to those of other cardioprotective treatments. The annual excess risk of myopathy with this regimen was about 0.01%. There were no significant adverse effects on cancer incidence or on hospitalization for any other non-vascular cause.

INTERPRETATION. Adding simvastatin to existing cardioprotective treatments safely produces substantial additional benefits for a wide range of high-risk patients, irrespective of their initial cholesterol concentrations. Allocation to 40 mg simvastatin daily reduced the rates of MI, of stroke, and of revascularization by about one-quarter. After making allowance for non-compliance, actual use of this regimen would probably reduce these rates by about one-third. Hence, among the many types of high-risk individual studied, five years of simvastatin would prevent about 70–100 people per 1000 from suffering at least one of these major vascular events (and longer treatment should produce further benefit). The size of the five-year benefit depends chiefly on such individuals' overall risk of major vascular events, rather than on their blood lipid concentrations alone.

Comment

The HPS was certainly large and it was certainly confirmatory of the benefit of statin therapy in various subgroups of patients for whom earlier trials had only indicated benefit by a process of extrapolation or meta-analysis. The HPS does not, however, provide any clear answer as to how to employ statin therapy in primary prevention. The annual CHD risk in the placebo group in HPS was 2.4%, similar to that in the secondary prevention trials, the Cholesterol and Recurrent Events study (CARE) |**5**| and LIPID |**3**|. Compare this with 1.5% in the West of Scotland Coronary Prevention Study (WOSCOPS) |**6**| and 1% in the Air Force/Texas Coronary Atherosclerosis Prevention Study (AFCAPS/TexCAPS) |**7**|. To enter, HPS patients had to have CHD already and/or other atherosclerosis (non-haemorrhagic stroke or transient cerebral ischaemia, peripheral (leg) arterial disease or carotid endarterectomy) and/or diabetes (type 1 or 2) and/or treated hypertension in men aged ≥65 years. Because these study categories overlap it is not possible to obtain from the report information about the effect of statin treatment in patients free of vascular disease at the time of randomization, such as those with diabetes alone. This is unfortunate because of the dichotomy of opinion about whether diabetes *per se* should be regarded as a coronary risk equivalent and thus diabetic patients generally be prescribed statin therapy regardless of their calculated CHD risk. Another limitation of HPS is that, although simvastatin 40 mg daily was compared with placebo, the placebo group became essentially a 'usual' care group as the results of other statin trials became available and physicians found it unethical not to prescribe statins to such high-risk patients whose cholesterol they considered to require lowering more than was being achieved in the trial. Thus by the fifth year of the trial, one-third of the placebo-allocated group was receiving statin medication. Therefore, although the dose of simvastatin employed in HPS was higher than the average of 27 mg daily in the earlier 4S study |**8**|, the cholesterol-lowering achieved in HPS led to a relative reduction in CHD risk of 27% which was less than in 4S.

So far I have dwelt on the limitations of HPS; what about its undoubted successes? Because of its size, it does establish beyond doubt the benefit of statin therapy in certain groups of patients who comprised only a small proportion of patients randomized in earlier trials. The most important of these are patients with cerebrovascular disease and with peripheral arterial disease who did not have clinical CHD at entry. This is particularly important because statin therapy was previously regarded by many as being of dubious benefit in both these categories of patients. Furthermore in the trial, simvastatin not only decreased the risk of CHD in these patients, but also of recurrent cerebrovascular events and the need for revascularization procedures. The authors of the HPS report concluded from this quite correctly that statins should be regarded as beneficial in atherosclerotic cardiovascular disease generally. They proposed that, in future, recommendations for statin use should be based on cardiovascular risk rather than CHD risk alone. The conclusion, however, requires some further consideration. It is clearly true in secondary prevention. HPS, however, does not directly provide evidence that it is also true in primary prevention. Also, in most recommendations for lipid-lowering drug therapy in primary prevention, CHD risk is calculated from the equations provided from the Framingham Study |**9**|. This CHD risk has a fairly fixed relationship with atherosclerotic cardiovascular risk in the ratio 3:4. Thus a 30% risk of CHD over 10 years equates generally to a cardiovascular risk of 40% over 10 years. It is thus quite reasonable to continue to base primary prevention recommendations on CHD risk as long as the threshold risk for intervention with statin therapy is lowered now that it is known that there are benefits beyond CHD prevention. Despite this it is likely, however, that because this will be confusing to some, most recommendations will increasingly be couched in terms of cardiovascular risk.

HPS also provides strong evidence that women can benefit from statin therapy. In addition, it contributes to the argument about whether statins should be used in the elderly. Consistent with earlier studies there was little evidence for any marked attenuation of the reduction in relative risk achieved by statin therapy with advancing age. Therefore because of the increase in the absolute cardiovascular risk with age, the number of elderly patients who must receive statin treatment to prevent one cardiovascular event is smaller than for younger patients. HPS is also very important in lowering the threshold of serum LDL cholesterol for statin therapy to be initiated which exists in some quarters. The entry serum cholesterol for participation in HPS was only 3.5 mmol/l (135 mg/dl) and the outcome was uninfluenced by the entry serum cholesterol or LDL cholesterol concentration. The trial, however, throws little light on what should be regarded as a target LDL cholesterol level for statin therapy. HPS provides only information about the LDL response to simvastatin before randomization. This had no effect on the relative reduction in vascular risk in the trial itself. From the results of GREACE, a larger LDL reduction might have been expected to produce a greater reduction in relative vascular risk. No doubt this will be food for speculation. Clearly the pre-randomization response may not have closely mirrored the in-trial response. Perhaps the range of responses with a fixed dose of statin was too small to detect differences in risk reduction. Perhaps some of the differences in statin potency or their pleotropic effects are important.

Finally HPS underscores the view that the degree of benefit of statin therapy, at least in terms of the number of people who must be treated to prevent one cardiovascular event, seems to be most strongly determined by absolute cardiovascular or CHD risk and appears to be largely uninfluenced by which risk factors lead to any particular level of risk.

MRC/BHF Heart Protection Study of antioxidant vitamin supplementation in 20 536 high-risk individuals: a randomised placebo-controlled trial.

Heart Protection Study Collaborative Group. *Lancet* 2002; **360**: 23–33.

BACKGROUND. It has been suggested that increased intake of various antioxidant vitamins reduces the incidence rates of vascular disease, cancer, and other adverse outcomes. A total of 20 536 UK adults (aged 40–80) with coronary disease, other occlusive arterial disease, or diabetes were randomly allocated to receive antioxidant vitamin supplementation (600 mg vitamin E, 250 mg vitamin C, and 20 mg β-carotene daily) or matching placebo. Intention-to-treat comparisons of outcome were conducted between all vitamin-allocated and all placebo-allocated participants. An average of 83% of participants in each treatment group remained compliant during the scheduled five-year treatment period. Allocation to this vitamin regimen approximately doubled the plasma concentration of alpha-tocopherol, increased that of vitamin C by one-third, and quadrupled that of β-carotene. Primary outcomes were major coronary events (for overall analyses) and fatal or non-fatal vascular events (for subcategory analyses), with subsidiary assessments of cancer and of other major morbidity. There were no significant differences in all-cause mortality (1446 [14.1%] vitamin-allocated *vs* 1389 [13.5%] placebo-allocated), or in deaths due to vascular (878 [8.6%] *vs* 840 [8.2%]) or non-vascular (568 [5.5%] *vs* 549 [5.3%]) causes. Nor were there any significant differences in the numbers of participants having non-fatal MI or coronary death (1063 [10.4%] *vs* 1047 [10.2%]), non-fatal or fatal stroke (511 [5.0%] *vs* 518 [5.0%]), or coronary or non-coronary revascularisation (1058 [10.3%] *vs* 1086 [10.6%]). For the first occurrence of any of these 'major vascular events', there were no material

differences either overall (2306 [22.5%] *vs* 2312 [22.5%]; event rate ratio 1.00 [95% CI 0.94–1.06]) or in any of the various subcategories considered. There were no significant effects on cancer incidence or on hospitalization for any other non-vascular cause.

Interpretation. Among the high-risk individuals that were studied, these antioxidant vitamins appeared to be safe. However, although this regimen increased blood vitamin concentrations substantially, it did not produce any significant reductions in the five-year mortality from, or incidence of, any type of vascular disease, cancer, or other major outcome.

Comment

The HPS had a factoralized design which allowed the effect of both simvastatin 40 mg daily and antioxidant vitamin supplementation to be investigated. It provides the last nail in the coffin of antioxidant vitamins as a potential means of CHD prevention. They were without any adverse or beneficial effect either alone or in combination with simvastatin. A previous meta-analysis by the Heart Outcomes Prevention Evaluation (HOPE) investigators which included results from three other trials as well as HOPE also showed no overall effect [10]. It is important that these findings are not regarded as a rejection of the hypothesis that oxidative modification of LDL is important in its atherogenicity. However, we should certainly consider the possibilities that LDL oxidation may be important much earlier in the progression of atheroma (when people are still at low risk), that chain-breaking antioxidants are not effective in preventing LDL oxidation in the arterial wall, that other systems for protecting LDL against oxidative modification, such as paraoxonase, may be more effective (see Part III Lipids and atherosclerosis by Michael Mackness) or that oxidation is not the most important modification of LDL rendering it atherogenic: glycation may, for example, be more important [11].

Early initiation of treatment with statins in acute coronary syndromes.

Olsson AG, Schwartz GG. *Ann Med* 2002; **34**: 37–41.

Background. Statins may act rapidly to reverse abnormalities of the arterial wall that may predispose to recurrent ischaemic events after ACSs. Such abnormalities are endothelial dysfunction, a local inflammatory response and an exaggerated thrombogenic tendency. In one study, almost 20 000 patients with first MI were studied with regard to statin treatment (28%) or not. Baseline characteristics were adjusted using multivariate regression analysis including propensity analysis. One-year mortality was 3.7/5.0% in statin/not statin groups, respectively, *P* = 0.001, relative risk 0.75. In another study of more than 20 000 patients, 18% were prescribed statin after an ACS and followed for six months. Propensity analysis was performed in this study as well. Deaths in statin/not statin groups were 1.7/3.5%, *P* <0.0001, relative

risk 0.48. In MIRACL, a double-blind randomized, placebo-controlled intervention study, 3086 patients with acute non-Q-wave coronary syndromes were allocated immediately in hospital to receive atorvastatin 80 mg daily or placebo for four months. No lower limit for plasma LDL cholesterol was used. Primary end-point was time to first occurrence of death, non-fatal MI, cardiac arrest, and worsening angina with objective evidence of ischaemia. This was significantly reduced compared to the placebo group by 2.4% (14.8 *vs* 17.2%, relative risk 0.84, *P* = 0.048). Atorvastatin also reduced significantly fatal or non-fatal strokes. Possible mechanisms behind these acute beneficial effects are discussed.

Interpretation. The studies highlight the importance of treatment with a statin in the early management of ACSs and the need to incorporate this therapeutic strategy in national guidelines and treatment recommendations.

Comment

This article summarizes the observations and clinical trial evidence that statin treatment introduced early after an ACS decreases the risk of further CHD events during the subsequent year. Randomized clinical trials of statin therapy have, for the most part not entered patients until some months after acute coronary events and benefit has not generally been apparent until the third year of the trial. On purely pragmatic grounds in clinical practice it is, however, generally wisest to introduce statin treatment during an admission for an ACS to ensure that it is not subsequently overlooked and thus the patient denied the long-term benefit of such treatment. Thus perhaps the most important clinical question is whether introducing statin therapy immediately after or during an acute coronary episode could do harm. The authors of this report certainly find no evidence of harm and report early benefit. They were both investigators in the randomized, placebo-controlled MIRACL study |**1**| and, given the vicissitudes of observational studies, this remains the strongest evidence for their conclusion.

Early statin initiation and outcomes in patients with acute coronary syndromes.

Newby LK, Kristinsson A, Bhapkar MV, *et al.*; SYMPHONY and 2nd SYMPHONY Investigators. *JAMA* 2002; **287**: 3087–95.

Background. The secondary prevention benefit of therapy with 3-hydroxy-3-methylglutaryl coenzyme A (HMG-CoA) reductase inhibitors (statins) has been clearly demonstrated; however, the role of early initiation of statins after ACSs is unknown. The aim of this study was to evaluate the association of early statin initiation (≤7 days) after ACS with 90-day and 1-year outcomes. The design was an observational cohort from databases of 2 randomized clinical trials, SYMPHONY (Sibrafiban *vs* Aspirin to Yield Maximum Protection from Ischaemic Heart Events Post-acute Coronary Syndromes) and 2nd SYMPHONY using data from 931 clinical

centres in 37 countries. The trial studied a total of 12 365 ACS patients randomized from August 1997 to August 1999 who were not taking statins prior to the index ACS and who either started statin therapy early (median, 2.0 [interquartile range, 1.0–3.1] days after ACS; *n* = 3952) or survived more than five days after ACS and never received statin therapy (*n* = 8413). The main outcome measures were 90-day incidence of death; death or MI; death, MI, or severe recurrent ischaemia; 1-year incidence of death. Ninety-day and 1-year unadjusted mortality comparison suggested early statin benefit (1.2% for early statins *vs* 2.1% for no statins; hazard ratio [HR] 0.58; 95% CI 0.42–0.81 for 90-day comparisons and 2.3% for early statins *vs* 4.4% for no statins; HR 0.52; 95% CI 0.40–0.68 for 1-year comparison). However, no benefit was evident for 90-day death or MI (6.5% *vs* 6.9%; HR 0.95; 95% CI 0.82–1.11) or death, MI, or severe recurrent ischaemia (9.2% *vs* 8.9%; HR 1.04; 95% CI 0.92–1.18). After propensity and covariate adjustment, there were no 90-day or 1-year differences between the early-statin group and the no-statin group. The 90-day adjusted HR for death was 1.08 (95% CI 0.75–1.56); for death or MI, 1.08 (95% CI 0.91–1.29); and for death, MI, or severe recurrent ischaemia, 1.15 (95% CI 0.99–1.34). One-year mortality-adjusted HR was 0.99 (95% CI 0.73–1.33). Among 2711 patients with core laboratory lipid analysis, early statin treatment was associated with higher adjusted risk for death or death or MI at cholesterol levels below treatment guidelines but was more favourable at higher levels.

Interpretation. In this study, there was no relationship between early initiation of statin therapy and improved outcomes although our subset analysis suggests that outcomes after early statin initiation may vary with cholesterol levels. Confirmation of early treatment effects of statins on outcomes awaits the results of adequately powered randomized clinical trials.

Comment

This is an observational study of patients whose physicians prescribed statin therapy early after an ACS and those who chose not to do so. No differences in early recurrence of CHD events were evident. The result is thus in contrast with the article by Schwartz and Olsson summarized immediately before this one. The lack of randomization to statin treatment means that some adverse confounding factor(s) present in the patients receiving statin therapy may have nullified its benefit, if any. Equally, of course, in observational studies, which apparently showed early benefit from statin therapy, the result could have been confounded by adverse confounding factors in patients not receiving statins. Thus MIRACL with its randomized design and small benefit is important |**1**|. Whilst confirmation of the benefit in further randomized trials is required, particularly in patients with lower LDL cholesterol levels and with lower doses of statins other than atorvastatin, none of the observational studies of early statin use have shown an adverse effect and so there is no reason to depart from the recommendations to prescribe statins during hospital admission with ACSs to ensure that the introduction of statins is not subsequently overlooked.

Early and sustained survival benefit associated with statin therapy at the time of percutaneous coronary intervention.

Chan AW, Bhatt DL, Chew DP, *et al. Circulation* 2002; **105**: 691–6.

BACKGROUND. Long-term administration of statin therapy has been shown to reduce major coronary events and cardiac mortality within randomized clinical trials. In addition to lowering lipids, statins favourably affect platelet adhesion, thrombosis, endothelial function, inflammation, and plaque stability, which may potentially improve outcome after percutaneous coronary intervention (PCI). Therefore, it was hypothesized that statin therapy has an early beneficial effect among patients undergoing PCI. Each year from 1993 to 1999, data were prospectively collected among the first 1000 patients undergoing PCI. Patients who presented with acute or recent MI or cardiogenic shock were excluded from the analysis. Baseline, procedural, and 6-month data of statin-treated and non-statin-treated patients were compared. Propensity score and multivariate survival analysis were used to adjust for heterogeneity between the two groups. Of 5052 patients who completed follow-up, 26.5% were treated with statin at the time of the procedure. Statin therapy was associated with a mortality reduction at 30 days (0.8% *vs* 1.5%; HR 0.53; $P = 0.048$) and at 6 months (2.4% *vs* 3.6%; HR 0.67; $P = 0.046$). After adjusting for the propensity to receive statin therapy before the procedure and other confounders, statin therapy remained an independent predictor for survival at six months after coronary intervention (HR 0.65; 95% CI 0.42–0.99; $P = 0.045$).

INTERPRETATION. In this large study cohort, statin therapy among PCI patients seems to be associated with a significant mortality advantage at early and intermediate-term follow-up.

Comment

This is another observational study, this time of the effect of statin therapy begun before PCI with a positive outcome on mortality during the first six months. The authors have been careful to adjust for confounding factors, although they concede that these still limit their conclusion. Nonetheless, they present an encouraging finding with no evidence of any adverse effect. Their evidence supports the use of these agents as soon as possible in high-risk patients because it is well established in randomized trials, for example in HPS, that they have a long-term advantage.

Cardiovascular prevention before admission reduces mortality following acute myocardial infarction in patients with diabetes.

Löndahl M, Katzman P, Nilsson A, Ljungdahl L, Prütz KG. *J Intern Med* 2002; **251**: 325–30.

BACKGROUND. Previous studies have shown that patients with diabetes mellitus have an increased mortality after suffering from acute myocardial infarction (AMI). Patients with diabetes have several risk factors for cardiovascular disease. The objective of this study was to quantify the prevalence of pharmacological cardiovascular prevention at admission and relate such treatment to short and long-term mortality following AMI in patients with and without diabetes. All patients discharged from the Department of Internal Medicine at Helsingborg Hospital in 1996 and 1997 with a principal diagnosis of AMI were included in the study. Patients were divided into two groups according to the presence or absence of diabetes. Cardiovascular risk factors, ongoing medication, type of ward following admission, peak creatine kinase MB mass (CKMB) and immediate treatment were registered. Information about death was obtained from the national register. Kaplan-Meier analysis was performed for life-expectancy. A total of 673 patients with AMI were registered, of whom 117 (17.4%) had diabetes. No differences in 30 days (17.1% *vs* 15.3%) or 1-year (24.8% *vs* 27.4%) mortality were seen between the diabetes and control groups, whereas the 2-year mortality was significantly higher in the diabetes group (40.2% *vs* 29.1%). Cardiovascular risk factors occurred more often in the diabetes group and the use of aspirin, angiotensin converting enzyme (ACE)-inhibitors, statins and diuretics was significantly more frequent. In patients treated with aspirin, in combination with either statin or ACE-inhibitor, or both, no differences were seen in 30-day, 1- or 2-year mortality between groups.

INTERPRETATION. In contrast to earlier studies, this study did not find an increased 30-day and 1-year mortality in patients with diabetes suffering from AMI. This discrepancy was linked to a higher frequency of pharmacological cardiovascular prevention, a finding supporting the hypothesis that survival of a diabetes patient after AMI could be affected by factors operating before the infarction.

Comment

This is a thought-provoking study, which despite its many limitations, suggests that the higher mortality following AMI in diabetes is diminished if cardioprotective drug therapy has been initiated before the infarction. The study thus strengthens the case for primary CHD prevention in diabetes with aspirin, statin and ACE-inhibitor therapy. It would be interesting to know whether there have been adequate numbers of diabetic patients experiencing ACSs in randomized, placebo-controlled studies of these agents to test the hypothesis suggested by this study in a pooled meta-analysis.

Therapy with hydroxymethylglutaryl coenzyme A reductase inhibitors (statins) and associated risk of incident cardiovascular events in older adults: evidence from the Cardiovascular Health Study.

Lemaitre RN, Psaty BM, Heckbert SR, Kronmal RA, Newman AB, Burke GL.
Arch Intern Med 2002; **162**: 1395–400.

BACKGROUND. Recommendations to treat older adults with hydroxymethylglutaryl coenzyme A reductase inhibitors (statins) for the primary prevention of CHD events are supported by a single clinical trial restricted to adults 73 years or younger with low levels of HDL cholesterol. The association of statin use with incident cardiovascular disease and all-cause mortality was investigated during up to 7.3 years' follow-up of 1250 women and 664 men from the Cardiovascular Health Study. Study participants were 65 years and older and free of cardiovascular disease at baseline. They received drug therapy to lower cholesterol levels at baseline or no treatment with a recommendation for therapy according to the NCEP guidelines. Use of these drugs was assessed annually. Proportional-hazards models were used to calculate HRs and 95% CIs, adjusted for confounding variables.

The study found 382 incident cardiovascular events (159 MIs, 159 strokes, and 64 deaths due to CHD) and 362 total deaths from 1 June 1989 to 31 May 1997. Compared with no use of drugs to lower cholesterol levels, statin use was associated with decreased risk of cardiovascular events (multivariate HR 0.44; 95% CI 0.27–0.71) and all-cause mortality (HR 0.56; 95% CI 0.36–0.88). Similar associations were observed among participants aged 74 years or older at baseline.

INTERPRETATION. Use of statins was associated with decreased risk of incident cardiovascular events among elderly adults. These findings lend support to the NCEP guidelines, which recommend therapy for the lowering of cholesterol levels for older adults with hypercholesterolaemia.

Comment

None of the randomized, placebo-controlled clinical event trials of statins has shown evidence of an attenuation of the reduction in relative benefit when the older patients in the trials are compared with the younger ones. From epidemiological studies, some attenuation might be anticipated because the steepness of the slope relating CHD incidence to serum cholesterol diminishes with age. In most recommendations, particularly those stemming from countries with a socialized health-care system eager to limit drug expenditure, the use of statins has been confined to people whose ages do not exceed those randomized in trials. However, because of the rise in absolute CHD event rates with increasing age, even if the statin effect on relative risk was quite markedly diminished, the number of elderly patients who must be treated to prevent a CHD event would still be fewer than for younger people. This study, despite its observational design, seems to support this conclusion.

A prospective study of Pravastatin in the elderly at risk (PROSPER): screening experience and baseline characteristics.

Ford I, Blauw G, Murphy MB, *et al. Curr Control Trials Cardiovasc Med* 2002; **3**: 8.

BACKGROUND. PROSPER was designed to investigate the benefits of treatment with pravastatin in elderly patients for whom a typical doctor might consider the prescription of statin therapy to be a realistic option. PROSPER is a randomized, double-blind, placebo-controlled trial to test the hypothesis that treatment with pravastatin (40 mg/day) will reduce the risk of CHD death, non-fatal MI, and fatal or non-fatal stroke in elderly men and women with pre-existing vascular disease or with significant risk of developing this condition. In Scotland, Ireland, and The Netherlands, 23 770 individuals were screened, and 5804 subjects (2804 men and 3000 women), aged 70 to 82 years (average 75 years) and with baseline cholesterol 4.0–9.0 mmol/l, were randomized. Randomized subjects had similar distributions with respect to age, blood pressure, and body mass index when compared to the entire group of those who were screened, but had a higher prevalence of smoking, diabetes, hypertension, and a history of vascular disease. The average total cholesterol level at baseline was 5.4 mmol/l (men) and 6.0 mmol/l (women).

INTERPRETATION. Compared with previous prevention trials of cholesterol-lowering drugs, the PROSPER cohort is significantly older and for the first time includes a majority of women. The study, having achieved its initial goal of recruiting more than 5500 elderly high-risk men and women, aims to complete all final subject follow-up visits in the first half of 2002 with the main results being available in the fourth quarter of 2002.

Comment

PROSPER will settle the issue about whether it is reasonable to limit the use of statin therapy to younger patients. Already there are strong indications that statin treatment may prove even more beneficial in elderly patients than younger ones. The outcome of PROSPER may seem to some a foregone conclusion, but, without a randomized, placebo-controlled trial, mainstream medical practice is unlikely to change and potentially huge numbers of elderly patients who could benefit from statin therapy will continue to be denied it.

Influence of low high-density lipoprotein cholesterol and elevated triglyceride on coronary heart disease events and response to simvastatin therapy in 4S.

Ballantyne CM, Olsson AG, Cook TJ, Mercuri MF, Pedersen TR, Kjekshus J. *Circulation* 2001; **104**: 3046–51.

BACKGROUND. Patients with low HDL cholesterol and elevated triglyceride had an increased risk for CHD events and received the greatest benefit with fibrate therapy in substudy analyses of the Helsinki Heart Study and the Bezafibrate Infarction Prevention Study. In this *post hoc* analysis of the 4S, which enrolled patients with elevated LDL cholesterol and CHD, subgroups defined by HDL cholesterol and triglyceride quartiles were compared to examine the influence of HDL cholesterol and triglyceride on CHD events and response to therapy. Patients in the lowest HDL cholesterol (<1.00 mmol/l [39 mg/dl]) and highest triglyceride (>1.80 mmol/l [159 mg/dl]) quartiles (lipid triad; n = 458) had increased proportions of other features of the metabolic syndrome (increased body mass index, hypertension, diabetes), men, prior MI, prior revascularization, and beta-blocker use than patients in the highest HDL cholesterol (>1.34 mmol/l [52 mg/dl]) and lowest triglyceride (<1.11 mmol/l [98 mg/dl]) quartiles (isolated LDL-C elevation; n = 545). The major coronary event rate was highest in lipid triad patients on placebo (35.9%), and this subgroup had the greatest event reduction (relative risk 0.48, 95% CI 0.33–0.69); a significant treatment-by-subgroup interaction (P = 0.03) indicated a greater treatment effect in the lipid triad subgroup than the isolated LDL cholesterol elevation subgroup.

INTERPRETATION. Patients with elevated LDL cholesterol, low HDL cholesterol, and elevated triglycerides were more likely than patients with isolated LDL cholesterol elevation to have other characteristics of the metabolic syndrome, had increased risk for CHD events on placebo, and received greater benefit with simvastatin therapy.

Comments

It is widely recognized that risk of CHD is more closely related to the ratio of serum cholesterol or LDL cholesterol to HDL cholesterol than to either serum or LDL cholesterol alone. This study clearly demonstrates that this applies also to patients who have established CHD and that in such patients the combination of raised serum cholesterol and low HDL cholesterol with moderately raised triglycerides (the lipid triad) carries the greatest risk for recurrent CHD events (Fig 12.2). It is of clinical importance that these patients with the highest risk also showed the greatest benefit from statin treatment. This triad is common in patients who have diabetes defined in terms of their fasting glucose concentration, but even when patients with diabetes were excluded, the greater risk and the greater benefit from statin therapy was still evident (Fig 12.3). Many of these non-diabetic patients with the lipid triad would develop diabetes were they to live long enough: it is known that low HDL and raised triglycerides may precede the development of diabetes by many

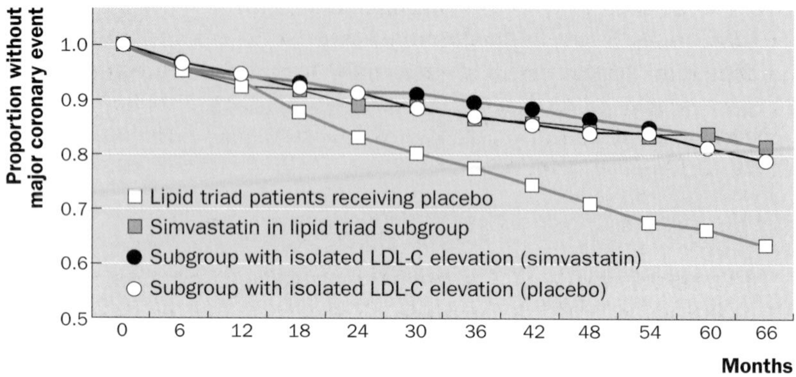

Fig. 12.2 Effect of simvastatin on event-free survival by subgroups defined by baseline HDL cholesterol and triglyceride. Event-free survival was significantly lower in lipid triad patients receiving placebo than in other treatment subgroups and improved more with simvastatin in lipid triad subgroup than subgroup with isolated LDL cholesterol elevation. Source: Ballantyne *et al.* (2001).

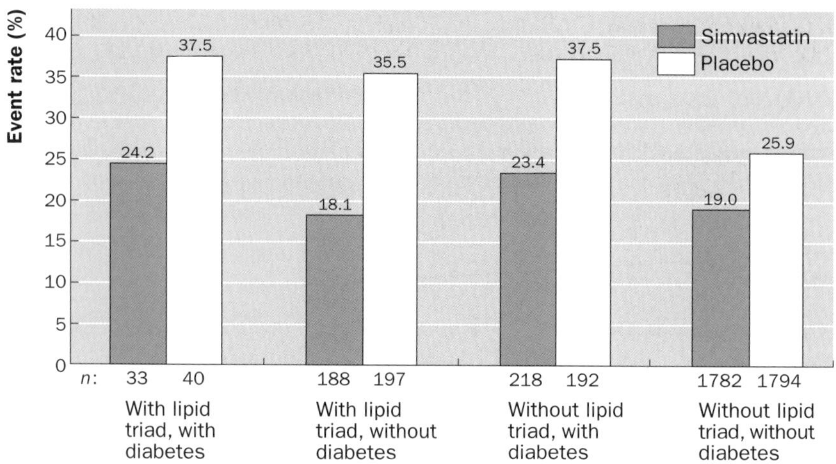

Fig. 12.3 Effect of simvastatin on major coronary events in patients with lipid triad, diabetes (either by history or glucose ≥126 mg/dL), or both. After exclusion of patients with diabetes, subgroup with lipid triad still received enhanced benefit with simvastatin, reducing major coronary events to a rate similar to that in patients without lipid triad and without diabetes. By comparison, event rates on placebo were similarly high in patients with lipid triad with or without diabetes and in diabetic patients without lipid triad. Source: Ballantyne *et al.* (2001).

years |**12**|. Thus they may develop CHD in advance of fulfilling the glycaemic criteria for diabetes. This explains the increased likelihood of patients with CHD developing diabetes and underlines the importance of being aware that the higher CHD risk stemming from diabetic dyslipidaemia may be present before diabetes is clinically present. Many such patients will at that stage have the metabolic syndrome, now recognized as important in planning how to intervene actively in CHD prevention in the NCEP III recommendations.

The relative role of low-density lipoprotein cholesterol and high-density lipoprotein cholesterol in coronary artery disease: evidence from large-scale statin and fibrate trials.

Sacks FM. *Am J Cardiol* 2001; **88**: 14–18.

BACKGROUND AND INTERPRETATION. Statin therapy has been remarkably successful in reducing coronary events and improving survival. Reduction in LDL cholesterol is the paramount mechanism for the benefit of statins. Although the relationship between reduction of risk of coronary artery disease and pre-treatment lipid levels has not been fully resolved, a feature of statin therapy emerging from previous large-scale trials is the diminished benefit when the baseline level of LDL cholesterol is low (e.g. <125 mg/dl). Increases in levels of HDL cholesterol with statin therapy are associated with mild reductions in coronary events, indicating that HDL cholesterol is an important element. However, the risk of low HDL cholesterol

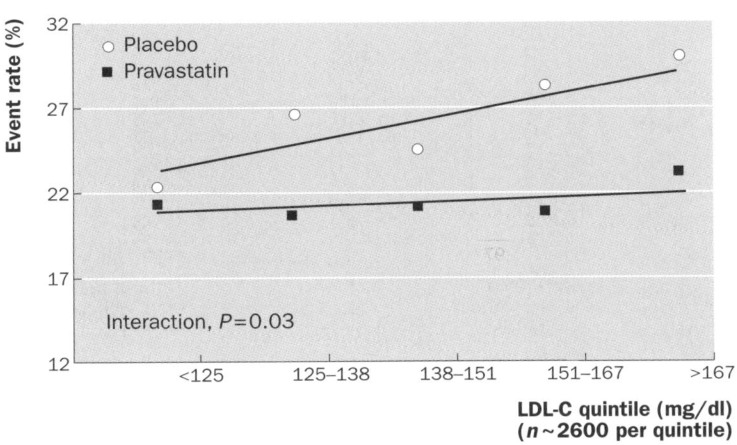

Fig. 12.4 Coronary event rates in the Cholesterol and Recurrent Events (CARE) and Long-Term Intervention with Pravastatin in Ischaemic Disease (LIPID) trials combined, in relation to LDL cholesterol (LDL-C). Source: Sacks (2001).

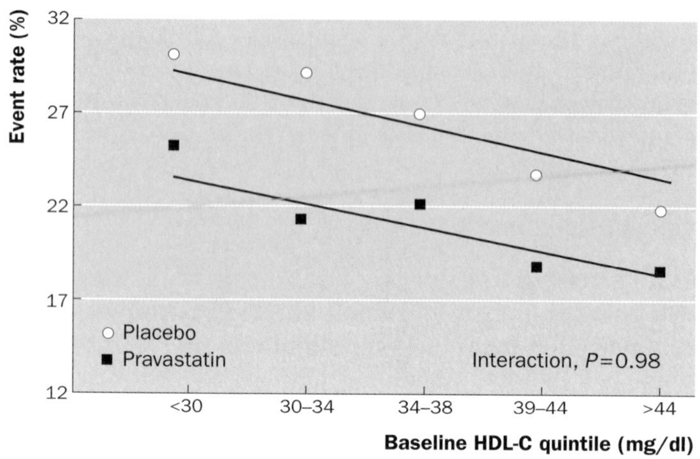

Fig. 12.5 Coronary event rates in the Cholesterol and Recurrent Events (CARE) and Long-Term Intervention with Pravastatin in Ischaemic Disease (LIPID) trials combined, in relation to high-density lipoprotein cholesterol (HDL-C). Source: Sacks (2001).

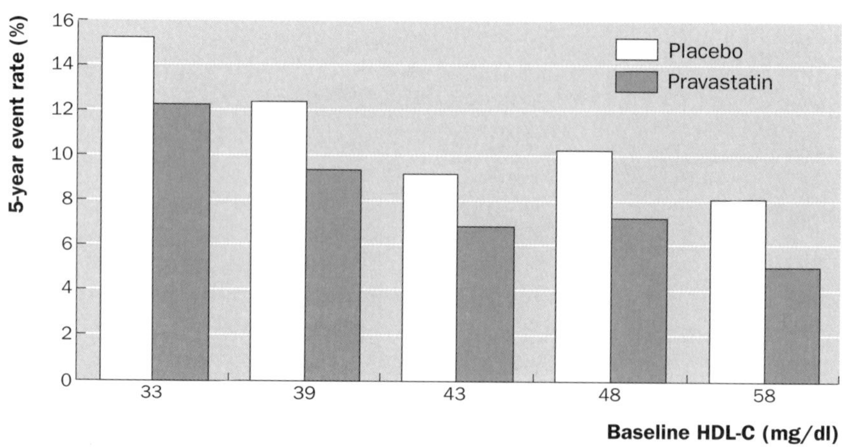

Fig. 12.6 Coronary event rates in the West of Scotland Coronary Prevention Study (WOSCOPS) in relation to high-density lipoprotein cholesterol (HDL-C). Source: Sacks (2001).

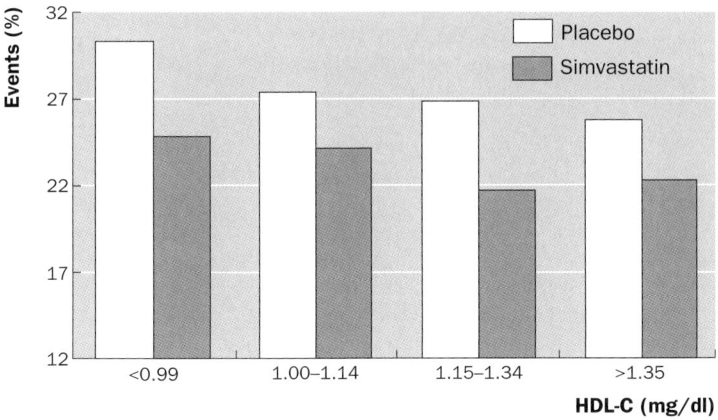

Fig. 12.7 Major event rates in the Scandinavian Simvastatin Survival Study (4S) in relation to high-density lipoprotein cholesterol (HDL-C). Source: Sacks (2001).

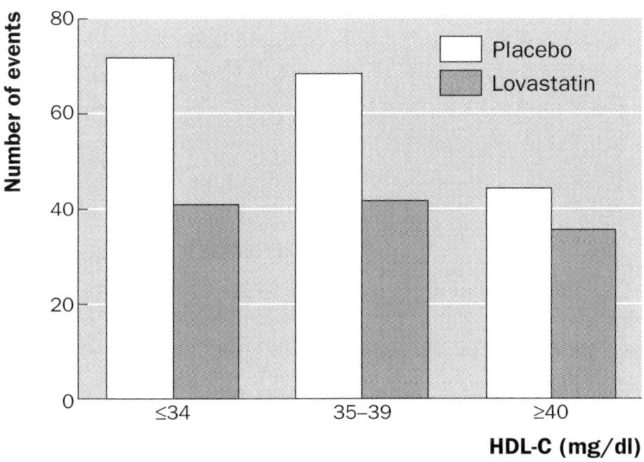

Fig. 12.8 Efficacy end-points with lovastatin in the Air Force/Texas Coronary Atherosclerosis Prevention Study (AFCAPS/TexCAPS) in relation to baseline high-density lipoprotein cholesterol (HDL-C). Source: Sacks (2001).

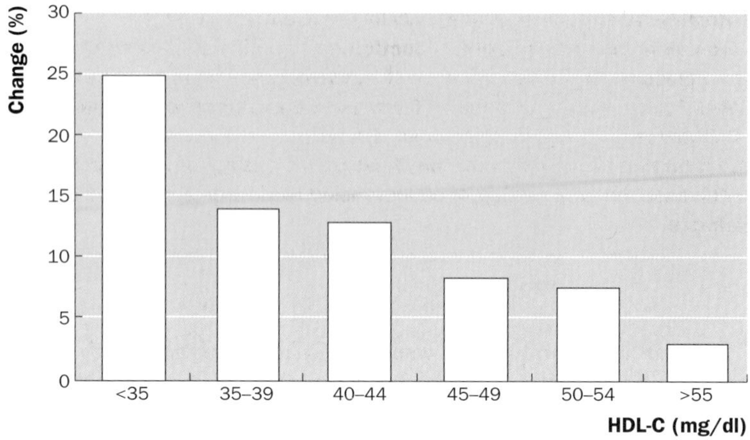

Fig. 12.9 Per cent change in high-density lipoprotein cholesterol (HDL-C) with gemfibrozil in relation to baseline HDL-C (Helsinki Heart Study). Source: Sacks (2001).

is not altered by statin therapy, in marked contrast with LDL cholesterol. Because most coronary events have not been prevented in the statin trials, non-statin lipid therapy, including fibrates, may be necessary to further reduce the unacceptably high rate of coronary artery disease.

Comment

The article brings together evidence that low HDL cholesterol compounds the CHD risk from LDL cholesterol (Figs. 12.4–12.9). In clinical trials, statins have a smaller effect on HDL than fibrate drugs and the author suggests that fibrate drugs in combination with statin therapy might be expected to produce a greater reduction in CHD risk than either can alone in patients with low HDL. Randomized, placebo-controlled trials of combined statin and fibrate therapy would be especially welcome to test this important hypothesis.

Targeting low high-density lipoprotein cholesterol for therapy: lessons from the Veterans Affairs High-density Lipoprotein Intervention Trial.

Robins SJ. *Am J Cardiol* 2001; **88**: 19–23.

Background and interpretation. Results of the Veterans Affairs High-Density Lipoprotein Intervention Trial (VA-HIT) showed that therapy with the fibric acid gemfibrozil significantly reduced the incidence of coronary artery disease events in

men with known coronary artery disease and a low level of HDL cholesterol. Coronary artery disease event reduction was inversely related to levels of HDL cholesterol achieved with gemfibrozil and, even at relatively low concentrations of HDL cholesterol, coronary artery disease event reduction with gemfibrozil was significantly greater than with placebo. Subjects with a low HDL cholesterol level who were recruited for VA-HIT had a high prevalence of features of the metabolic syndrome with obesity, type 2 diabetes, and hyperinsulinaemia. These individuals especially benefited from gemfibrozil, and thus, fibrate therapy may have an increasing role in reducing coronary artery disease risk in a setting of an increasing worldwide prevalence of obesity and diabetes.

Comment

The VA-HIT |**2**| was a randomized, placebo-controlled trial of gemfibrozil in patients with CHD whose LDL cholesterol levels were low, but who also had low HDL cholesterol levels. There was a 22% reduction in coronary events in the actively treated patients. We now know from HPS that patients of this type would also have experienced benefit from statin therapy. It is not possible to determine whether fibrate or statin therapy should be the first line of treatment in such patients, although the greater body of evidence for both the efficacy and safety of statins mean that most authorities would recommend statin first. Again the issue is raised about whether the greatest benefit would accrue from combined therapy.

High-density lipoprotein and coronary heart disease: lessons from recent intervention trials.

Bloomfield Rubins H. *Prev Cardiol* 2000; **3**: 33–9.

BACKGROUND AND INTERPRETATION. Epidemiological studies have consistently implicated low plasma HDL cholesterol as an important, independent risk factor for the development of CHD. However, clinical trials specifically designed to evaluate the role of lipid therapy in patients with low HDL cholesterol have only been recently reported. They include two trials with angiographic end-points, the Lipid Coronary Angiography Trial and the Bezafibrate Coronary Atherosclerosis Intervention Trial, and three clinical end-points trials, AFCAPS/TexCAPS, VA-HIT, and the Bezafibrate Infarction Prevention study. These and other trials clearly indicate that persons with CHD and high LDL cholesterol (>130 mg/dl [3.36 mmol/l]), with or without low HDL cholesterol, benefit from statin therapy. AFCAPS/TexCAPS showed that persons at high risk of CHD but without known disease, who have moderate levels of LDL cholesterol as well as low levels of HDL cholesterol, also appear to benefit from statin therapy although the cost-effectiveness of this approach is unclear. The results from VA-HIT provide convincing evidence that patients without high LDL cholesterol and with established CHD and low HDL cholesterol benefit from gemfibrozil. This drug may be particularly beneficial for patients who, in addition to low HDL cholesterol, present with other features of the metabolic syndrome, such as obesity, glucose intolerance, and high triglycerides. Whether other fibrates, niacin, or statins lower CHD risk in persons with

low HDL cholesterol in the absence of high or moderately high LDL cholesterol is unknown.

Comment

If we are to use fibrates in CHD prevention generally, we need to know more about their safety in combination with statins. Also despite the fact that fibrates have been extant in clinical practice far longer than statins there are still fewer data from clinical trials about their efficacy. In both the Helsinki Heart Study |**13**|, a primary prevention trial, and VA-HIT |**2**|, a secondary prevention trial, CHD risk was reduced. Both these studies employed gemfibrozil. Bezafibrate, in the Bezafibrate Infarction Prevention study |**14**|, a secondary prevention trial, did not significantly decrease risk overall, although it did appear to do so in the subgroup with raised triglycerides. Similar trials, such as Fenofibrate Intervention and Event Lowering in Diabetes (FIELD) using fenofibrate, with statistical power to give details on clinical events are still to report.

The role of high-density lipoprotein (HDL) cholesterol in the prevention and treatment of coronary heart disease: Expert Group recommendations.

Sacks FM. *Am J Cardiol* 2002; **90**: 139–43.

Background. Ample evidence supports the importance of HDL cholesterol for CHD risk. Raising low HDL cholesterol levels should be considered important, as is lowering LDL cholesterol, to prevent CHD, and increased prominence should be given to HDL cholesterol as an intervention target.

Interpretation. This report proposes the following recommendations:

1. HDL cholesterol $\geq$40 mg/dl ($\geq$1.0 mmol/l) be recommended as a goal for patients with cardiovascular disease and those without clinical cardiovascular disease but at high risk, and especially in those with type 2 diabetes or features of the metabolic syndrome, most particularly abdominal obesity with high fasting insulin values.
2. Lifestyle changes that include smoking cessation, weight loss, a diet moderate in unsaturated fat rather than low fat, and regular exercise be encouraged to reach this HDL cholesterol goal.
3. Consideration be given to fibrate therapy for the previously mentioned patients with a low HDL cholesterol and low-risk LDL cholesterol, defined as below the threshold for LDL-lowering drug treatments according to recognized guidelines (e.g. NCEP $<$130 mg/dl for secondary prevention). Niacin may also be considered in appropriate patients.
4. These recommendations would apply to patients who do not require statin therapy to reduce LDL cholesterol, and to those who are being treated with statins in accordance with current national or international guidelines.

Comment

These recommendations raise more questions than they answer. They do not really address the issue that statin therapy will be regarded by many as the most appropriate first-line therapy in patients with low HDL because they will benefit from these drugs and they are known to be safe. As has previously been discussed, they may derive an even greater decrease in CHD risk from combining the statin with a fibrate, such as gemfibrozil. But what are the long-term hazards of such a combination? Perhaps in high-risk patients these are risks worth taking, but how high should the CHD risk be to make it acceptable? Introducing niacin as an alternative to fibrates is not particularly helpful because the medication has many side-effects and its safety in combination with statins is even more dubious than that of fibrates. Added to this, placebo-controlled trails of niacin are not even possible because of the unpleasant flushing which virtually everyone compliant with this medication experiences.

Cardiovascular disease outcomes during 6.8 years of hormone therapy: Heart and Estrogen/progestin Replacement Study follow-up (HERS II).

Grady D, Herrington D, Bittner V, *et al.* HERS Research Group. *JAMA* 2002; **288**: 49–57.

BACKGROUND. HERS found no overall reduction in risk of CHD events among post-menopausal women with CHD. However, in the hormone group, findings did suggest a higher risk of CHD events during the first year, and a decreased risk during years 3 to 5. The aim of this study was to determine if the risk reduction observed in the later years of HERS persisted and resulted in an overall reduced risk of CHD events with additional years of follow-up. This was a randomized, blinded, placebo-controlled trial of 4.1 years' duration (HERS) and subsequent unblinded follow-up for 2.7 years (HERS II) conducted at outpatient and community settings at 20 US clinical centres. Participants were a total of 2763 post-menopausal women with CHD and average age of 67 years at enrolment in HERS; 2321 women (93% of those surviving) consented to follow-up in HERS II. Participants were randomly assigned to receive 0.625 mg/d of conjugated oestrogens and 2.5 mg of medroxyprogesterone acetate (n = 1380), or placebo (n = 1383) during HERS; open-label hormone therapy was prescribed at personal physicians' discretion during HERS II. The proportions with at least 80% adherence to hormones declined from 81% (year 1) to 45% (year 6) in the hormone group, and increased from 0% (year 1) to 8% (year 6) in the placebo group. The primary outcome was non-fatal MI and CHD death. Secondary cardiovascular events were coronary revascularization, hospitalization for unstable angina or congestive heart failure, non-fatal ventricular arrhythmia, sudden death, stroke or transient ischaemic attack, and peripheral arterial disease. There were no significant decreases in rates of primary CHD events or secondary cardiovascular events among women assigned to the hormone group compared with the placebo group in HERS,

HERS II, or overall. The unadjusted relative hazard (RH) for CHD events in HERS was 0.99 (95% CI 0.81–1.22); HERS II, 1.00 (95% CI 0.77–1.29); and overall, 0.99 (0.84–1.17). The overall RHs were similar after adjustment for potential confounders and differential use of statins between treatment groups (RH, 0.97; 95% CI 0.82–1.14), and in analyses restricted to women who were adherent to randomized treatment assignment (RH, 0.96; 95% CI 0.77-1.19).

Interpretation. Lower rates of CHD events among women in the hormone group in the final years of HERS did not persist during additional years of follow-up. After 6.8 years, hormone therapy did not reduce risk of cardiovascular events in women with CHD. Post-menopausal hormone therapy should not be used to reduce risk for CHD events in women with CHD.

Comment

The result represents the abandonment of hope for those devotees of combined HRT as a means of secondary CHD prevention. The suggestion that in the longer-term than the original HERS such therapy might have been associated with diminished CHD risk, was not borne out in this longer follow-up period.

Statin therapy, cardiovascular events and total mortality in the Heart and Estrogen/Progestin Replacement Study (HERS).

Herrington DM, Vittinghoff E, Lin F, *et al.* for the HERS Study Group. *Circulation* 2002; **105**: 2962–7.

Background. Although effects of statins on cardiovascular outcomes are well established in men, fewer data exist for women, Furthermore, the effects of statins plus HRT on cardiovascular outcome are uncertain. This study examined statin use, cardiovascular events, and total mortality in HERS, a randomized clinical trial of oestrogen plus progestin *versus* placebo in post-menopausal women with heart disease (n =2763). A non-randomized comparison of statin users and non-users revealed lower rates of the primary outcome, non-fatal MI or CHD death (RH = 0.79, 95% CI 0.63–0.99, P = 0.04), and total mortality (RH = 0.67, 95% CI 0.51–0.87, P = 0.003). Rates of venous thromboembolic events were also lower among statin users (RH = 0.45, 95% CI 0.23–0.88, P = 0.02). HRT resulted in a significant increase in early risk for primary events in women who did not use statins (RH = 1.75, 95% CI 1.02–3.03, P = 0.04), but not in statin users (RH = 1.34, 95% CI 0.63–2.86, P = 0.45). Adjustment for post-randomization statin use showed no effect of HRT on risk for the primary outcome (RH = 0.96, 95% CI 0.77–1.29, P = 0.72).

Interpretation. In HERS, statin use was associated with lower rates of cardiovascular event, venous thromboembolic events and total mortality. These data provide strong support for statin use in eligible women with coronary disease.

Comment

HERS was a secondary prevention trial and the women participating were not randomized to statin therapy so the apparent benefit of statins on CHD risk in this trial could have been confounded by, for example, treatment indication. None the less, the finding is similar to that reported for women in randomized placebo-controlled secondary prevention trials of statin treatment. Only in the HPS, however, was the study deliberately powered to give a definitive answer in women. It is of practical interest that the increase in CHD events in HERS associated with treatment with 0.625 mg conjugated equine oestrogens and medroxyprogesterone 2.5 mg daily was statistically significant only in the women not receiving statin. However, this tells us nothing about the mechanism by which this type of HRT increases the risk of CHD events. The decreased risk of venous thromboembolism in the statin-treated women was also interesting, but could be confounded and not due to statin treatment itself.

Risks and benefits of estrogen plus progestin in healthy post-menopausal women: principal results from the Women's Health Initiative randomized controlled trial.

Writing Group for the Women's Health Initiative Investigators. *JAMA* 2002; **288**: 321–33.

Background. Despite decades of accumulated observational evidence, the balance of risks and benefits for hormone use in healthy post-menopausal women remains uncertain. The aim of this study was to assess the major health benefits and risks of the most commonly used combined hormone preparation in the United States. The study examined the oestrogen plus progestin component of the Women's Health Initiative, a randomized controlled primary prevention trial (planned duration, 8.5 years) in which 16 608 post-menopausal women aged 50–79 years with an intact uterus at baseline were recruited by 40 US clinical centres in 1993–1998. Participants received conjugated equine oestrogens, 0.625 mg/d, plus medroxyprogesterone acetate, 2.5 mg/d, in one tablet (*n* = 8506) or placebo (*n* = 8102). The primary outcome was CHD (CHD non-fatal MI and CHD death), with invasive breast cancer as the primary adverse outcome. A global index summarizing the balance of risks and benefits included the two primary outcomes plus stroke, pulmonary embolism (PE), endometrial cancer, colorectal cancer, hip fracture, and death due to other causes.

On 31 May 2002, after a mean of 5.2 years of follow-up, the data and safety monitoring board recommended stopping the trial of oestrogen plus progestin *v.s* placebo because the test statistic for invasive breast cancer exceeded the stopping boundary for this adverse effect and the global index statistic supported risks exceeding benefits. This report includes data on the major clinical outcomes through to 30 April 2002. Estimated HRs (nominal 95% CIs) were as follows: CHD, 1.29 (1.02–1.63) with 286 cases; breast cancer, 1.26 (1.00–1.59) with 290 cases; stroke, 1.41 (1.07–1.85) with 212 cases; PE, 2.13 (1.39–3.25) with 101 cases; colorectal

cancer, 0.63 (0.43–0.92) with 112 cases; endometrial cancer, 0.83 (0.47–1.47) with 47 cases; hip fracture, 0.66 (0.45–0.98) with 106 cases; and death due to other causes, 0.92 (0.74–1.14) with 331 cases. Corresponding HRs (nominal 95% CIs) for composite outcomes were 1.22 (1.09–1.36) for total cardiovascular disease (arterial and venous disease), 1.03 (0.90–1.17) for total cancer, 0.76 (0.69–0.85) for combined fractures, 0.98 (0.82–1.18) for total mortality, and 1.15 (1.03–1.28) for the global index. Absolute excess risks per 10 000 person-years attributable to oestrogen plus progestin were seven more CHD events, eight more strokes, eight more PEs, and eight more invasive breast cancers, while absolute risk reductions per 10 000 person-years were six fewer colorectal cancers and five fewer hip fractures. The absolute excess risk of events included in the global index was 19 per 10 000 person-years.

Interpretation. Overall health risks exceeded benefits from use of combined oestrogen plus progestin for an average 5.2-year follow-up among healthy post-menopausal US women. All-cause mortality was not affected during the trial. The risk–benefit profile found in this trial is not consistent with the requirements for a viable intervention for primary prevention of chronic diseases, and the results indicate that this regimen should not be initiated or continued for primary prevention of CHD.

Comment

This is a disturbing result which will be debated in many places. As the authors conclude, combined HRT can no longer be regarded as a means for CHD prevention. The risks of such therapy are small but are they acceptably small? Many women will find the relief of menopausal symptoms they experience with HRT a compelling reason to take it but we need an objective comparison of the values of such symptomatic relief relative to the hazards of HRT. As far as the prevention of osteoporosis is concerned, alternative non-hormonal therapies are increasingly available but we do not know what effect they have on CHD risk. That is true even of calcium supplements. What are the hazards of oestrogen therapy alone in women who have undergone hysterectomy, but not oophorectomy? Should we re-evaluate oestrogen alone even in those who have undergone hysterectomy and oophorectomy? Should we also do so in women with spontaneous premature menopause? Are the adverse effects as great with other oestrogens, other progestogens or with tibolone?

Early protection against sudden death by n-3 polyunsaturated fatty acids after myocardial infarction: time-course analysis of the results of the Gruppo Italiano per lo Studio della Sopravvivenza nell'Infarto Miocardico (GISSI)-Prevenzione.

Marchioli R, Barzi F, Bomba E, *et al.*; GISSI-Prevenzione Investigators. *Circulation* 2002; **105**: 1897–903.

Background. The purpose of this study was to assess the time course of the benefit of omega–3 polyunsaturated fatty acids (PUFAs) on mortality documented by

the GISSI-Prevenzione trial in patients surviving a recent (<3 months) MI. In this study, 11 323 patients were randomly assigned to supplements of omega–3 PUFAs, vitamin E (300 mg/d), both, or no treatment (control) on top of optimal pharmacological treatment and lifestyle advice. Intention-to-treat analysis adjusted for interaction between treatments was carried out. Early efficacy of omega–3 PUFA treatment for total, cardiovascular, cardiac, coronary, and sudden death; non-fatal MI; total CHD; and cerebrovascular events was assessed by right-censoring follow-up data 12 times from the first month after randomization up to 12 months. Survival curves for omega–3 PUFA treatment diverged early after randomization, and total mortality was significantly lowered after three months of treatment (relative risk [RR] 0.59; 95% CI 0.36–0.97; $P = 0.037$). The reduction in risk of sudden death was specifically relevant and statistically significant already at four months (RR 0.47; 95% CI 0.219–0.995; $P = 0.048$). A similarly significant, although delayed, pattern after six to eight months of treatment was observed for cardiovascular, cardiac, and coronary deaths.

Interpretation. The early effect of low-dose (1 g/d) n-3 PUFAs on total mortality and sudden death supports the hypothesis of an antiarrhythmic effect of this drug. Such a result is consistent with the wealth of evidence coming from laboratory experiments on isolated myocytes, animal models, and epidemiological and clinical studies.

Comment

Omacor, the drug employed in the GISSI study, will soon be available in many parts of the world including the USA and Europe. Despite the decrease in sudden CHD deaths with omega–3 fatty acids in GISSI (Fig. 12.10) its licence in the USA will be to provide a means of lowering serum triglycerides whereas in Europe it will be licensed both for this purpose and for secondary cardioprotection. The difference is because the FDA require two randomized placebo-controlled clinical event trials to establish the indication for secondary CHD prevention. Whether this is reasonable is debatable. However, Omacor, albeit in larger doses than those used in GISSI, does provide an alternative to fibrates or niacin in decreasing serum triglycerides even in patients already receiving statin therapy |**15**|. This is important now that hypertriglyceridaemia, at least as part of the metabolic syndrome, is recognized as a target in CHD prevention by the NCEPIII recommendations. If in Europe, Omacor at a dose of 1 g daily becomes part of the cardioprotective regimen for patients with established CHD, then it will be relatively easy for clinicians to use a somewhat larger dose in patients with persisting hypertriglyceridaemia despite statin therapy. Ideally, a randomized, placebo-controlled trial of triglyceride-lowering doses of Omacor in such patients to establish whether this does produce greater benefit than 1 g daily or placebo should be undertaken. This would also help to settle the issue about whether raising HDL cholesterol is as important as lowering triglycerides, because omega–3 fatty acids have less effect on HDL than fibrates or niacin.

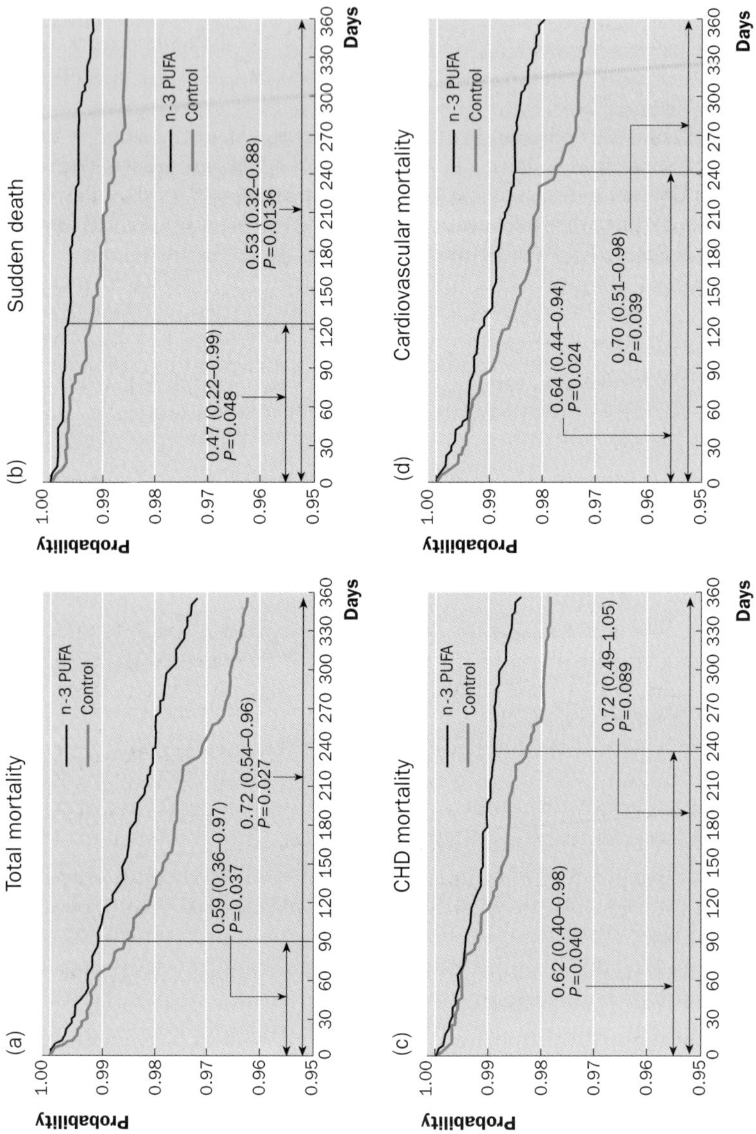

Fig. 12.10 Early benefit of n-3 PUFA therapy. (a) Total mortality; (b) sudden death; (c) coronary heart disease mortality; and (d) cardiovascular mortality. Source: Marchioli *et al.* (2002).

References

1. Schwartz GG, Olsson AG, Ezekowitz MD, *et al.* Effects of atorvastatin on early recurrent ischemic events in acute coronary syndromes: the MIRACL study. A randomized controlled trial. *JAMA* 2001; **285**: 1711–18.
2. Rubins HB, Robins SJ, Collins D, Fye CL, Anderson JW, Elam MB, Faas FH, Linares E, Schaeffer EJ, Schectman G, Wilt TJ, Wittes J, for the Veterans Affairs High-Density Lipoprotein Cholesterol Intervention Trial Study Group. Gemfibrozil for the secondary prevention of coronary heart disease in men with low levels of high-density lipoprotein cholesterol. *N Engl J Med* 1999; **341**: 410–18.
3. The LIPID Study Group. Prevention of cardiovascular events and death with pravastatin in patients with coronary heart disease and a broad range of initial cholesterol levels. *N Engl J Med* 1998; **339**: 1349–57.
4. Pedersen TR, Wilhelmsen L, Faergeman O, *et al.* Follow-up study of patients randomized in the Scandinavian Simvastatin Survival Study (4S) of cholesterol lowering. *Am J Cardiol* 2000; **86**: 257–62.
5. Sacks FM, Pfeffer MA, Moye LA, Rouleau JL, Rutherford JD, Cole TG, Brown L, Warnica JW, Arnold JMO, Wun C-C, Davis BR, Braunwald E, for the Cholesterol and Recurrent Events Trial Investigators. The effect of pravastatin on coronary events after myocardial infarction in patients with average cholesterol levels. *N Engl J Med* 1996; **335**: 1001–9.
6. Shepherd J, Cobbe SM, Ford I, Isles CG, Lorimer AR, MacFarlane PW, McKillop JH, Packard C, *et al.* for the West of Scotland Coronary Prevention Study Group. Prevention of coronary heart disease with pravastatin in men with hypercholesterolaemia. *N Engl J Med* 1995; **333**: 1301–7.
7. Downs GR, Clearfield M, Weiss S, Whitney E, Shapiro DR, Beere PA, *et al.* Primary prevention of acute coronary events with lovastatin in men and women with average cholesterol levels: results of the AFCAPS/TexCAPS (Air Force/Texas Coronary Atherosclerosis Prevention Study). *JAMA* 1998; **279**: 1615–22.
8. Scandinavian Simvastatin Survival Study Group. Randomized trial of cholesterol lowering in 4444 patients with coronary heart disease; the Scandinavian Survival Study. *Lancet* 1994; **344**: 1383–9.
9. Anderson KM, Odell PM, Wilson PWF, Kannel WB. Cardiovascular disease risk profiles. *Am Heart J* 1990; **121**: 293–8.
10. The Heart Outcomes Prevention Evaluation Study Investigators. Vitamin E supplementation and cardiovascular events in high-risk patients. *N Engl J Med* 2000; **342**: 154–60.
11. Tames FJ, Mackness MI, Arrol S, Laing I, Durrington PN. Non-enzymatic glycation of apolipoprotein B in the sera of diabetic and non-diabetic subjects. *Atherosclerosis* 1992; **93**: 237–44.
12. Haffner SM, Stern MP, Hazuda HP, Mitchell BD, Patterson JK. Cardiovascular risk factors in confirmed prediabetic individuals: does the clock for coronary disease start ticking before the onset of clinical diabetes? *JAMA* 1990; **263**: 2893–8.

13. Frick MH, Elo O, Haapa K, Heinonen OP, Heinsalmi P, Helo P, Huttunen JK, *et al.* Hensinki Heart Study: primary-prevention trial with gemfibrozil in middle-aged men with dyslipidaemia. Safety of treatment, changes in risk factors, and incidence of coronary heart disease. *N Engl J Med* 1987; **317**: 1237–45.

14. The BIP Study Group. Secondary prevention by raising HDL cholesterol and reducing triglycerides in patients with coronary artery disease. The Bezafibrate Infarction Prevention (BIP) Study. *Circulation* 2000, **102**: 21–7.

15. Durrington PN, Bhatnagar D, Mackness MI, Morgan J, Julier K, Khan MA, France M. The effect of an omega–3 polyunsaturated fatty acid concentrate administered for one year to simvastatin treated patients with established coronary heart disease and persisting hypertriglyceridaemia. *Heart* 2001; **85**: 544–8.

13

Clinical trials in the context of epidemiology and clinical practice

Introduction

This year saw the publication of the most thorough and far-reaching national recommendations for the treatment of dyslipidaemia in the shape of the National Cholesterol Education Program (NCEP) Adult Treatment Panel (ATP) III report. These are discussed in this chapter in some detail.

Several studies this year addressed how well clinical trial evidence about coronary heart disease (CHD) prevention is being translated into clinical practice. Not too well it would appear even in the relatively uncontroversial area of secondary CHD prevention as revealed by European Action on Secondary Prevention by Intervention to Reduce Events (EUROASPIRE) II. Studies by Reid *et al.*, Whincup *et al.*, Benner *et al.*, and Jackevicius *et al.* also indicated that statins are underprescribed in older patients who themselves also comply poorly. This was the case even in older patients in the age ranges embraced by the clinical trials of statins where their use is undisputed. This represents a major failure in CHD prevention.

Failure of clinicians to achieve satisfactory low-density lipoprotein (LDL)-lowering with statin treatment has many causes. One is a reluctance to titrate statin doses upwards. One solution to this problem is to employ more potent statins which require dose titrations less frequently. This was discussed in an article by Stein. Resistance to statins can also be a factor and this year two articles were published on this subject. Miettinen and Gylling reported that resistant patients tend to absorb more cholesterol from their intestines and thus to have lower rates of hepatic cholesterol biosynthesis and Scanu and Hinman pointed out that apparent resistance to LDL cholesterol-lowering by statins can result from high levels of lipoprotein (a) which is included in measurements of LDL cholesterol, but not lowered by statins.

Clinicians are fearful of inducing myositis in their patients with statin therapy and frequently are unaware of how rare genuine cases of myositis are in patients receiving statin treatment. The issue was discussed by Evans and Rees this year. In this chapter I have also reviewed how little we understand the phenomenon of statin-induced myositis as we enter an era when combination therapy is increasingly advocated. Thus, whilst statin drugs which have been tested in randomized, placebo-controlled trials have been shown to be safe, greater insight into rare instances of severe myositis would be welcome. Another rare complication of statin therapy is peripheral neuropathy. Gaist and colleagues have this year provided better evidence for this than hitherto existed.

Executive Summary of the Third Report of the National Cholesterol Education Program (NCEP) Expert Panel on Detection, Evaluation, and Treatment of High Blood Cholesterol in Adults (Adult Treatment Panel III).

Expert Panel on Detection, Evaluation, and Treatment of High Blood Cholesterol in Adults. *JAMA* 2001; **285**: 2486–97.

BACKGROUND. The third adult treatment panel (ATP III) report updates the existing recommendations for clinical management of high blood cholesterol. The NCEP periodically produces ATP clinical updates as warranted by advances in the science of cholesterol management. Each of the guideline reports – ATP I, II, and III – has a major thrust. ATP I outlined a strategy from primary prevention of CHD in persons with high levels of LDL cholesterol (≥ 160 mg/dl) or those with borderline high LDL cholesterol (130–159 mg/dl) and multiple (2+) risk factors. ATP II affirmed the importance of this approach and added a new feature: the intensive management of LDL cholesterol in persons with established CHD. For patients with CHD, ATP II set a new, lower LDL cholesterol goal of ≤100 mg/dl. ATP III adds a call for more intensive LDL-lowering therapy in certain groups of people, in accord with recent clinical trial evidence, but its core is based on ATP I and ATP II. Some of the important features shared with previous reports are shown in Table 13.1. While ATP III maintains attention to intensive treatment of patients with CHD, its major new feature is a focus on primary prevention in persons with multiple risk factors. Many of these persons have a relatively high risk of CHD and will benefit from more intensive LDL-lowering treatment than recommended in ATP II. Table 13.2 shows the new features of ATP III. (Note: to convert cholesterol to mmol/l, divide values by 38.7)

Comment

NCEP III recommends that in adults aged 20 years or older, fasting serum cholesterol, LDL cholesterol, HDL cholesterol and triglyceride should be measured every five years. If the patient is non-fasting serum cholesterol and HDL cholesterol can still be measured and if these are >200 mg/dl (5.2 mmol/l) or <40 mg/dl (1.03 mmol/l) respectively, the full fasting lipoprotein profile should be obtained.

The essential features of the recommendations are outlined in Tables 13.1 and 13.2 and Fig. 13.1. A major new development in NCEP III is the treatment of diabetes as a coronary risk equivalent. In primary CHD prevention in non-diabetic people, assessment of risk is recommended in some patients using a scoring system based on the results of the Framingham Heart Study (Table 13.3 and Table 13.4). When I checked this against the first risk equation published from

Framingham |**1**|, which is the one used in many other recommendations such as those in Europe and Australia and New Zealand, I found that the new scoring system gave lower levels of risk for many patients. Until more information is available about this discrepancy, the NCEP III scoring system *should not be used in combination with guidelines based on the earlier equation.* The Joint British Societies Coronary Risk Prediction charts |**2**| published in last year's *The Year in Dyslipidaemia 2002* are more suitable for use outside the USA.

In primary prevention, NCEP III recognises that most patients with genetic hypercholesterolaemia, such as familial hypercholesterolaemia will have LDL cholesterol levels ≥190 mg/dl (4.91 mmol/l). Drug treatment is recommended at these levels of LDL cholesterol, if unresponsive to lifestyle measures, regardless of calculated CHD risk. An intermediate level of LDL cholesterol between 160 and 189 mg/dl (4.13–4.9 mmol/l) even in the presence of no other risk factors or only one risk factor is regarded as optional for lipid-lowering drug therapy. Lipid-lowering drugs are, however, recommended with these levels of LDL cholesterol if two or

Table 13.1 Shared features of ATP III and ATP II*

- Continued identification of LDL cholesterol lowering as the primary goal of therapy
- Consideration of high LDL cholesterol (≥160 mg/dl) as a potential target for LDL-lowering drug therapy, specifically as follows:
  - For persons with multiple risk factors whose LDL levels are high (≥160 mg/dl) after dietary therapy, consideration of drug therapy is recommended
  - For persons with 0–1 risk factor, consideration of drug therapy (after dietary therapy) is optional for LDL 160–189 mg/dl and recommended for LDL ≥190 mg/dl
- Emphasis on intensive LDL-lowering therapy in persons with established CHD
- Identification of 3 categories of risk for different LDL goals and different intensities of LDL-lowering therapy
  - CHD and CHD risk equivalents† (other forms of clinical atherosclerotic disease)
  - Multiple (2+) risk factors‡
  - 0–1 risk factor
- Identification of subpopulations, besides middle-aged men, for detection of high LDL cholesterol (and other lipid risk factors) and for clinical intervention. These include:
  - Young adults
  - Post-menopausal women
  - Older persons
- Emphasis on weight loss and physical activity to enhance risk reduction in persons with elevated LDL cholesterol

* ATP, Adult Treatment Panel; LDL, low-density lipoprotein; CHD, coronary heart disease.
† A CHD risk equivalent is a condition that carries an absolute risk for developing new CHD equal to the risk for having recurrent CHD events in persons with established CHD.
‡ Risk factors that continue to modify the LDL goal include cigarette smoking, hypertension, low HDL cholesterol, family history of premature CHD, age (male ≥45 years and female ≥55 years), and diabetes (in ATP III diabetes is regarded as a CHD risk equivalent).

Source: Expert Panel on Detection, Evaluation, and Treatment of High Blood Cholesterol in Adults (2001).

more risk factors are present. Risk assessment is recommended in non-diabetic patients without CHD or other atherosclerosis with two or more risk factors whose LDL cholesterol is in the range 100–160 mg/dl (2.58–4.13 mmol/l).

The LDL cholesterol goals of lipid-lowering drug therapy are shown in Fig. 13.1. As can be seen they vary according to the patient's category of risk. It is not clear what the scientific basis for this is.

In addition to the primary target of NCEP III which is LDL cholesterol, triglyceride levels are now recognized as a secondary target. Normal triglycerides are defined as <150 m/dl (<1.69 mmol/l) and borderline high triglycerides 150–199 mg/dl (1.69–2.25 mmol/l). For high triglycerides in the range 200–499 mg/dl (2.26–5.63 mmol/l), non-HDL cholesterol becomes a secondary target of therapy. Aside from weight reduction and increased physical activity (also recommended for borderline high triglycerides), drug therapy can be considered in high-risk people to achieve a non-HDL cholesterol goal (i.e. serum cholesterol minus

Table 13.2 New features of ATP III*

Focus on multiple risk factors

- Raises persons with diabetes without CHD, most of whom have multiple risk factors, to the risk level of CHD risk equivalent
- Uses Framingham projections of 10-year absolute CHD risk (i.e. the per cent probability of having a CHD event in 10 years) to identify certain patients with multiple (2+) risk factors for more intensive treatment
- Identifies persons with multiple metabolic risk factors (metabolic syndrome) as candidates for intensified therapeutic lifestyle changes

Modifications of lipid and lipoprotein classification

- Identifies LDL cholesterol ≤100 mg/dL as optimal
- Raises categorical low HDL cholesterol from <35 mg/dL to <40 mg/dL because the latter is a better measure of depressed HDL
- Lowers the triglyceride classification cutpoints to give more attention to moderate elevations

Support for implementation

- Recommends a complete lipoprotein profile (total, LDL, and HDL cholesterol and triglycerides as the preferred initial test, rather than screening for total cholesterol and HDL alone
- Encourages use of plant stanols/sterols and viscous (soluble) fibre as therapeutic dietary options to enhance lowering of LDL cholesterol
- Presents strategies for promoting adherence to therapeutic lifestyle changes and drug therapies
- Recommends treatment beyond LDL lowering for persons with triglycerides ≥200 mg/dL

* ATP, Adult Treatment Panel; CHD, coronary heart disease; LDL, low-density lipoprotein; HDL, high density lipoprotein.

Source: Expert Panel on Detection, Evaluation, and Treatment of High Blood Cholesterol in Adults (2001).

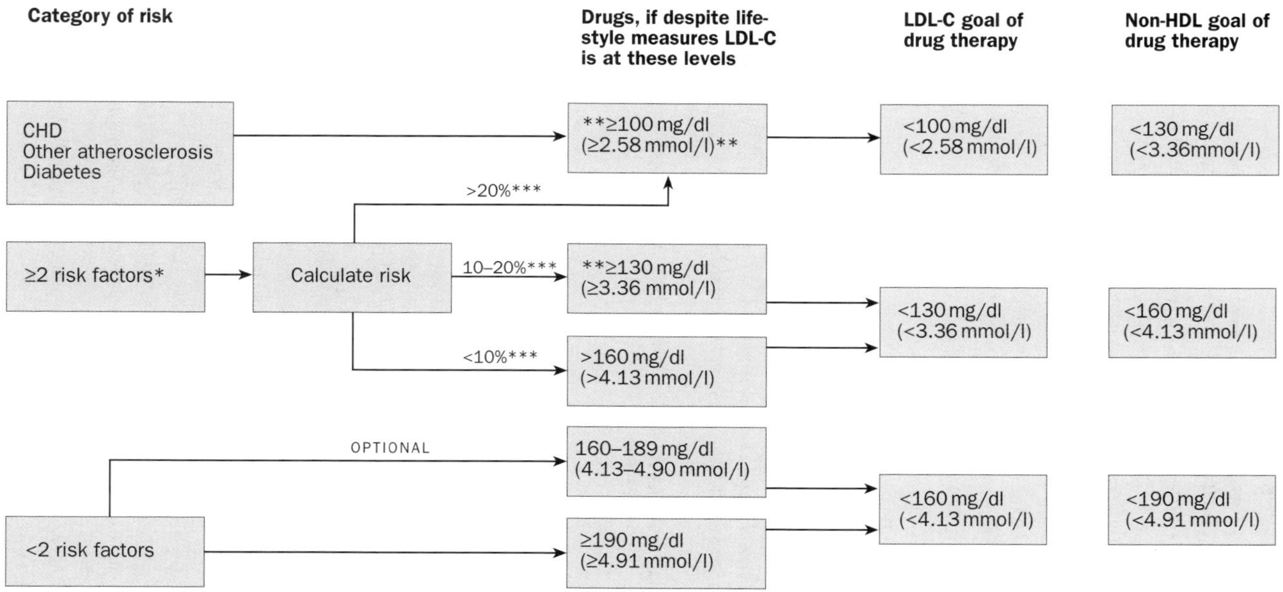

Fig. 13.1 The author's synopsis of the Third Report of the National Cholesterol Education Program (NCEP) Expert Panel on Detection Evaluation and Treatment of High Blood Cholesterol in Adults

* Cigarette-smoking, HDL cholesterol <40 mg/dl (<1.0mmol/l) (remove a risk factor, if HDL cholesterol >60 mg/dl [>1.5mmol/l]), family history (CHD in 1st degree male relative aged <55 years or female <65 years) age (men ≥45 years and women ≥55 years).
** Commencing lipid-lowering drugs at ≥ 130 mg/dl (≥3.36 mmol/l) is actually recommended with 100–130 mg/dl (2.58–3.36 mmol/l) being optional. However, this is illogical because the statin trials show greater benefit (absolute risk reductions) in patients in this category of risk with LDL cholesterol levels as low as 100 mg/dl (2.58 mmol/l) than in some of the other categories in which lipid-lowering drugs are definitely recommended.
*** CHD risk over next 10 years calculated using method based on Framingham findings (NCEPIII provides a scoring system for this).

Source: Expert Panel on Detection, Evaluation, and Treatment of High Blood Cholesterol in Adults (2001).

Table 13.3 Method of estimating 10-year risk of CHD for **Women** (Framingham Point Scores) provided in the Third Report of the National Cholesterol Education Program (NCEP) Expert Panel on Detection, Evaluation, and Treatment of High Blood Cholesterol in Adults (Adult Treatment Panel III)

Age, yr	Points	Age, yr	Points
20–34	–7	55–59	8
35–39	–3	60–64	10
40–44	0	65–69	12
45–49	3	70–74	14
50–54	6	75–79	16

Total cholesterol mg/dl	Points				
	Age 20–39 yr	Age 40–49 yr	Age 50–59 yr	Age 60–69 yr	Age 70–79 yr
<160	0	0	0	0	0
160–199	4	3	2	1	1
200–239	8	6	4	2	1
240–279	11	8	5	3	2
≥280	13	10	7	4	2

	Points				
	Age 20–39 yr	Age 40–49 yr	Age 50–59 yr	Age 60–69 yr	Age 70–79 yr
Non-smoker	0	0	0	0	0
Smoker	9	7	4	2	1

HDL, mg/dl	Points
≥60	–1
50–59	0
40–49	1
<40	2

Systolic BP, mmHg	If untreated	If treated
<120	0	0
120–129	1	3
130–139	2	4
140–159	3	5
≥160	4	6

Point total	10-year risk, %	Point total	10-year risk, %
<9	<1	18	6
9	1	19	8
10	1	20	11
11	1	21	14
12	1	22	17
13	2	23	22
14	2	24	27
15	3	≥25	≥30
16	4		
17	5		

Source: Expert Panel on Detection, Evaluation, and Treatment of High Blood Cholesterol in Adults (2001).

Table 13.4 Method of estimating 10-year risk of CHD for **Men** (Framingham Point Scores) provided in the Third Report of the National Cholesterol Education Program (NCEP) Expert Panel on Detection, Evaluation, and Treatment of High Blood Cholesterol in Adults (Adult Treatment Panel III)

Age, yr	Points	Age, yr	Points
20–34	–9	55–59	8
35–39	–4	60–64	10
40–44	0	65–69	11
45–49	3	70–74	12
50–54	6	75–79	13

Total cholesterol mg/dl	Points				
	Age 20–39 yr	Age 40–49 yr	Age 50–59 yr	Age 60–69 yr	Age 70–79 yr
<160	0	0	0	0	0
160–199	4	3	2	1	0
200–239	7	5	3	1	0
240–279	9	6	4	2	1
≥280	11	8	5	3	1

	Points				
	Age 20–39 yr	Age 40–49 yr	Age 50–59 yr	Age 60–69 yr	Age 70–79 yr
Non-smoker	0	0	0	0	0
Smoker	8	5	3	1	1

HDL, mg/dl	Points
≥60	–1
50–59	0
40–49	1
<40	2

Systolic BP, mmHg	If untreated	If treated
<120	0	0
120–129	0	1
130–139	1	2
140–159	1	2
≥160	2	3

Point total	10-year risk, %	Point total	10-year risk, %
<0	<1	9	5
0	1	10	6
1	1	11	8
2	1	12	10
3	1	13	12
4	1	14	16
5	2	15	20
6	2	16	25
7	3	≥17	≥30
8	4		

Source: Expert Panel on Detection, Evaluation, and Treatment of High Blood Cholesterol in Adults (2001).

HDL cholesterol as opposed to LDL cholesterol – this will thus include very low-density lipoprotein (VLDL) cholesterol also lowered by statins) (Fig. 13.1). There are two approaches to drug therapy. First, the non-HDL cholesterol goal can be achieved by intensifying therapy with an LDL-lowering drug, such as a statin; second, nicotinic acid (niacin) or fibrate can be added, if used with appropriate caution to achieve the non-HDL cholesterol goal by further lowering VLDL cholesterol. In rare cases in which triglycerides are very high (≥500 mg/dl (5.64 mmol/l)), the initial aim of therapy is to prevent acute pancreatitis through triglyceride-lowering. This approach requires very low fat diets (≤15% of calorie intake), weight reduction, increased physical activity, and usually a triglyceride-lowering drug (fibrate or nicotinic acid). Only after triglyceride levels have been lowered to <500 mg/dl (<5.64 mmol/l) should attention turn to LDL lowering to reduce risk for CHD. The logic of this latter recommendation is difficult to appreciate, because some patients whose triglycerides persist above 500 mg/dl do have elevated LDL cholesterol levels and they could surely benefit from statin therapy which besides lowering LDL cholesterol will also often have some triglyceride-reducing effect. Bile acid sequestrating agents would certainly not be helpful in this because these agents exacerbate hypertriglyceridaemia.

NCEP III does not specify a goal for HDL raising, because clinical trial evidence is insufficient to specify one. The importance of low HDL (40 mg/dl [1.03 mmol/l]) is, however recognized. Weight reduction and increased physical activity to raise it is recommended. Realistically it is appreciated that currently available drugs do not 'robustly raise HDL cholesterol'. Many patients with low HDL cholesterol will, however, have increased triglyceride and treatment aimed at non-HDL cholesterol goals outlined above is then appropriate. In patients without raised triglycerides (<200 mg/dl [2.26 mmol/l]), but low HDL cholesterol, drugs for HDL raising (fibrates on nicotinic acid) can be considered. However, treatment for this isolated low HDL is mostly reserved for people with CHD and CHD risk equivalents.

NCEP III provides a definition of the metabolic syndrome (Table 13.5) and

Table 13.5 Three or more of these features constitutes metabolic syndrome as defined in the Third Report of the National Cholesterol Education Program (NCEP) Expert Panel on Detection, Evaluation and Treatment of High Blood Cholesterol in Adults

1)	Abdominal obesity:waist circumference	
		Men >102 cm (>40 ins)
		Women >88 cm (>35 ins)
2)	Serum triglycerides	≥1.7 mmol/l (≥150 mg/dl)
3)	HDL cholesterol	
		Men <1.0 mmol/l (<40 mg/dl)
		Women <1.3 mmol/l (<50 mg/dl)
4)	Blood pressure	≥130/≥85 mmHg
5)	Fasting glucose	≥6.1 mmol/l (≥110 mg/dl)

Source: Expert Panel on Detection, Evaluation, and Treatment of High Blood Cholesterol in Adults (2001).

recommends weight reduction, increased physical activity and treatment of hypertension, use of aspirin whilst recognizing that guidelines for its use in primary prevention are not firmly established, and treatment of elevated triglycerides and low HDL cholesterol as discussed above.

The NCEP III recommendations are the most thorough and comprehensive national guidelines for the management of dyslipidaemia ever produced. Whilst I have drawn attention to a few points which could receive further clarification, they are of greater practical value and take much more account of the full range of evidence available than any other national recommendations. They represent a great step forward.

Clinical reality of coronary prevention guidelines: a comparison of EUROASPIRE I and II in nine countries. EUROASPIRE I and II Group. European Action on Secondary Prevention by Intervention to Reduce Events.

EUROASPIRE I and II Group; European Action on Secondary Prevention by Intervention to Reduce Events. *Lancet* 2001; **357**: 995–1001.

BACKGROUND. Patients with CHD are the top priority for preventive cardiology. The first EUROASPIRE survey among patients with established CHD in nine countries in 1995–96 showed substantial potential for risk reduction. A second survey (EUROASPIRE II) was carried out in 1999–2000 in the same countries to see whether preventive cardiology had improved since the first. It compared the proportion of patients in both studies who achieved the lifestyle, risk-factor, and therapeutic goals recommended by the Joint European Societies report on coronary prevention. The surveys were undertaken in the same selected geographical areas and hospitals in the Czech Republic, Finland, France, Germany, Hungary, Italy, The Netherlands, Slovenia, and Spain. Consecutive patients (men and women ≤70 years of age) were identified after coronary-artery bypass graft or percutaneous transluminal coronary angioplasty, or a hospital admission with acute myocardial infarction or ischaemia, and were interviewed at least six months later.

A total of 3569 and 3379 patients were interviewed in the first and second surveys, respectively. The prevalence of smoking remained almost unchanged at 19.4% *vs* 20.8%. The prevalence of obesity (body mass index ≥30 kg/m^2) increased substantially from 25.3% to 32.8%. The proportion with high blood pressure (≥140/90 mmHg) was virtually the same (55.4% *vs* 53.9%), whereas the prevalence of high total cholesterol concentrations (≥5.0 mmol/l) decreased substantially from 86.2% to 58.8%. Aspirin or other antiplatelet therapy was as widely used in the second survey as the first (83.9% overall), and reported use of beta-blockers, angiotensin-converting-enzyme inhibitors, and especially lipid-lowering drugs increased.

INTERPRETATION. The adverse lifestyle trends among European CHD patients are a cause for concern, as is the lack of any improvement in blood-pressure management, and the fact that most CHD patients are still not achieving the cholesterol goal of less than 5 mmol/l. There is a collective failure of medical practice in Europe to achieve the substantial potential among patients with CHD to reduce the risk of recurrent disease and death.

Comment

Clinical practice frequently does not make the best use of available evidence. Patients may not be compliant, but often too there is reluctance on the part of clinicians to use medication in doses which achieve desirable targets, most regrettably in the case of secondary CHD prevention in the management of blood pressure and cholesterol. Frequently medical and nursing provision is inadequate after the acute event to fully instigate preventive therapy. Yet half of patients dying from CHD are already known to have the disease. Smoking cessation and reduction in obesity are difficult to achieve, particularly when social policies often conflict. There can, however, be little excuse for any confusion about the aims of blood pressure and cholesterol management which are exclusively the province of health-care providers. The targets examined in this study were simple ones; blood pressure less than 140/90 mmHg and serum cholesterol <5 mmol/l (200 mg/dl). There seems no reason why clinicians and patients alike should not be aware of them.

Use of statins in the secondary prevention of coronary heart disease: is treatment equitable?

Reid FD, Cook DG, Whincup PH. *Heart* 2002; **88**: 15–19.

BACKGROUND. The purpose of this study was to investigate possible inequities in the use of statins for people with CHD according to a wide range of social and clinical factors. It used a cross-sectional analysis of data from the Health Survey for England 1998, a population-based survey. Subjects were 760 adults with CHD. Only 19.9% of subjects with CHD were receiving lipid-lowering drugs (151 of 760; 95% confidence interval [CI] 17.0%–22.7%). The likelihood of receiving statins was greatly reduced for older age groups: compared with those aged less than 65 years, the odds of receiving statin treatment were 0.53 (95% CI 0.35–0.80) for subjects aged 65–74 years, and 0.11 (95% CI 0.06–0.21) for subjects aged 75 years and over. Statins were given less often to current cigarette smokers than to non-smokers (odds ratio 0.55, 95% CI 0.32–0.96), and to subjects with angina compared with those with a previous myocardial infarction (odds ratio 0.63, 95% CI 0.43–0.93). Lower levels of statin use were also seen with increasing time since diagnosis (*P* = 0.12). No clear associations were observed with social measures.

INTERPRETATION. Important inequalities were found in the use of statins among people with CHD, which could not be justified by evidence from the large statin trials. Proactive policies are required to ensure that the vast majority of (if not all) patients with CHD are receiving statins, regardless of age, sex, social class, smoking status, type of CHD, or time since diagnosis.

Low prevalence of lipid lowering drug use in older men with established coronary heart disease.

Whincup PH, Emberson JR, Lennon L, Walker M, Papacosta O, Thomson A. *Heart* 2002; **88**: 25–9.

BACKGROUND. This study was carried out to determine the prevalence and correlates of lipid-lowering drug use among older British men with established CHD. The design was a cross-sectional survey within a cohort study (British regional heart study) carried out at 20 years of follow-up in 1998–2000. Participants were 3689 men aged 60–75 years (response rate 76%) from general practices in 24 British towns. Main outcome measures were diagnoses of myocardial infarction and angina based on detailed review of general practice records. Lipid-lowering drug use and blood cholesterol concentrations were ascertained at 20 year follow-up examination. Among 286 men with definite myocardial infarction, 102 (36%) were taking a lipid-lowering drug (93 [33%] a statin); among 360 men with definite angina without myocardial infarction, 84 (23%) were taking a lipid-lowering drug (78 [21%] a statin). Most men with documented CHD who were not receiving a lipid-lowering drug had a total cholesterol concentration of 5.0 mmol/l or more (87% of those with myocardial infarction, 82% with angina). Fewer than half of men with CHD receiving a statin had a total cholesterol concentration below 5.0 mmol/l (45% of those with myocardial infarction and 47% of those with angina). Only one-third of the men taking a statin were receiving trial validated dosages. Among men with CHD, a history of revascularization, more recent diagnosis, and younger age at diagnosis were associated with a higher probability of receiving lipid-lowering drug treatment.

INTERPRETATION. Among patients with established CHD, the prevalence of lipid-lowering drug use remains low and statin regimens suboptimal. Major improvements in secondary prevention are essential if the benefits of statins are to be realized.

Long-term persistence in use of statin therapy in elderly patients.

Benner JS, Glynn RJ, Mogun H, Neumann PJ, Weinstein MC, Avorn J. *JAMA* 2002; **288**: 455–61.

BACKGROUND. Knowledge of long-term persistence with 3-hydroxy-3-methylglutaryl coenzyme A reductase inhibitor (statin) therapy is limited because previous studies have observed patients for short periods of time, in closely monitored clinical trials, or in other unrepresentative settings. The aim of this study was to describe the patterns and predictors of long-term persistence with statin therapy in an elderly US population. It was a retrospective cohort study including 34 501 subjects enrolled in the New Jersey Medicaid and Pharmaceutical Assistance to the Aged and Disabled programs who were 65 years of age and older, initiated statin treatment between

1990 and 1998, and who were followed up until death, disenrolment, or 31 December 1999.

The main outcome measures were the proportion of days covered (PDC) by a statin in each quarter during the first year of therapy and every six months thereafter; predictors of suboptimal persistence during each interval (PDC <80%) were identified using generalized linear models for repeated measures. The mean PDC was 79% in the first three months of treatment, 56% in the second quarter, and 42% after 120 months. Only one patient in four maintained a PDC of at least 80% after five years. The proportion of patients with a PDC less than 80% increased in a log-linear manner, comprising 40%, 61%, and 68% of the cohort after 3, 12, and 120 months, respectively. Independent predictors of poor long-term persistence included non-white race, lower income, older age, less cardiovascular morbidity at initiation of therapy, depression, dementia, and occurrence of CHD events after starting treatment. Patients who initiated therapy between 1996 and 1998 were 21% to 25% more likely to have a PDC of at least 80% than those who started in 1990.

Interpretation. Persistence with statin therapy in older patients declines substantially over time, with the greatest drop occurring in the first six months of treatment. Despite slightly better persistence among patients who began treatment in recent years, long-term use remains low. Interventions are needed early in treatment and among high-risk groups, including those who experience CHD events after initiating treatment.

Adherence with statin therapy in elderly patients with and without acute coronary syndromes.

Jackevicius CA, Mamdani M, Tu JV. *JAMA* 2002; **288**: 462–7.

Background. Landmark clinical trials have demonstrated the survival benefits of statins, with benefits usually starting after one to two years of treatment. Research prior to these trials of older lipid-lowering agents demonstrated low levels of one-year adherence. The aim of this study was to compare two-year adherence following statin initiation in three cohorts of patients: those with recent acute coronary syndrome (ACS), those with chronic coronary artery disease (CAD), and those without coronary disease (primary prevention). It was a cohort study using linked population-based administrative data from Ontario. All patients aged 66 years or older who received at least one statin prescription between January 1994 and December 1998 and who did not have a statin prescription in the previous year were followed up for two years from their first statin prescription. There were 22 379 patients in the ACS, 36 106 in the chronic CAD, and 85 020 in the primary prevention cohorts. The main outcome measures were adherence to statins, defined as a statin being dispensed at least every 120 days after the index prescription for two years. Two-year adherence rates in the cohorts were only 40.1% for ACS, 36.1% for chronic CAD, and 25.4% for primary prevention. Relative to the ACS cohort, non-adherence was more likely among patients receiving statins in the chronic CAD (relative risk [RR], 1.14; 95% CI, 1.11–1.16) and primary prevention cohorts (RR, 1.92; 95% CI, 1.87–1.96).

Interpretation. Elderly patients with and without recent ACS have low rates of adherence to statins. This suggests that many patients initiating statin therapy may receive no or limited benefit from statins because of premature discontinuation.

Comment on the last four papers

Age is the major risk factor for coronary death. It seems, however, that in both Europe and North America there is a reluctance by clinicians to initiate and to continue statin treatment in older patients and a disinclination on the part of elderly people to take it. Increasingly elderly patients are taking multiple medication and it may be easy for patients to neglect those which do not provide any immediate symptomatic relief. There is also an ambivalence on the part of health services, perhaps fuelled by issues of cost, to provide therapy which can extend the lives of elderly people and thus to save the expense of their various therapies. Some recommendations for the use of statins have stuck rigidly to the age ranges at randomization in clinical trials of statins, but this seems unjustified when there is no evidence from these that the benefits decline with age. However, this cannot explain the findings of these studies which show a decrease in statin use even at the age of 65 years which is well within the age range included in trials. There is great scope here for improvement.

Identification and treatment of individuals at high risk of coronary heart disease.

Stein EA. *Am J Med* 2002; **112**(Suppl 8A): 3S–9S.

Background and interpretation. LDL cholesterol reduction remains the cornerstone of CHD prevention. The most dramatic and consistent reductions in LDL cholesterol and CHD risk are achieved with statin therapy. Identification of individuals at high CHD risk is important, not only for initiating appropriate treatment and minimizing morbidity and mortality but also for optimizing the cost-effectiveness of such treatment. A simple method for identifying high-risk individuals is to identify those with pre-existing atherosclerotic disease, diabetes, or familial hypercholesterolaemia (FH). Treatment options for achieving LDL cholesterol goals in high-risk patients include statins, bile acid sequestrants, niacin, and plant stanols. Statin therapy should be instituted at a dose likely to result in achievement of LDL cholesterol goals based on average response; it should then be aggressively titrated if the goals are not achieved. If LDL cholesterol goals are not achieved with maximal statin therapy, combination with a bile acid sequestrant, niacin, and/or stanols should be considered. Options likely to be available in the near future include more efficacious statins with greater potential for reducing LDL cholesterol in all patients but especially in high-risk patients, such as those with FH, enabling a greater proportion to achieve LDL cholesterol goals. Other options that may soon be available as additive agents to statins to achieve greater LDL cholesterol reductions include bile acid transport inhibitors and cholesterol absorption inhibitors.

Table 13.6 Effect of statin therapy on CHD in clinical events trials

Trial	Baseline LDL cholesterol (mg/dl)	LDL cholesterol level achieved, mg/dl (% reduction)	Drug group event rate (%)	Placebo group event rate (%)	Relative risk reduction (%)	Absolute risk reduction (%)	No. of patients needed to treat to prevent event
Secondary prevention							
4S	188	122 (35)	19.4	28.0	34	8.5	12
CARE	139	98 (32)	10.2	13.2	24	3.0	34
LIPID	150	112 (25)	12.3	15.7	23	3.4	30
Primary prevention							
WOSCOPS	192	159 (26)	5.3	7.5	29	2.2	46
AFCAPS/TexCAPS	150	115 (25)	3.5	5.5	37	2.0	50

AFCAPS/TexCAPS, Air Force/Texas Coronary Atherosclerosis Prevention Study; CARE, Cholesterol and Recurrent Events trial; LDL, low-density lipoprotein; LIPID, Long-Term Intervention with Pravastatin in Ischemic Disease trial; 4S, Scandinavian Simvastatin Survival Study; WOSCOPS, West of Scotland Coronary Prevention Study.

Source: Stein (2002).

Table 13.7 Summary of findings on National Cholesterol Education Program Adult Treatment Panel II low-density lipoprotein cholesterol goal attainment: rosuvastatin *versus* atorvastatin, simvastatin, and pravastatin

	% of patients achieving goals	
Agent (dose)	**All categories**	**High-risk group***
Rosuvastatin (5 mg)	87	42
Rosuvastatin (10 mg)	84	67
Atorvastatin (10 mg)	73	19
Simvastatin (20 mg)	64	19
Pravastatin (20 mg)	53	7

*Low-density lipoprotein cholesterol goal <100 mg/dl.
Source: Stein (2002).

Comment

Statin therapy is highly effective in preventing CHD events. The author summarizes trials which preceded the GREek Atorvastatin and Coronary-heart-disease Evaluation (GREACE) trial and the Heart Protection Study (HPS) (Table 13.6). One reason for the failure of many patients to achieve LDL cholesterol goals with statin therapy may be the reluctance of clinicians to titrate doses upwards. Also patients with particularly high levels of LDL cholesterol, such as those with FH, may not achieve target levels even at the highest doses. The article provides tables comparing the efficacy of various statins in achieving LDL goals in patients with LDL cholesterol levels typical of those who will receive statin treatment (Table 13.7) and patients with FH (Table 13.8).

Ineffective decrease of serum cholesterol by simvastatin in a subgroup of hypercholesterolemic coronary patients.

Miettinen TA, Gylling H. *Atherosclerosis* 2002; **164**: 147–52.

Background and interpretation. Serum cholesterol concentrations and synthesis markers (e.g. serumlathosterol to cholesterol ratio), and absorption markers (e.g. serum campesterol to cholesterol ratio) of cholesterol were measured in 319 good responders (GR; dose 20 mg up to 1 year) and in 115 poor responders (PR; dose increased at six weeks to 40 mg) among Finnish participants in the Scandinavian Simvastatin Survival Study at baseline, six weeks and one year of the simvastatin treatment. The baseline cholesterol level and the ratios of the absorption markers were higher and those of the synthesis markers lower in PR than GR. The ratios of the precursor sterols were negatively related to the baseline cholesterol in GR only (*P* = 0.003). The cholesterol levels, and the ratios of the precursor sterols were decreased and those of the absorption marker sterols increased less consistently in PR than GR by 20 mg, the group differences being only slightly lessened by the dose

Table 13.8 Effects of rosuvastatin 20/40/80 mg *versus* atorvastatin 20/40/80 mg in patients with familial hypercholesterolaemia at 18 weeks

	% change from baseline							% of patients achieving NCEP ATP II LDL cholesterol goal	
	LDL cholesterol	HDL cholesterol	Total cholesterol	Triglycerides	ApoB	ApoA-I	LDL cholesterol: HDL cholesterol	All categories	High-risk group
Rosuvastatin (n = 435)	−58*	+12*	−46*	−28	−50*	+6*	−62*	61	24 (n = 134)
Atorvastatin (n = 187)	−50	+3	−42	−32	−44	−2	−51	46	3 (n = 63)

Apo, apolipoprotein; ATP II, Adult Treatment Panel II; HDL, high-density lipoprotein; LDL, low-density lipoprotein; NCEP, National Cholesterol Education Program.

Source: Stein (2002).

addition to 40 mg. One-year differences were still frequently significant. The baseline cholesterol concentrations were negatively related to the reduction of the precursor sterol ratios in GR, the change of cholesterol being positively related to those of the synthesis markers and negatively to those of the absorption markers only in PR. Thus, patients needing large statin doses for cholesterol normalization have high absorption and low synthesis of cholesterol, yet baseline synthesis is inversely related to cholesterol level only in GR. The synthesis rate is less markedly reduced by the large than by the small statin dose in the PR, and the reduction is related, in contrast to that in the GR, to lowering of cholesterol.

Comment

It is undoubtedly more difficult to lower cholesterol with statins in some patients than others. This study provides evidence that more resistant patients tend to have higher rates of intestinal cholesterol absorption and probably as a consequence lower rates of cholesterol biosynthesis, rendering inhibition of synthesis less effective. With the advent of cholesterol absorption inhibitors it will be of great interest to see how effectively the problem can be overcome.

Issues concerning the monitoring of statin therapy in hypercholesterolemic subjects with high plasma lipoprotein(a) levels.

Scanu AM, Hinman J. *Lipids* 2002; **37**: 439–44.

BACKGROUND AND INTERPRETATION. Most studies on the topic have shown that statin therapy decreases plasma LDL levels but not those of lipoprotein(a) (Lp(a)). This specificity of action, although previously noted, has not been systematically investigated. In the current study this problem was approached by monitoring LDL- and Lp(a) cholesterol in 80 hypercholesterolaemic subjects with high Lp(a) levels, at entry and 8 months after initiation of statin therapy. We found that commonly used direct and indirect LDL cholesterol assays gave an LDL cholesterol value that comprised both true LDL- and Lp(a) cholesterol. We estimated these two analytes from the values of Lp(a) protein determined by ELISA (enzyme linked immunosorbent assay) and from knowledge of the Lp(a) chemical composition, complemented by data from immunochemical and ultracentrifugal analyses. Statin therapy, while not affecting plasma Lp(a) protein levels (21.710.4, before, and 22.010.1 mg/dl, after), caused a decrease in the estimated or true LDL cholesterol (*P* <0.0001) to values in some cases as low as 10 mg/dl. This drop in true LDL was validated by the decrease in the LDL band in the ultracentrifugation profiles, and its magnitude was proportional to the degree of total cholesterol-lowering and to the pre-treatment true LDL/Lp(a) cholesterol weight ratio. We conclude that true LDL but not Lp(a) cholesterol is affected by statin therapy and that this specific response cannot be monitored by current LDL cholesterol assays and must, rather, rely on estimates of these two analytes.

Comment

Lp(a) is an LDL-like particle in which apolipoprotein B, the principal protein component of LDL, is disulphide-linked to another protein, apolipoprotein (a), which is a member of the plasminogen gene family. Lp(a) levels are frequently high in FH and renal disease. There is also considerable individual variation in Lp(a) levels in the general population, largely explicable on a genetic basis with those families in which smaller molecular weight isoforms of Lp(a) are present having the highest levels. The cholesterol composition of Lp(a) is similar to LDL and in some people Lp(a) levels can exceed those of LDL. Statins do not lower Lp(a).

This interesting study reveals that many laboratory methods for measuring LDL cholesterol include Lp(a) cholesterol. This is another explanation therefore of why some patients appear resistant to LDL cholesterol-lowering by statins.

Effects of HMG-CoA reductase inhibitors on skeletal muscle: are all statins the same?

Evans M, Rees A. *Drug Saf* 2002; **25**: 649–63.

BACKGROUND AND INTERPRETATION. The 3-hydroxy-3-methyl coenzyme A (HMG-CoA) reductase inhibitors or statins, specifically inhibit the enzyme HMG-CoA in the liver, thereby inhibiting the rate limiting step in cholesterol biosynthesis and so reducing plasma cholesterol levels. Numerous studies have consistently demonstrated that cholesterol-lowering with statin therapy reduces morbidity and mortality from CHD, whilst recent evidence has demonstrated that benefits of statin therapy may also extend into stroke prevention. Since hypercholesterolaemia is a chronic condition, the long-term safety and tolerability of these agents is an important issue. Numerous large-scale clinical trials have consistently demonstrated a positive safety and tolerability profile for statins. Hepatic, renal and muscular systems are rarely affected during statin therapy, with adverse reactions involving skeletal muscle being the most common, ranging from mild myopathy to myositis and occasionally to rhabdomyolysis and death. Post-marketing data supports the positive safety and tolerability profile of statins, with an overall adverse event frequency of less than 0.5% and a myotoxicity event rate of less than 0.1%. The recent withdrawal of cerivastatin from the world market due to deaths from rhabdomyolysis has, however, focused attention on the risk of adverse events and in particular myotoxicity associated with statins. Indeed, initial clinical trial data supports post-marketing data, demonstrating a higher incidence of myotoxicity associated with cerivastatin, particularly when used in combination with fibrates. The potential mechanisms underlying statin-induced myotoxicity are complex with no clear consensus of opinion. Candidate mechanisms include intracellular depletion of essential metabolites and destabilization of cell membranes, resulting in increased cytotoxicity. Cytochrome P450 3A4 is the main isoenzyme involved in statin metabolism. Reduced activity of this enzyme due to either reduced expression or inhibition by other drugs prescribed concomitantly such as cyclosporin or itraconazole may increase drug bioavailability and the risk of myotoxicity. Such factors may partly account for

the inter-individual variability in susceptibility to statin-induced myotoxicity, although other, as yet unclarified, genetic factors may also be involved. The risk of rhabdomyolysis is increased with combination fibrate-statin therapy, with initial evidence suggesting that gemfibrozil–statin combination may particularly increase the risk of myotoxicity, with pharmacodynamic as well as pharmacokinetic mechanisms being involved.

Comment

In large randomized, placebo-controlled trials of statin therapy such as the 4S |**3**|, the Cholesterol and Recurrent Events Study (CARE) |**4**|, the Long-term Intervention with Pravastatin in Ischaemic Disease (LIPID) |**5**|, the West of Scotland Coronary Prevention Study (WOSCOPS) |**6**|, the Air Force/Texas Coronary Atherosclerosis Prevention Study (AFCAPS/Tex CAPS) |**7**| and HPS, lasting up to six years, there were no deaths or other clinical episodes related to severe myositis (rhabdomyolysis) and very few incidences of asymptomatic marked increases in creatine kinase (CK) activity. These studies have involved simvastatin, pravastatin and lovastatin. The same was also true of atorvastatin in the Myocardial Ischaemia Reduction with Aggressive Cholesterol Lowering study (MIRACL) |**8**|, Atorvastatin Versus Revascularization Treatment (ADVERT) |**9**| and the GREACE trial, over shorter time intervals of up to three years. It was disconcerting therefore that cerivastatin was withdrawn because of fatal episodes of rhabdomyolsis. These were reported in association with larger doses, particularly 0.8 mg daily and undoubtedly some were in combination with gemfibrozil, but many were not. More details of these episodes have yet to emerge. Meanwhile we have no clear understanding of the mechanism by which statins induce myositis, although it is more likely to occur in combination with drugs which compete for elimination and with drugs, such as fibrates, which themselves can cause myositis (albeit it rarely and by presumably different mechanisms).

For the clinician this creates difficulties. The statin effect in preventing CHD is emerging as a class effect, although the evidence that the extent to which CHD is prevented is simply related to changes in LDL cholesterol is unclear. Whether other effects of statins on lipoprotein metabolism or their other potential antiatherogenic actions are important and whether the statins differ to any great extent in these regards remains controversial. Furthermore statins were, at least until the withdrawal of cerivastatin, increasingly being viewed as a safe group of drugs. Now we are less certain that their safety is a class effect. Certainly in the doses employed in clinical trials they would appear to be safer than many other commonly used drugs, including aspirin. Some statins are now available, however, in doses higher than those used in clinical trials, new statins are being introduced and increasingly for certain patients their use in combination with fibrates, bile-acid sequestrating agents, nicotinic acid, omega–3 fatty acids and ezetimibe is advocated. Fibrate therapy does undoubtedly increase the likelihood of myositis, but probably rarely and it has not been often encountered in the many small, short-term trials published so far. Most reported cases of myositis in combination with statin use have been with gemfibrozil, but is this because the combination has been most widely used with gemfibrozil in, for example, the USA where other fibrates are used less frequently? The evidence that fibrates decrease CHD is currently stronger for gemfibrozil than for other fibrates as a result of the Helsinki Heart Study |**10**| and the Veterans Affairs Cooperative HDL Cholesterol Intervention Trial (VA-HIT) |**11**|, so its use in combination with statin therapy is more logical. Therefore it would be important to know whether gemfibrozil really does have a greater potential to cause myositis in combination with statins.

In the clinical trials patients were closely scrutinized for symptoms of myositis or significant CK elevations, but does this have any bearing on the low rates of myositis in these trials? Many patients reported muscle aching in these trials as they do in clinical practice where they are also encouraged to report such symptoms. Muscle aching is, however, extremely common and in the trials there were no clear differences in the frequency of this symptom in patients on active treatment and those on placebo. CK monitoring also revealed that minor elevations just outside the normal range were common both on active treatment and on placebo. So what is the value of detecting small increases up to two or three times the upper limit of normal? Does this really predict the rises up to several thousand IU/l associated with myositis? We are increasingly being encouraged to advocate physical exercise and this complicates the picture further. Also how rapid was the onset of severe myositis and rhabdomyolysis in cases reported outside clinical trials? We do not know whether CK measurement every six months or so would have predicted these episodes. It is hard to comprehend why patients did not report symptoms to their clinicians before their condition had become so serious as to involve fatality or was it so fulminating there was no time? Most of the cases of non-statin-induced myositis one encounters in clinical practice recover spontaneously: fatal cases must be the tip of the iceberg. So unless there is something odd about statin-induced myositis, for each fatal case one would expect a huge number of non-fatal but clinically obvious cases to occur with statins. Yet they do not. It would be very helpful to have answers to some of this questions as soon as possible.

Statins and risk of polyneuropathy: a case-control study.

Gaist D, Jeppesen U, Andersen M, Garcia Rodriguez LA, Hallas J, Sindrup SH. *Neurology* 2002; **58**: 1333–7.

Background. Several case reports and a single epidemiologic study indicate that use of statins occasionally may have a deleterious effect on the peripheral nervous system. The authors therefore performed a population-based study to estimate the relative risk of idiopathic polyneuropathy in users of statins. The authors used a population-based patient registry to identify first-time-ever cases of idiopathic polyneuropathy registered in the five-year period 1994 to 1998. For each case, validated according to pre-defined criteria, 25 control subjects were randomly

selected among subjects from the background population matched for age, sex, and calendar time. The authors used a prescription register to assess exposure to drugs and estimated the odds ratio of use of statins (ever and current use) in cases of idiopathic polyneuropathy compared with control subjects. The authors verified a diagnosis of idiopathic polyneuropathy in 166 cases. The cases were classified as definite (35), probable (54), or possible (77). The odds ratio linking idiopathic polyneuropathy with statin use was 3.7 (95% CI 1.8–7.6) for all cases and 14.2 (5.3–38.0) for definite cases. The corresponding odds ratios in current users were 4.6 (2.1–10.0) for all cases and 16.1 (5.7–45.4) for definite cases. For patients treated with statins for two or more years the odds ratio of definite idiopathic polyneuropathy was 26.4 (7.8–45.4).

INTERPRETATION. Long-term exposure to statins may substantially increase the risk of polyneuropathy.

Comment

This is a well-designed, case–control study which considerably improves on earlier reports suggesting a small excess risk of peripheral neuropathy associated with statin therapy. The risk was estimated to be one case for every 2200 person-years of statin use. I have occasionally encountered patients in whom symptoms of peripheral neuropathy have occurred on taking statins which have largely resolved on discontinuation of treatment, but hitherto it has been difficult to be certain that they were due to the therapy because, of course, idiopathic peripheral neuropathy which resolves spontaneously does occur. In clinical practice the clinician will have to decide whether the cardiovascular risk justifies rechallenge with a different statin. The cases encountered in this study were receiving simvastatin, pravastatin, lovastatin and fluvastatin. The absence of atorvastatin may have reflected limited use of atorvastatin during the time the cases were diagnosed because this medication was also absent in the controls. Peripheral neuropathy may thus be a rare class effect.

References

1. Anderson KM, Odell PM, Wilson PWF, Kannel WB. Cardiovascular disease risk profiles. *Am Heart J* 1990; **121**: 293–8.
2. British Cardiac Society, British Hyperlipidaemia Association, British Hypertension Society, British Diabetic Association. Joint British guidelines on prevention of coronary heart disease in clinical practice: summary. *Br Med J* 2000; **320**: 705–8.
3. Scandinavian Simvastatin Survival Study Group. Randomized trial of cholesterol lowering in 4444 patients with coronary heart disease; the Scandinavian Survival Study. *Lancet* 1994; **344**: 1383–9.

4. Sacks FM, Pfeffer MA, Moye LA, Rouleau JL, Rutherford JD, Cole TG, Brown L, Warnica JW, Arnold JMO, Wun C-C, Davis BR, Braunwald E, for the Cholesterol and Recurrent Events Trial Investigators. The effect of pravastatin on coronary events after myocardial infarction in patients with average cholesterol levels. *N Engl J Med* 1996; **335**: 1001–9.

5. The Long-Term Intervention with Pravastatin in Ischaemic Disease (LIPID) Study Group. Prevention of cardiovascular events and death with pravastatin in patients with coronary heart disease and a broad range of initial cholesterol levels. *N Engl J Med* 1998; **339**: 1349–57.

6. Shepherd J, Cobbe SM, Ford I, Isles CG, Lorimer AR, MacFarlane PW, McKillop JH, Packard C, *et al.* for the West of Scotland Coronary Prevention Study Group. Prevention of coronary heart disease with pravastatin in men with hypercholesterolaemia. *N Engl J Med* 1995; **333**: 1301–7.

7. Downs GR, Clearfield M, Weiss S, Whitney E, Shapiro DR, Beere PA, *et al.* Primary prevention of acute coronary events with lovastatin in men and women with average cholesterol levels: results of the AFCAPS/TexCAPS (Air Force/Texas Coronary Atherosclerosis Prevention Study). *JAMA* 1998; **279**: 1615–22.

8. Schwartz GG, Olsson AG, Ezekowitz MD, *et al.* Effects of atorvastatin on early recurrent ischemic events in acute coronary syndromes: the MIRACL study. A randomized controlled trial. *JAMA* 2001; **285**: 1711–18.

9. Pitt B, Waters D, Brown WV, von Boven AJ, Schwarts L, Title LM, Eisenberg D, Shurzinske L, McCormick LS, for the Atorvastatin versus Revascularisation Treatment (ADVERT) Investigators. Aggressive lipid-lowering therapy compared with angioplasty in stable coronary disease. *N Engl J Med* 1999; **341**: 70–6.

10. Frick MH, Elo O, Haapa K, Heinonen OP, Heinsalmi P, Helo P, Huttunen JK, *et al.* Hensinki Heart Study: primary-prevention trial with gemfibrozil in middle-aged men with dyslipidaemia. Safety of treatment, changes in risk factors, and incidence of coronary heart disease. *N Engl J Med* 1987; **317**: 1237–45.

11. Rubins HB, Robins SJ, Collins D, Fye CL, Anderson JW, Elam MB, Faas FH, Linares E, Schaeffer EJ, Schectman G, Wilt TJ, Wittes J, for the Veterans Affairs High-Density Lipoprotein Cholesterol Intervention Trial Study Group. Gemfibrozil for the secondary prevention of coronary heart disease in men with low levels of high-density lipoprotein cholesterol. *N Engl J Med* 1999; **341**: 410–18.

14

Trials of therapy to modify serum lipids and lipoproteins

Introduction

This year Gagné and colleagues have reported on the first of the intestinal cholesterol absorption inhibitors, ezetimibe, in homozygous familial hypercholesterolemia (FH). Ezetimibe seems to produce a decrease in low-density lipoprotein (LDL) cholesterol of around 20% in this and most other conditions in which it has been tested. Doubling the dose of a statin induces a decrement in LDL cholesterol of around 6% so the additional decrease of 20% when ezetimibe is added to statin therapy is of great interest.

We now have so much clinical trial evidence about lipid-lowering drugs that we could propose the ideal profile for such a drug. It would be even more potent in lowering LDL than currently available statins and would lower triglycerides and raise high-density lipoprotein (HDL) cholesterol by at least as much as a fibrate. As each new statin struggles from its pupa to become available for general prescription, it must match itself against this ideal. This year Davidson *et al.* reported on rosuvastatin which looks promising. Guerin *et al.* studied the effect of different doses of atorvastatin, the most potent statin in terms of its cholesterol-lowering capacity currently available. They showed many effects on the atherogenicity of lipoproteins particularly with large doses which would not be evident from the cholesterol measurements performed in routine clinical practice or in clinical trials. Partly for this reason and because it may predict the outcome of statin therapy better than LDL cholesterol, Miremadi and colleagues have proposed that serum apolipoprotein B (apoB) should be measured to monitor the effectiveness of statin therapy. This has much to advocate it |**1**|. The reduction in triglyceride levels achieved with statins has attracted much interest this year. Forster *et al.* showed that the effect is due to increased catabolism of triglyceride-rich lipoproteins. A study by Roglass showed that triglyceride biosynthetic enzymes in rat liver are actually induced by statins again strongly suggesting their triglyceride-lowering effect is catabolic and also suggesting one mechanism limiting the decrease in triglycerides which statins can achieve. The triglyceride-lowering effect of statins can, of course, be enhanced with a fibrate or nicotinic acid in combination. Athyros and coworkers found that gemfibrozil in combination with statin was more effective at raising HDL whereas ciprofibrate combined with statin was more effective in lowering triglyceride and fibrinogen levels. The triglyceride-lowering effects of statins can be augmented not only with fibrates and nicotinic acid, but also, as was shown in a study by Chan *et al.*, with concentrated omega–3 fatty acids.

The age at which to introduce statin therapy in heterozygous FH is controversial. Although many centres have begun to employ statins in childhood, no national recommendations advocate this except perhaps in very exceptional circumstances. The time-honoured approach is to use bile acid sequestrating agents in children. However, compliance with these agents is poor and there is no evidence that such therapy is more effective than no drug treatment. McCrindle *et al.* this year reported on the effect of bile acid sequestrating agents alone and smaller doses of these agents in combination with statin in children.

Efficacy and safety of ezetimibe coadministered with atorvastatin or simvastatin in patients with homozygous familial hypercholesterolemia.

Gagné C, Gaudet D, Bruckert E; the Ezetimibe Study Group. *Circulation* 2002; **105**: 2469–75.

BACKGROUND. Patients with homozygous familial hypercholesterolaemia (HoFH) have a high incidence of cardiovascular morbidity and mortality from premature atherosclerosis, and the efficacy of pharmacological therapy has been limited. This study evaluated the efficacy, safety, and tolerability of ezetimibe, a novel cholesterol absorption inhibitor, in a multicentre, double-blind, randomized trial of HoFH patients receiving atorvastatin or simvastatin. Fifty patients with a diagnosis of HoFH on the National Cholesterol Education Program Step 1 or stricter diet and taking open-label atorvastatin 40 mg/d or simvastatin 40 mg/d (statin-40) with (n = 25) or without (n = 25) concomitant LDL-apheresis were randomized to one of three double-blind treatments: atorvastatin or simvastatin 80 mg/d (statin-80, n = 17); ezetimibe 10 mg/d plus atorvastatin or simvastatin 40 mg/d (n = 16); or ezetimibe 10 mg/d plus atorvastatin or simvastatin 80 mg/d (n = 17) for 12 weeks. The primary end-point was mean percentage change in LDL cholesterol (LDL-C) from statin-40 baseline to the end-point for patients receiving statins alone (statin-80) *versus* patients receiving ezetimibe plus atorvastatin or simvastatin at either dose (ezetimibe plus statin-40/80). Ezetimibe plus statin-40/80 significantly reduced LDL-C levels compared with statin-80 (–20.7% *versus* –6.7%, P = 0.007). In the high-dose statin cohorts, ezetimibe plus statin-80 reduced LDL-C by an additional 20.5% (P = 0.0001) *versus* statin-80. Similar significant reductions in LDL-C concentrations were observed for patients with genotype-confirmed HoFH (n = 35). Ezetimibe was safe and well tolerated.

INTERPRETATION. Ezetimibe coadministered with atorvastatin or simvastatin in patients with HoFH produced clinically important LDL-C reductions compared with best current therapy. Ezetimibe provides a new, complementary pharmacological approach for this high-risk population.

Comment

Ezetimibe is the first of a new class of cholesterol absorption inhibitors that impedes dietary and biliary cholesterol absorption from the brush border of the intestine. It is rapidly conjugated to glucuronide in the intestine and circulated enterohepatically. In a dose of 10 mg daily in primary hypercholesterolaemia it decreased LDL cholesterol by about 20% either alone or in combination with statin therapy |**2**|. In the study by Gagné and colleagues it produced an additional decrease of LDL cholesterol also of 20% in patients with homozygous FH. Ezetimibe is thus of potential value both in HoFH and in other resistant hypercholesterolaemia. Dietary cholesterol has only a small effect on serum cholesterol, but the reabsorption of cholesterol secreted into the intestine has a more substantial effect and this therefore underlies the effectiveness of ezetimibe.

Comparison of effects on low-density lipoprotein cholesterol and high-density lipoprotein cholesterol with rosuvastatin versus atorvastatin in patients with type IIa or IIb hypercholesterolemia.

Davidson M, Ma P, Stein EA, *et al.* *Am J Cardiol* 2002; **89**: 268–75.

Background and interpretation. This randomized, double-blind, placebo-controlled trial was conducted in 52 centres in North America to compare the effects of the new, highly effective statin, rosuvastatin, with atorvastatin and placebo in hypercholesterolaemic patients. After a 6-week dietary run-in, 516 patients with LDL cholesterol ≥4.14 mmol/l (160 mg/dl) and <6.47 mmol/l (250 mg/dl) and triglycerides ≤4.52 mmol/l (400 mg/dl) were randomized to 12 weeks of once-daily placebo (*n* = 132), rosuvastatin 5 mg (*n* = 128), rosuvastatin 10 mg (*n* = 129), or atorvastatin 10 mg (*n* = 127). The primary efficacy end-point was percentage change in LDL cholesterol. Secondary efficacy variables were achievement of National Cholesterol Education Program Adult Treatment Panel II (ATP II), ATP III, and European Atherosclerosis Society LDL cholesterol goals and percentage change from baseline in HDL cholesterol, total cholesterol, triglycerides, non-HDL cholesterol, apoB, and apolipoprotein A-I (apoA-I). Rosuvastatin 5 and 10 mg compared with atorvastatin 10 mg were associated with greater LDL cholesterol reductions (–40% and –43% *vs* 35%; *P* <0.01 and *P* <0.001, respectively) and HDL cholesterol increases (13% and 12% *vs* 8%, *P* <0.01 and *P* <0.05, respectively). Total cholesterol and apoB reductions and apoA-I increases were also greater with rosuvastatin; triglyceride reductions were similar. Rosuvastatin 5 and 10 mg were associated with improved achievement in ATP II (84% in both rosuvastatin groups *vs* 73%) and ATP III (84% and 82% *vs* 72%) LDL cholesterol goals, and rosuvastatin 10 mg was more effective than atorvastatin in achieving European Atherosclerosis Society LDL cholesterol goals. Both treatments were well tolerated.

Comment

Rosuvastatin is a new statin which in clinical trials shows greater potential to lower serum LDL cholesterol than existing statins |**3,4**|. It could therefore reduce the need for dose titration in achieving LDL targets in patients with typical LDL cholesterol levels encountered in CHD prevention and achieve greater reductions in patients with higher levels than are currently possible. Many issues concerning whether evidence for a class effect on CHD prevention with established statins can be applied to new statins and long-term safety will need to be settled before its place in clinical practice is finally decided, but early results are encouraging.

Dose-dependent action of atorvastatin in type IIB hyperlipidemia: preferential and progressive reduction of atherogenic apoB-containing lipoprotein subclasses (VLDL-2, IDL, small dense LDL) and stimulation of cellular cholesterol efflux.

Guerin M, Egger P, Soudant C, *et al. Atherosclerosis* 2002; **163**: 287–96.

BACKGROUND. Type IIB hyperlipidaemia is associated with premature vascular disease, an atherogenic lipoprotein phenotype characterized by elevated levels of triglyceride-rich very low-density lipoprotein (VLDL) and small dense LDL, together with subnormal levels of HDL. The dose-dependent and independent effects of a potent HMG-CoA reductase inhibitor, atorvastatin, at daily doses of 10 and 40 mg, were evaluated on triglyceride-rich lipoprotein subclasses (VLDL-1, VLDL-2 and IDL), on the major LDL subclasses (light LDL, LDL-1+LDL-2, D: 1.019–1.029 g/ml; intermediate LDL, LDL-3, D: 1.029–1.039 g/ml and small dense LDL, LDL-4+LDL+5, D: 1.039–1.063 g/ml), on cholesteryl ester transfer protein (CETP)-mediated cholesteryl ester transfer from HDL to apoB-containing lipoproteins, on phospholipid transfer protein activity and on plasma-mediated cellular cholesterol efflux in patients ($n = 10$) displaying type IIB hyperlipidaemia. Plasma concentrations of triglyceride-rich lipoprotein subclasses (TRL: VLDL-1, Sf 60-400; VLDL-2, Sf 20-60 and IDL, Sf 12-20) and of LDL (D: 1.019–1.063 g/ml) were markedly diminished after six weeks of statin treatment at 10 mg per day (–31 and –36%, respectively; $P < 0.002$) and by 42 and 51%, respectively at the 40 mg per day dose. Increasing doses of atorvastatin progressively normalized both the quantitative and qualitative features of the LDL subclass profile, in which dense LDL predominated at baseline. Indeed, dense LDL levels were reduced by up to 57% at the 40 mg dose, leading to a shift in the peak of the density profile towards larger, buoyant LDL particles typical of normolipidaemic subjects. In addition, marked reduction in numbers of apoB-100-containing particle acceptors led to a 30% decrease ($P < 0.02$) in CETP-mediated cholesteryl ester transfer from HDL. Finally, a significant dose-dependent statin-mediated elevation (+15% at 10 mg; $P = 0.0003$ and +35% at 40 mg; $P < 0.0001$ compared to baseline) in the capacity of plasma from type IIB subjects to mediate free cholesterol efflux from Fu5AH hepatoma cells was observed. Moreover, atorvastatin (40 mg per day) significantly increased plasma apoA-I levels (+24%; $P < 0.05$), thereby suggesting that

this statin enhances production of apoA-I and with it, formation of nascent pre-beta HDL particles. Plasma PLTP (phospholipid transfer protein) activity was not affected by either dose of atorvastatin.

Interpretation. It was concluded that increasing the dose of atorvastatin leads to dose-dependent, preferential and progressive reduction in particle numbers of atherogenic VLDL-2, IDL and dense LDL, and concomitantly, to enhanced cellular cholesterol efflux in type IIB dyslipidaemia, thereby diminishing the atherosclerotic burden in subjects characterized by high cardiovascular risk.

Comment

The risk of hypercholesterolaemia is increased if hypertriglyceridaemia is also present. Triglycerides do not themselves appear to participate directly in the process of atherogenesis. They do, however, increase the concentration of other lipoproteins which are believed to be directly involved in the events in the arterial wall leading to atherosclerosis. In the presence of hypertriglyceridaemia there are increases in the concentration of intermediate density lipoprotein (IDL) which is formed in the conversion of VLDL to LDL and small dense LDL. Small dense LDL is formed when serum triglycerides exceed about 1.5 mmol/l (135 mg/dl) from LDL which might otherwise be cleared from the circulation by the LDL-receptors. The transfer of triglyceride to LDL and VLDL by CETP in exchange for cholesteryl ester when triglyceride levels are high appears to be important in the formation of small dense LDL, which is cholesteryl ester depleted. Statins are well known to decrease the more buoyant type of LDL between IDL and small dense LDL, but there is increasing evidence that whilst this produces a decrease in serum cholesterol which is pleasing to the clinician, it is not as anti-atherogenic as decreasing IDL and small dense LDL. This may also explain some of the pleiotropic effects of statins in which changes in serum cholesterol and LDL cholesterol do not fully explain the effect of statins in reducing CHD events.

The study is therefore valuable in demonstrating that higher doses of atorvastatin decrease CETP, IDL and small dense LDL in combined hyperlipidaemia.

Can measurement of serum apolipoprotein B replace the lipid profile monitoring of patients with lipoprotein disorders?

Miremadi S, Sniderman A, Frohlich J. *Clin Chem* 2002; **48**: 484–8.

Background. Current clinical guidelines require that five indices (total cholesterol, LDL cholesterol, HDL cholesterol, triglycerides, and the total/HDL cholesterol ratio) be measured or calculated to assess the lipid-related risk of vascular disease. All five are also targets of therapy and therefore all must be measured initially and at follow-up. Considerable evidence indicates that apoB is a better index of reaching or not

reaching treatment targets than total or LDL cholesterol. The objective of this study was to examine whether measurement of a single marker (apoB) led to the same categorization of risk as the traditional five indices (lipid profile). If both apoB and lipid profile indicated that the patient was either within or outside their respective treatment targets, the indices were considered concordant. If not, the indices were considered discordant. Concordance/discordance was examined in 215 patients at their first and last clinic visit. Concordance was high in both higher (88% at the first and 92% at the last clinic visit) and lower (76% at the first and 78% at the last clinic visit) risk groups at both the initial and final visits. Discordance was virtually restricted to the group with hypertriglyceridaemia with normal concentrations of apoB, a group in which little independent evidence points to any substantially increased risk of vascular disease.

INTERPRETATION. These data raise the possibility that at least for high-risk patients treated with statins, follow-up could be simplified and expenses reduced if only apoB were measured. They also raise the possibility that outcome might be improved if the therapeutic algorithm were simplified.

Comment

As was discussed for the previous abstract, serum cholesterol and LDL cholesterol do not necessarily directly represent the decrease in the most atherogenic subclasses of LDL. However, because the cholesterol-depleted, small dense LDL continues to contain one molecule of apoB per LDL particle, the measurement of serum apoB may more closely reflect the antiatherogenic action of statins. In this article Miremadi and colleagues argue that apoB should be employed in monitoring the clinical response to statins and review evidence that apoB assays are now sufficiently robust to be used for this purpose and that the outcome of clinical event trials is more accurately predicted by apoB than LDL cholesterol.

Influence of atorvastatin and simvastatin on apolipoprotein B metabolism in moderate combined hyperlipidemic subjects with low VLDL and LDL fractional clearance rates.

Forster LF, Stewart G, Bedford D, *et al. Atherosclerosis* 2002; **164**: 129–45.

BACKGROUND AND INTERPRETATION. Subjects with moderate combined hyperlipidaemia (*n* = 11) were assessed in an investigation of the effects of atorvastatin and simvastatin (both 40 mg per day) on apoB metabolism. The objective of the study was to examine the mechanism by which statins lower plasma triglyceride levels. Patients were studied on three occasions, in the basal state, after eight weeks on atorvastatin or simvastatin and then again on the alternate treatment. Atorvastatin produced significantly greater reductions than simvastatin in

LDL cholesterol (49.7 *vs* 44.1% decrease on simvastatin) and plasma triglyceride (46.4 *vs* 39.4% decrease on simvastatin). ApoB metabolism was followed using a tracer of deuterated leucine. Both drugs stimulated direct catabolism of large VLDL(1) apoB (4.52 ± 3.06 pools per day on atorvastatin; 5.48 ± 4.76 pools per day on simvastatin *vs* 2.26 ± 1.65 pools per day at baseline (both *P* <0.05)) and this was the basis of the 50% reduction in plasma VLDL(1) concentration; apoB production in this fraction was not significantly altered. On atorvastatin and simvastatin the fractional transfer rates (FTR) of VLDL(1) to VLDL(2) and of VLDL(2) to IDL were increased significantly, in the latter instance nearly two-fold. IDL apoB direct catabolism rose from 0.54 ± 0.30 pools per day at baseline to 1.17 ± 0.87 pools per day onatorvastatin and to 0.95 + 0.43 pools per day on simvastatin (both *P* <0.05). Similarly, the FTR for IDL to LDL conversion was enhanced 58–84% by statin treatment (*P* <0.01) LDL apoB fractional catabolic rate (FCR) which was low at baseline in these subjects (0.22 ± 0.04 pools per day) increased to 0.44 ± 0.11 pools per day on atorvastatin and 0.38 ± 0.11 pools per day on simvastatin (both *P* <0.01). ApoB-containing lipoproteins were more triglyceride-rich and contained less free cholesterol and cholesteryl ester on statin therapy. Further, patients on both treatments showed marked decreases in all LDL subfractions. In particular the concentration of small dense LDL (LDL-III) fell 64% on atorvastatin and 45% on simvastatin.

Interpretation. It was concluded that in patients with moderate combined hyperlipidaemia who initially have a low FCR for VLDL and LDL apoB, the principal action of atorvastatin and simvastatin is to stimulate receptor-mediated catabolism across the spectrum of apoB-containing lipoproteins. This leads to a substantial, and approximately equivalent, percentage reduction in plasma triglyceride and LDL cholesterol.

Comment

Statins are well known to decrease circulating LDL by increasing its catabolism. This is a consequence of inhibition of cholesterol biosynthesis in the liver which decreases the intrahepatic levels of cholesterol which is a signal for increased hepatic LDL-receptor expression and thus increased removal of LDL from the circulation. Statins do decrease serum triglyceride concentrations, particularly in patients with combined hyperlipidaemia. This study examined the mechanism by which simvastatin and atorvastatin lower serum triglyceride in combined hyperlipidaemia. Neither statin reduced hepatic VLDL production, although both decreased serum triglyceride. The mechanism for this decrease was increased removal from the circulation of VLDL1 (large VLDL), and IDL. The statins also decreased the atherogenic LDLIII (small-dense LDL) fraction. LDLIII is not a good ligand for the LDL-receptor so it is possible that it was reduced because of the enhanced removal of the more buoyant LDL which is its precursor. It is also possible that the increase in CETP reported by Guerin *et al.* with a statin in similar patients was at least partly responsible.

High doses of atorvastatin and simvastatin induce key enzymes involved in VLDL production.

Roglans N, Verd JC, Peris C, *et al. Lipids* 2002; **37**: 445–54.

Background and interpretation. Treatments with high doses of 3-hydroxy-3-methylglutaryl-coenzyme A (HMG-CoA) reductase inhibitors may induce the expression of sterol regulatory element binding protein (SREBP)-target genes, causing different effects from those attributed to the reduction of hepatic cholesterol content. The aim of this study was to investigate the effects of high doses of statins on the key enzymes involved in VLDL production in normolipidaemic rats. To examine whether the effects caused by statin treatment are a consequence of HMG-CoA reductase inhibition, the effect of atorvastatin on these enzymes in mevalonate-fed rats was tested. Atorvastatin and simvastatin enhanced not only HMG-CoA reductase but also the expression of the SREBP-2 gene itself. As a result of the over-expression of SREBP-2 caused by the statin treatment, genes regulated basically by SREBP-1, as FA synthase and acetyl-coenzyme A carboxylase, were also induced and their messenger ribonucleic acid (mRNA) levels increased. DAG acyltransferase and microsomal triglyceride transfer protein mRNA levels as well as phosphatidate phosphohydrolase activity were increased by both statins. Simvastatin raised liver cholesterol content, ACAT mRNA levels, and CTP:phosphocholine cytidylyltransferase activity, whereas it reduced liver DAG and phospholipid content. Mevalonate feeding reversed all changes induced by the atorvastatin treatment. These results show that treatment with high doses of statins induces key enzymes controlling rat liver lipid synthesis and VLDL assembly, probably as a result of SREBP-2 overexpression. Despite the induction of the key enzymes involved in VLDL production, both statins markedly reduced plasma triglyceride levels, suggesting that different mechanisms may be involved in the hypotriglyceridaemic effect of statins at high or low doses.

Comment

This study in rats revealed that although statins lower both serum LDL cholesterol and serum triglyceride, the effect of inhibiting HMG-CoA reductase in the liver is to induce this enzyme and also enzymes involved in triglyceride and VLDL production. This effect is probably due to the decrease in mevalonate which induces SREBP-2. Effectively, induction of these enzymes will tend to oppose both the LDL and triglyceride-lowering effects of statins, both of which were shown in the previous abstract to be due to enhanced VLDL, IDL and LDL catabolism. This is therefore one effect causing resistance to the lipid-lowering action of statins. Development of agents which can overcome SREBP-2 induction may therefore enhance the efficacy of statin treatment.

Atorvastatin versus four statin–fibrate combinations in patients with familial combined hyperlipidaemia.

Athyros VG, Papageorgiou AA, Athyrou VV, Demitriadis DS, Pehlivanidis AN, Kontopoulos AG. *J Cardiovasc Risk* 2002; **9**: 33–9.

BACKGROUND. Statin–fibrate combinations are very effective in the treatment of familial combined hyperlipidaemia (FCH). Nonetheless, they have not been extensively used because of the fear of side-effects. Thus, a therapeutic alternative is required for this lipid disorder. The aim of this study was to compare the long-term (one-year) efficacy of atorvastatin monotherapy with that of four statin–fibrate combinations in 675 FCH patients. Patients were randomly assigned to atorvastatin monotherapy (A; 20 mg/day) *n* = 134, or pravastatin (P; 20 mg/day) + gemfibrozil (G; 1200 mg/day) *n* = 135, simvastatin (S; 20 mg/day) + G (1200 mg/day) *n* = 137, P (20 mg/day) + ciprofibrate (C; 100 mg/day) *n* = 135, and S (20 mg/day) + C (100 mg/day) *n* = 134. Twelve patients on statin–fibrate combinations were withdrawn from the study because of side-effects: three because of creatine kinase (CK) elevation, two because of myalgia and seven due to increase in serum transaminase levels. One patient on A was withdrawn because of persistent epigastric discomfort. Atorvastatin reduced LDL cholesterol and apoprotein B more than all four combinations (–45% *vs* maximum –40% of S + C, and –39% *vs* maximum –32% of the same combination, respectively, *P* <0.001 for both), but had a lesser effect on triglycerides (–38% *vs* maximum –53% of S + C, *P* = 0.0002) and HDL cholesterol (6% vs maximum 21% of S + G, *P* = 0.0003). The effect of A on plasma fibrinogen was analogous to that of G combinations (–8% *vs* –9% of P + G and –11% of S + G, *P* = NS *vs* baseline and among each other) and inferior to that of the ciprofibrate combinations (–8% *vs* –24% of P + C, *P* = 0.0002 and –26% S + C, *P* = 0.0001). A had a lower treatment cost and better patient compliance, *P* = 0.04 *vs* C combinations and *P* = 0.02 *vs* G combinations.

INTERPRETATION. The data suggest that statin–fibrate combinations have a beneficial effect on all lipid parameters. Atorvastatin monotherapy has a better effect on LDL-C and apoprotein B than statin–fibrate combinations, but a lesser effect on HDL-C, triglycerides and in the case of ciprofibrate combinations, fibrinogen. The clinical significance of these findings should be tested in a large, long-term survival study.

Comment

This is an interesting study revealing that statin combined with ciprofibrate had a greater effect in reducing triglyceride and fibrinogen levels, and statin with gemfibrozil produced the greater increase in HDL cholesterol compared to atorvastatin alone or other gemfibrozil and ciprofibrate combinations. Given the greater LDL-lowering with atorvastatin alone it would have been interesting to know how this statin combined with fibrates affected triglyceride, HDL cholesterol and fibrinogen. Clearly more studies of this type will be important if we are to increase our use of combined statin and fibrate therapy.

Factorial study of the effects of atorvastatin and fish oil on dyslipidaemia in visceral obesity.

Chan DC, Watts GF, Mori TA, Barrett PH, Beilin LJ, Redgrave TG. *Eur J Clin Invest* 2002; **32**: 429–36.

Background. Dyslipidaemia may account for increased risk of cardiovascular disease in central obesity. Pharmacotherapy is often indicated in these patients, but the optimal approach remains unclear. This study investigated the effects of atorvastatin and fish oil on plasma lipid and lipoprotein levels, including remnant-like particle-cholesterol and apolipoprotein C-III, in dyslipidaemic men with visceral obesity. The authors carried out a 6-week randomized, placebo-controlled, 2 $\times$ 2 factorial intervention study of atorvastatin (40 mg/day) and fish oil (4 g/day) on plasma lipids and lipoproteins in 52 obese men (age 53.1 years, body mass index 33.7 $\pm$ 0.55 kg/m^2) with dyslipidaemia and insulin resistance. Treatment effects were analyzed by general linear modelling. Atorvastatin had significant main effects in decreasing triglycerides (–0.38 0.02 mmol/l, *P* = 0.002), total cholesterol (–1.89 0.17 mmol/l, *P* = 0.001), LDL cholesterol (–1.78 0.14 mmol/l, *P* = 0.001), remnant-like particle-cholesterol (–0.08 0.04 mmol/l, *P* = 0.035), apoB (–49.4 mg/dl, *P* = 0.001), apolipoprotein C-III (–12.6 6.1 mg/l, *P* = 0.044) and in increasing HDL cholesterol (+0.10 0.04 mmol/l, *P* = 0.007). Fish oil had significant main effects in decreasing triglycerides (–0.38 0.11 mmol/l, *P* = 0.002) and in increasing HDL cholesterol (+0.07 0.04 mmol/l, *P* = 0.041). There were no significant changes in weight or insulin resistance during the study.

Interpretation. Atorvastatin and fish oil have independent and additive effects in correcting dyslipidaemia in viscerally obese men. Improvement in abnormalities in remnant lipoproteins may occur only with use of atorvastatin. Combination treatment with statin and fish oil may, however, offer an optimal therapeutic approach for globally correcting dyslipidaemia in obesity.

Comment

This is a study which shows the effectiveness of a concentrated omega–3 fatty acid preparation in combination with a statin in decreasing triglyceride levels more than with statin alone. Chylomicron remnants, which may be atherogenic, were decreased by atorvastatin but not by omega–3 fatty acids which may be explained because these fatty acids contribute to chylomicron formation. However, it is not known whether remnants of omega–3-containing chylomicrons are atherogenic; therefore it is possible that the other benefits of triglyceride-lowering on IDL and small dense LDL are still present with omega–3-induced decreases in triglycerides.

A randomized crossover trial of combination pharmacologic therapy in children with familial hyperlipidemia.

McCrindle BW, Helden E, Cullen-Dean G, Conner WT. *Pediatr Res* 2002; **51**: 715–21.

BACKGROUND. This investigation sought to determine whether a low-dose combination of a bile acid-binding resin (colestipol) with an HMG-CoA reductase inhibitor (pravastatin) would result in improved acceptability, compliance, and effectiveness in lipid-lowering compared with conventional therapy with a higher dose of a bile acid-binding resin only, with fewer side-effects. The authors performed a randomized, crossover open-label clinical trial with two 18-week medication regimens separated by an 8-week washout period in 36 children and adolescents with FH or FCH. The regimens included colestipol 10 g/d (10 pills) *vs* a combination of colestipol 5 g/d with pravastatin 10 mg/d (6 pills). All patients were maintained on a fat-reduced diet. Acceptability was better with the combination regimen. Mean compliance was similar and suboptimal (approximately 60%) with all medication components. Mean relative LDL cholesterol-lowering was significantly better with the combination regimen (–17 ± 16% *vs* –10 ± 13%; *P* = 0.045), although insufficient to achieve recommended target values in the majority of patients on either regimen. Both regimens were equally free of adverse effects, with no important effect on chemistry or haematologic values. Patient-reported adverse effects were more common with the conventional-dose colestipol-only regimen.

INTERPRETATION. Compliance with medication regimens using the bile acid-binding resins is suboptimal, although combination with a low dose of a statin may result in better lipid-lowering.

Comment

The age at which to introduce statin treatment in young people with heterozygous FH is controversial. It is effective in lowering LDL cholesterol, but evidence that it has no long-term harmful effects during growth and development does not exist. There are reasonable data to suggest that bile acid sequestrants are safe in childhood. If compliance is good, these agents effectively lower LDL cholesterol to a similar degree to statins. However, compliance is not good and children hate the medication. Parental attitudes may also be bad towards bile acid sequestrating agents which the parents themselves may have also endured before the advent of statins or in their own childhood. With the concern that atherosclerosis is rapidly developing during childhood, the use of statins at an earlier age than at present appears attractive. The alternative is to give bile acid sequestrants or nothing, if statins are not used. Essentially, giving bile acid sequestrants is often about as effective as giving nothing because of poor compliance. Furthermore to give such an unpleasant medication to children is to encourage resentment and outrage which finds its full expression during the rebellious years of adolescence. This frequently leads to the alienation of the patient from medical services in adolescence and young adulthood, exactly at the age when the patient should be accessible to start statin therapy. Giving no drug therapy in childhood has the attraction that there is no need for frequent blood testing, another factor encouraging bad attitudes to medical contact. Keeping the patients in a receptive mood to receive statins as early as considered safe, has much to commend it. In clinical trials much of the benefit of statin therapy is achieved within three years (although of course these trials were not in young people) so maybe introducing statins in boys with FH, whose prognosis is worse than that of girls, at around 18 years and in girls in their mid-twenties and avoiding bile acid sequestrating agents altogether is a valid approach. It is the one I adopt except in those rare families where CHD is manifest in the late twenties or early thirties. It would be very welcome to have clinical trial evidence to test this approach and some of the non-invasive methods for assessing early atherosclerosis might permit testing of this and other approaches to therapy. Meanwhile this study is particularly thought-provoking.

References

1. Durrington PN. Can measurement of apolipoprotein B replace the lipid profile in the follow-up of patients with lipoprotein disorders? *Clin Chem* 2002; **48**: 401–2.

2. Bays H, Moore PB, Drehobl MA, *et al.* Effectiveness and tolerability of ezetimibe in patients with primary hypercholesterolaemia: pooled analysis of two phase II studies. *Clin Ther* 2001; **23**: 1209–30.

3. Olsson AG, Pears J, McKellar J, Mizan J, Raza A. Effect of rosuvastatin on low-density lipoprotein cholesterol in patients with hypercholesterolaemia. *Am J Cardiol* 2001; **88**: 504–8.

4. Karlipara K. Statin therapy: rationale for a new agent, rosuvastatin. *Int J Clin Pract* 2002; **56**: 379–87.

15

Trials of lipid-lowering drugs on non-lipid risk factors for coronary heart disease and other diseases

Introduction

This chapter focuses on recent publications concerning the effects of lipid-lowering drugs either indirectly related or unrelated to their effects on serum lipoproteins. This year highly-sensitive C-reactive protein (CRP) has occupied centre stage as a potentially important means of identifying individual atherosclerosis risk as was highlighted by the article by Devaray. Whether the effect of statins in reducing CRP is directly antiatherogenic or whether CRP is simply a means of identifying individuals who are more likely to benefit from statin therapy is unsettled, but in either case it has enormous potential. This year Riesen *et al.* showed that atorvastatin like other statins diminishes CRP and Blake and colleagues argued for the use of highly-sensitive CRP measurement to augment existing methods for the identification of people who can benefit from statin therapy.

Of great interest too are reports which are emerging that statins will diminish proteinuria in a variety of clinical conditions. Some evidence suggests that their effect is to limit renal damage, although thus far there is no evidence that they can halt the decline in glomerular fitration rates in primary or secondary renal disease. It is also fascinating that in some diseases producing proteinuria, particularly diabetes, evidence is now emerging that the increased transcapillary escape of macromolecules is not confined to the kidneys and may increase atherogenesis due, for example, to the greater rate of entry of low-density lipoproteins (LDL) into the artery wall. Perhaps this process too can be ameliorated by statins. This year Lee *et al.* showed that pravastatin decreases proteinuria associated with hypertension. Nakamara and colleagues found a similar effect in chronic glomerulonephritis and Danesh *et al.* reported a direct effect of statins on glucose-induced renal mesangial cell proliferation in tissue culture. Kornerup *et al.* found increased transvascular escape of LDL in diabetes.

Statins do not reduce serum fibrinogen levels, another risk factor for atherosclerosis which has for many years had its champions. On the other hand fibrates, in which there is renewed interest as antiatherogenic agents, do decrease fibrinogen, which this year formed the basis of an article by Maison *et al.* The fibrate effect on fibrinogen may therefore augment the decrease in triglycerides and small dense LDL and the increase in high-density lipoprotein (HDL) cholesterol that they induce.

This year evidence for the proposed decrease in osteoporosis with statin therapy has receded somewhat and with it one of the strands of evidence for its pleiotropic effects. The Heart Protection Study (HPS) showed no decrease in bone fracture rates with simvastatin treatment. Whilst a study of bone density by Funkhouser and coworkers was positive, that by Sirola and colleagues showed no statin effect and both studies were unrandomized.

Inflammation and atherosclerosis: the value of the high-sensitivity C-reactive protein assay as a risk marker.

Jialal I, Devaraj S. *Am J Clin Pathol* 2001; **116**(Suppl): S108–15.

Background and interpretation. CRP is a prototypic marker of inflammation. Numerous prospective studies in healthy volunteers have confirmed that high-sensitivity CRP (hsCRP) predicts cardiovascular events (CVEs), and hsCRP seems additive to an elevated total cholesterol level and a total/HDL cholesterol ratio in men and women in predicting risk. In smokers and people with metabolic syndrome, hsCRP levels are elevated; in elderly people, there seems to be a relationship between hsCRP and CVEs and mortality. Several properties of CRP make it proatherogenic; however; pending further studies, it should be considered as a risk marker. In people with acute coronary syndromes, hsCRP measurement may be valuable. Elevated levels in the highest quantile seem to predict greater mortality and poorer prognosis in patients with unstable angina and myocardial infarction (MI). While hsCRP is a strong independent predictor of risk of future MI, stroke, peripheral arterial disease, and vascular death, the validity of hsCRP as a risk marker needs to be assessed in all populations. Weight loss, statin drugs, aspirin, and high-dose alpha tocopherol therapy could affect hsCRP. It has its greatest validity as an adjunctive measure in the primary prevention of cardiovascular disease.

Comment

This report summarizes evidence that hsCRP is a risk marker for atherosclerosis which could improve on existing methods for risk prediction.

Short-term effects of atorvastatin on C-reactive protein.

Riesen WF, Engler H, Risch M, Korte W, Noseda G. *Eur Heart J* 2002; **23**: 794–9.

Background. The aim of this investigation was to study the short-term effect of atorvastatin on CRP in patients with or at risk for coronary heart disease. One hundred and fifty-five randomly selected patients from the Swiss Intervention Trial for lowering

Cholesterol (SWITCH) were assessed for hsCRP, total cholesterol, LDL cholesterol, HDL cholesterol and triglycerides at baseline, and after one and three months of treatment with atorvastatin at various doses to reach pre-defined lipid target values. The median decrease of cholesterol was 28% after one month and 35% after three months. LDL cholesterol was decreased by 37% and 45%, HDL cholesterol was increased by 7% and 8%, respectively. Patients with a low CRP baseline concentration (lowest quartile <1.34 mg/ l displayed no significant change, whereas patients in the other quartiles showed a significant decrease, of 22% to 40% (*P* <0.05 to <0.001) at one month and of 32% to 36% after three months compared to baseline. The decrease in CRP-lowering was thus fully established by one month and this response was independent of lipid and lipoprotein changes as well as atorvastatin doses.

INTERPRETATION. Atorvastatin significantly decreases CRP concentrations after four weeks of therapy. These results may be important with respect to the early benefit of statin therapy.

Comment

This is a study showing the reduction in highly-sensitive CRP with atorvastatin, adding to the existing literature that this is a general effect of statins.

Projected life-expectancy gains with statin therapy for individuals with elevated C-reactive protein levels.

Blake GJ, Ridker PM, Kuntz KM. *J Am Coll Cardiol* 2002; **40**: 49–55.

BACKGROUND. Persons with LDL cholesterol levels below current treatment guidelines and elevated CRP levels are at increased risk of cardiovascular disease and may benefit from statin therapy. This study sought to estimate the potential gains in life expectancy achieved with statin therapy for individuals without overt hyperlipidaemia but with elevated CRP levels. The authors constructed a decision-analytic model to estimate the gains in life expectancy with statin therapy for individuals without overt hyperlipidaemia but with elevated CRP levels. The annual risks of MI and stroke, as well as the efficacy of statin therapy, were based on evidence from randomized trials. Estimates of prognosis after MI or stroke were derived from population-based studies. It was estimated that 58-year-old men and women with CRP levels ≥0.16 mg/dl but LDL cholesterol <149 mg/dl would gain 6.6 months and 6.4 months of life expectancy, respectively, with statin therapy. These gains were similar to those for patients with LDL cholesterol ≥149 mg/dl (6.7 months for men and 6.6 months for women). In sensitivity analyses, the baseline risk of MI and the efficacy of statin therapy for preventing MI were identified as the most important factors in determining the magnitude of benefit with statin therapy.

INTERPRETATION. These results suggest that individuals with elevated CRP levels, many of whom do not meet current National Cholesterol Education Program guidelines for drug treatment, may receive a substantial benefit from statin therapy. This analysis supports a crucial need for direct intervention trials aimed at subjects with elevated CRP levels.

Comment

These authors have been in the forefront of the work suggesting that highly-sensitive CRP can improve our ability to predict CHD and other atherosclerotic disease. Here they combine their predictive models with the degree of reduction in CRP with statins to estimate its potential benefit and suggest that some patients who would not receive statins on current recommendations, but who have raised CRP, could benefit. Currently what is missing is evidence that CRP is causally involved in atherosclerosis or as to whether it is a response to the development of atherosclerosis at a stage when it is not clinically identifiable. Such evidence would be hard to come by until there is a drug which lowers CRP without lowering LDL cholesterol. In the meantime we already know that the number of patients who must receive statin treatment to prevent a CVE is inversely related to their risk no matter how this risk is identified. Highly-sensitive CRP appears to provide an additional means of assessing this risk

Effect of pravastatin on proteinuria in patients with well-controlled hypertension.

Lee TM, Su SF, Tsai CH. *Hypertension* 2002; **40**: 67–73.

Background and interpretation. Proteinuria is an important risk factor for cardiovascular and renal morbidity and mortality. The effects of HMG-CoA reductase inhibitor (statin) therapy on proteinuria in normolipidaemic patients with well-controlled hypertension have not been studied. A total of 63 normolipidaemic (total cholesterol <240 mg/dl) and proteinuric (300 to 3000 mg/d) patients with well-controlled blood pressure (<140/90 mmHg) were randomized to receive either placebo ($n = 32$) or pravastatin (10 mg/d; $n = 31$) after a three-month placebo period. Pravastatin lowered proteinuria after six months by 54% (P <0.0001). Creatinine clearance was stable throughout the study in the two groups. Despite unchanged plasma endothelin-1 levels throughout the study, urinary excretion of the peptide was decreased and significantly correlated with improvement in urinary protein excretion in pravastatin-treated patients ($r = 0.64$, $P = 0.001$). The urinary excretion of retinol-binding protein decreased after pravastatin administration, probably reflecting an improvement in tubular function. In contrast, the urinary excretion of IgG did not change significantly throughout the study in either group. Multivariate analysis revealed that proteinuria was only significantly correlated with statin use (P <0.0001, $R^2 = 0.66$). Linear regression analysis in the statin-treated group did not show any correlation between changes in lipid profiles and proteinuria regression. Thus, in addition to their primary function of antilipidaemia, the addition of pravastatin to treatment for well-controlled hypertension may have an additive effect on reducing proteinuria independent of haemodynamics and lipid-lowering effects, possibly through inhibiting renal endothelin-1 synthesis and improving tubular function.

Comment

Several reports have appeared suggesting that statin therapy will reduce proteinuria in primary renal disease and secondary to hypertension and diabetes. The effect as in this study may be independent of antihypertensive agents such as angiotensin converting enzyme (ACE) inhibitors. It is difficult to know whether this is a direct effect of statins on renal tissue or whether it is mediated through the correction of dyslipidaemia which may itself be nephrotoxic. In either event it is another potential benefit of statin therapy. Whether in the longer term it will ameliorate the decline in glomerular filtration as well as diminishing proteinuria is as yet unknown.

Effect of cerivastatin on proteinuria and urinary podocytes in patients with chronic glomerulonephritis.

Nakamura T, Ushiyama C, Hirokawa K, *et al. Nephrol Dial Transplant* 2002; **17**: 798–802.

BACKGROUND. We previously reported urinary podocytes to be a marker of glomerular injury. The aim of the present study was to determine whether cerivastatin, a newly developed, potent synthetic statin, affects proteinuria and urinary podocyte excretion in patients with chronic glomerulonephritis (CGN). The authors randomly assigned 40 normotensive hypercholesterolaemic patients with CGN to receive either cerivastatin 0.15 mg/day (n = 20) or placebo (n = 20). Subjects were 24 men and 16 women, with a mean age of 40.8144 years; 27 had IgA nephropathy and 13 had non-IgA proliferative glomerulonephritis. Treatment was continued for six months. Plasma total cholesterol, HDL cholesterol, LDL cholesterol and triglycerides, urinary protein excretion and the number of podocytes were measured before treatment and at three and six months after treatment. After 6 months, a significant reduction in total cholesterol (P <0.001), LDL cholesterol (P <0.001) and triglycerides (P <0.05), and a significant increase in HDL cholesterol (P <0.001) were observed in the group treated with cerivastatin. Urinary protein excretion decreased from 1.806 to 0.804 g/day, (P <0.01) in this group, and urinary podocyte excretion decreased from 1.606 to 0.904 cells/ml (P <0.01). However, placebo showed little effect on these lipid levels, urinary protein excretion and urinary podocyte excretion. The differences between the cerivastatin group and the placebo group were significant (cholesterol, P <0.001; LDL cholesterol, P <0.001; triglycerides, P <0.05; HDL cholesterol, P <0.001; urinary protein, P <0.01; and urinary podocytes, P <0.01).

INTERPRETATION. Statins such as cerivastatin may be beneficial for restoration of injured podocytes in patients with CGN and hypercholesterolaemia.

Comment

This study reveals that statin therapy decreased urinary protein excretion in patients with primary chronic glomerulonephritis. The decrease in urinary podocytes suggested that the effect, however it was mediated, was due to a decrease in renal damage.

3-hydroxy-3-methylglutaryl CoA reductase inhibitors prevent high glucose-induced proliferation of mesangial cells via modulation of Rho GTPase/p21 signaling pathway: Implications for diabetic nephropathy.

Danesh FR, Sadeghi MM, Amro N, *et al. Proc Nat Acad Sci USA* 2002; **99**: 8301–5.

Background and interpretation. Inhibitors of HMG-CoA reductase, also known as statins, are lipid-lowering agents widely used in the prevention of coronary heart disease. Recent experimental and clinical data, however, indicate that the overall benefits of statin therapy may exceed its cholesterol-lowering properties. These investigators postulate that statins may ameliorate the detrimental effects of high glucose (HG)-induced proliferation of mesangial cells (MCs), a feature of early stages of diabetic nephropathy, by preventing Rho isoprenylation. Rat MCs cultured in HG milieu were treated with and without simvastatin, an HMG-CoA reductase inhibitor. Simvastatin inhibited HG-induced MC proliferation as measured by [(3)H]thymidine incorporation. This inhibitory effect was reversed with geranylgeranyl pyrophosphate, an isoprenoid intermediate of the cholesterol biosynthetic pathway. At the cell-cycle level, the HG-induced proliferation of MCs was associated with a decrease in cyclin dependent kinase (CDK) inhibitor p21 protein expression accompanied by an increase in CDK4 and CDK2 kinase activities. Simvastatin reversed the downregulation of p21 protein expression and decreased CDK4 and CDK2 kinase activities. Exposure of MCs to HG was associated with an increase in membrane-associated Ras and Rho GTPase protein expression. Cotreatment of MCs with simvastatin reversed HG-induced Ras and Rho membrane translocation. Immunofluorescence microscopy revealed that the overexpression of the dominant-negative RhoA led to a significant increase in p21 expression. Our data suggest that simvastatin represses the HG-induced Rho GTPase/p21 signalling in glomerular MCs. Thus, this study provides a molecular basis for the use of statins, independently of their cholesterol-lowering effect, in early stages of diabetic nephropathy.

Comment

This study suggests that statin therapy may directly ameliorate the process of diabetic nephropathy.

Transvascular low-density lipoprotein transport in patients with diabetes mellitus (type 2): a noninvasive *in vivo* isotope technique.

Kornerup K, Nordestgaard BG, Feldt-Rasmussen B, Borch-Johnsen K, Jensen KS, Jensen JS. *Arterioscler Thromb Vasc Biol* 2002; **22**: 1168–74.

BACKGROUND. The increased risk of atherosclerosis associated with diabetes cannot be explained by conventional cardiovascular risk factors alone. It was hypothesized that transvascular lipoprotein transport may be increased in patients with diabetes, possibly explaining increased intimal lipoprotein accumulation and, thus, atherosclerosis. An *in vivo* method was developed for measurement of transvascular transport of LDL and it was applied in 16 patients with maturity-onset diabetes (type 2) and 29 healthy control subjects. Autologous ^{131}I-labelled LDL was reinjected intravenously in addition to ^{125}I-labelled albumin, and the one-hour fractional escape rates were taken as indices of transvascular transport. Both parameters were normally distributed, and they were tightly correlated ($R^2 = 0.69$, $P < 0.0001$). Transvascular LDL transport was 5.42.9%/h and 4.11.5%/h in patients with diabetes and control subjects, respectively ($P < 0.05$); equivalent values for albumin were 6.52.5%/h and 5.31.6%/h ($P < 0.05$). This difference most likely was not caused by altered hepatic LDL-receptor expression, glycosylation of LDL, small LDL size, nephropathy, statin use, or different plasma insulin levels in diabetic patients.

INTERPRETATION. Transvascular LDL transport may be increased in patients with type 2 diabetes. This suggests that lipoprotein flux into the arterial wall is increased in people with diabetes, possibly explaining the accelerated development of atherosclerosis.

Comment

The notion that capillaries in diabetes may be generally more permeable to macromolecules such as LDL and albumin, rather than simply those in the kidney, has been abroad for some while. These well-designed experiments do much to lend support to it. Now that it is emerging that statins can decrease proteinuria it will be fascinating to discover whether they affect this more generalized increase in transcapillary filtration associated with diabetes.

A pharmacoepidemiological assessment of the effect of statins and fibrates on fibrinogen concentration.

Maison P, Mennen L, Sapinho D, *et al.*; DESIR Study Group. *Atherosclerosis* 2002; **160**: 155–60.

BACKGROUND AND INTERPRETATION. Plasma fibrinogen is strongly associated with cardiovascular morbi-mortality. Within the large cohort of the DESIR (data from an epidemiological study on the insulin resistance syndrome) study, the relationship

between change in fibrinogen concentration over a three-year follow-up of fibrate and statin use was investigated. Fibrinogen concentrations were higher at baseline among individuals treated with statins (*n* = 130) compared to those treated with fibrates (*n* = 251), even after adjustment for confounding factors (including total cholesterol, HDL cholesterol and triglycerides) (mean [SD]: 2.8 [0.6] *vs* 3.1 [0.6], *P* <0.001). The change in fibrinogen concentrations was compared at three years of follow-up, between individuals who started fibrate (*n* = 126) or statin (*n* = 127) treatment during the follow-up and individuals (*n* = 3906) who stayed without treatment during this period. After adjustment for baseline fibrinogen level, age, sex and changes in total cholesterol, triglycerides and alcohol intake, fibrinogen concentration decreased after fibrate treatment, while it increased after statin treatment and in those not using lipid-lowering drugs (–0.07 [0.54] *vs* 0.10 [0.54] *vs* 0.08 [0.52] g/l respectively, *P* = 0.01). No differences were observed between different statins or different fibrates.

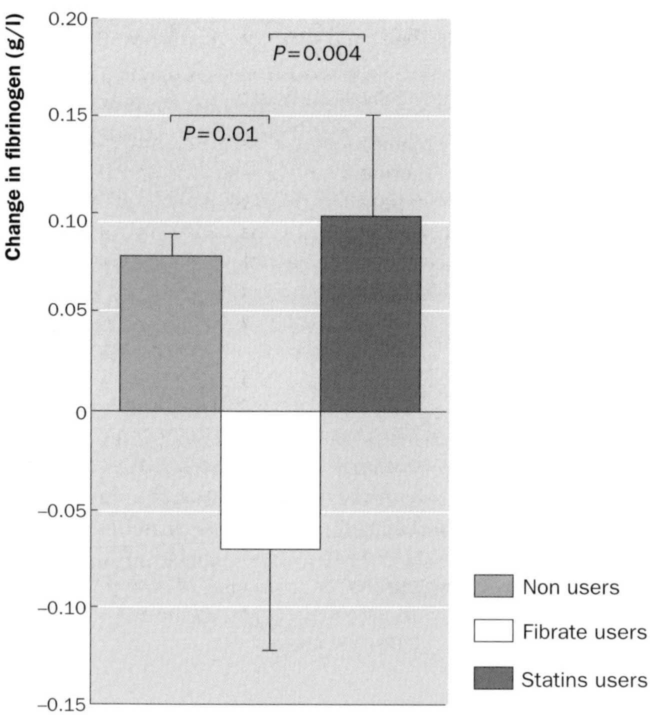

Fig. 15.1 Mean (SD) change in fibrinogen concentration after 3 years in subjects who started fibrate (*n* = 126) or statin treatment (*n* = 127) during the follow-up period and those who stayed without lipid-lowering treatment (*n* = 3906), adjusted for age, sex, baseline fibrinogen concentration and changes in total cholesterol, triglycerides and alcohol intake. The DESIR study. Source: Maison *et al.* (2002).

INTERPRETATION. Fibrates in contrast with statins may combine lipid-lowering with a beneficial effect on fibrinogen. This effect is independent of changes in cholesterol and triglyceride concentrations.

Comment

Fibrate therapy has previously been reported to decrease serum fibrinogen levels. The effect is most conspicuous with fibrates which are true fibric acid derivatives, such as clofibrate, bezafibrate and fenofibrate rather than gemfibrozil which is a fibric acid analogue. In this study there is a clear difference in fibrinogen levels between patients receiving fibrates and statins or no lipid-lowering therapy (Fig. 15.1). Whilst the patients were not randomized to therapy the result accords well with earlier randomized trials of individual statins or fibrates.

Effect of HMG-CoA reductase inhibitors (statins) on bone mineral density.

Funkhouser HL, Adera T, Adler RA. *J Clin Densitom* 2002; **5**: 151–8.

BACKGROUND AND INTERPRETATION. Recent studies have suggested that HMG-CoA reductase inhibitors (statins) can increase the bone mineral density (BMD). The objective of this study was to determine if patients on statin drugs were more likely to have a greater BMD and lower risk of osteoporosis than patients not taking these drugs. A computerized pharmacy system provided complete medication dispensing records for the 983 patients (697 men and 286 women) referred for BMD testing at a single Veterans Affairs Medical Centre. In an analysis of covariance model that adjusted for age, body mass index, race, and vitamin use, men using statin drugs were more likely to have a greater BMD of the spine (P <0.005). The mean difference (effect size) was 0.05 g/cm^2 (95% confidence interval [CI] 0.02–0.09), about 5.3% greater BMD. In women, the association was not significant. The risk of osteoporosis (defined as a T-score ≤ -2.5) was determined using logistic regression analysis after adjustment for potential confounding variables. Although not statistically significant, men who received statin drugs for more than two years were approximately half as likely to develop osteoporosis (odds ratio [OR] = 0.55, 95% CI = 0.28–1.08). A similar effect was observed in women taking statins for any length of time (OR = 0.36, 95% CI = 0.12–1.07). This study suggests that statin drugs may decrease osteoporosis risk, warranting a randomized controlled trial.

Comment

There have been earlier reports that statin use is associated with diminished risk of bone fracture. This observational study suggests that bone mineral mass may be preserved in statin users. The investigators call for a randomized, placebo-controlled clinical trial. It was published before the HPS. In that trial which was randomized and involved over 20 000 patients no effect on bone fracture rates was observed between active simvastatin treatment and placebo. Is this a case of an elegant hypothesis cruelly destroyed by cold fact or should a trial be conducted in patients at high risk of osteoporotic fracture?

Relation of statin use and bone loss: a prospective population-based cohort study in early postmenopausal women.

Sirola J, Sirola J, Honkanen R, *et al. Osteoporos Int* 2002; **13**: 537–41.

BACKGROUND AND INTERPRETATION. Recent experimental and epidemiological studies have suggested that the lipid-lowering drugs, statins, may have bone-protective effects. This investigation studied the effects of statin use on the change in BMD in a prospective 4.5-year cohort study based on subjects from the Kuopio Osteoporosis Risk Factor and Prevention (OSTPRE) Study, Finland. Six-hundred-and-twenty women aged 53–64 years were divided into four groups: 55 women reported continuous and 63 women occasional statin use during the follow-up; 142 non-users of statins reported hypercholesterolaemia whereas 360 non-users did not. Spinal and femoral BMDs were measured by dual-energy X-ray densitometry in 1995–1996 and 1999–2000 and the BMD changes of the four groups were compared. Characteristics of the study population were obtained with postal inquiries. The mean annual spinal and femoral BMD changes of the study groups were 0.29% and –0.50% for the continuous statin users, 0.19% and –0.57% for the occasional statin users, 0.52% and –0.29% for the hypercholesterolaemic non-users of statins, and 0.39% and –0.33% for the non-users of statins without hypercholesterolaemia, ($P = 0.398$ and $P = 0.404$) respectively. The corresponding BMD changes adjusted for age, years since menopause, body mass index, BMD at baseline, calcium intake, oestrogen and cortisone therapy, duration of follow-up and statin use before the baseline were –0.20% and –0.47%, 0.19% and –0.54%, 0.54% and –0.32%, 0.47% and –0.33% ($P = 0.134$ and $P = 0.628$), respectively. These results suggest that statins do not protect from early postmenopausal bone loss. Randomized trials are needed to confirm these results.

Comment

This is a negative observational study of the effect of statin use in early postmenopausal women on bone mineral mass. It accords well the negative findings of HPS with respect to the effect of simvastatin on bone fracture rates.

16

Gender and coronary heart disease

Review of the literature

One of the most striking epidemiological observations relating to coronary heart disease (CHD) is the greater risk in men as opposed to women. The incidence of coronary artery disease in men continues to be higher even in advanced old age. The greater longevity of women compared to men means that in old age there are more women with CHD, but its incidence continues to be greater in the surviving men. It is true that the greatest disparity in CHD incidence is evident in younger age groups, but in women it never catches up with that of their male counterparts, even after the menopause. Furthermore there is no abrupt acceleration in CHD incidence in women following the menopause: their risk accelerates smoothly with age. In men the rate of acceleration does diminish in older age, although the rate itself, of course, continues to increase |**1**|.

Even when there is a major risk factor for CHD present, women are frequently still at lesser risk than men. This is most clearly the case in familial hypercholesterolaemia (FH) |**2**|, but it is also evident for cigarette-smoking and hypertension. In diabetes mellitus and insulin resistance it is often stated that women lose their relative immunity from CHD, but even then their risk is generally still somewhat lower than that of diabetic or insulin-resistant men |**3**|. Diabetes and insulin resistance, are particularly interesting in this context and will be discussed later. Let us, however, return to FH because this is the one condition which in its homozygous form is associated with CHD in childhood. It thus allows us to explore how early the gender difference in CHD risk arises. In homozygous FH, CHD frequently presents before puberty. Some years ago a report appeared which sought to demonstrate the effectiveness of extracorporeal removal of low-density lipoprotein (LDL) in delaying death from CHD in homozygous FH |**4**|. Inevitably the patients studied were children or adolescents, because at the time few patients survived beyond their third decade. The authors of the report were able to find families in which there were two siblings with homozygous FH with only the younger sibling being alive at the time that plasmapheresis was introduced. They compared the longevity of the sibling receiving plasmapheresis with the sibling who had not done so. There was a statistically significant improvement in survival, but the finding could have been confounded by gender differences. The authors therefore reviewed the reported series of homozygous FH, none of whom had received plasmapheresis, to establish whether there was any gender difference in longevity. None was evident, suggesting that boys and girls are at equal CHD risk. This is not, however, the case in heterozygous FH in which CHD occurs later in life. In heterozygous FH, women are clearly at an advantage with about half the men succumbing to CHD before the age of 60 whereas only about 15% of women will have done so |**2**|.

In family studies of heterozygous FH, the affected male first-degree relatives develop symptomatic CHD on average about nine years before their affected female first degree relatives |**5**|. This is an important observation for two reasons. First, it overcomes the confounding effect of the severity of the particular LDL-receptor mutation because within a family the same mutation is running. Secondly, it establishes that some event must occur between childhood and early adulthood which clearly increases the CHD risk in men or which protects the women. Obviously the most important event to consider is puberty. At puberty differences in CHD risk factors between men and women become evident. We must not discount differences in rates of cigarette-smoking and hypertension |**6**|, but it is the changes in lipoprotein metabolism which occur most generally. We can examine these in great detail because of a remarkable series of population studies undertaken in the USA as an extension of the Lipid Research Clinics Prevelance Program |**7,8**|. The major findings of these are shown in Figs. 16.1–16.4. At puberty there is a profound

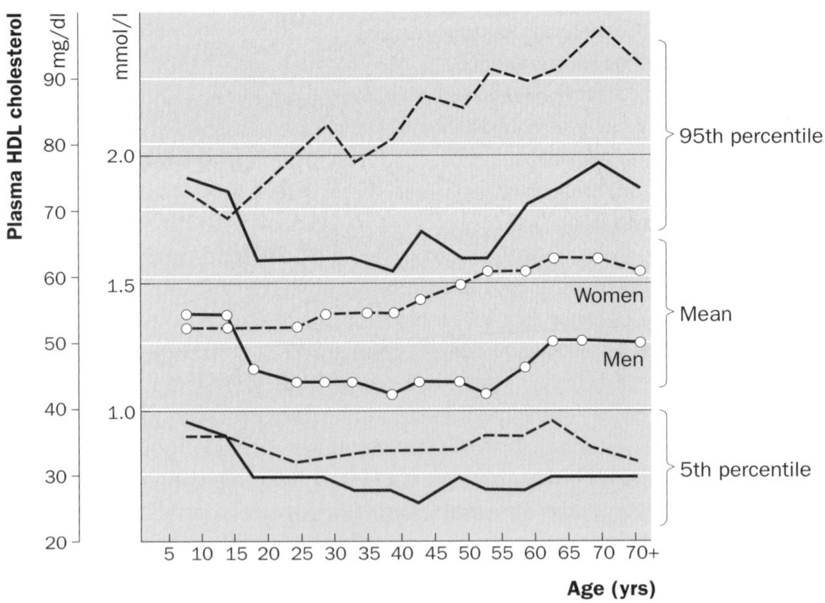

Fig. 16.1 Plasma high-density lipoprotein (HDL) cholesterol in men and women as a function of age. Redrawn with permission from Durrington PN. *Hyperlipidaemia. Diagnosis and Management*, 2nd edn. Butterworth Heinemann, 1995.

decrease in high-density lipoprotein (HDL) cholesterol in men. This difference persists throughout life. There is no decline in HDL cholesterol in women at the menopause. Indeed, it is not until later around the late 50s that average HDL cholesterol levels in men and women become closer and then it is not because of a decline in women but because of a tendency for levels to drift upwards in men. This effect could be due to a higher CHD mortality in men with lower HDL cholesterol levels, but the percentage of men dying prematurely from CHD is probably not high enough to make this the major cause. More likely is that some change in men has occurred in their late 50s which allows their HDL cholesterol to rise.

The plasma cholesterol and LDL cholesterol levels are higher in men compared to women from puberty and continue to rise in them until around the age of 40 after which they plateau. In women, cholesterol and LDL cholesterol also rise, but it is not until the menopause that they reach levels in excess of men and they remain higher on average thereafter.

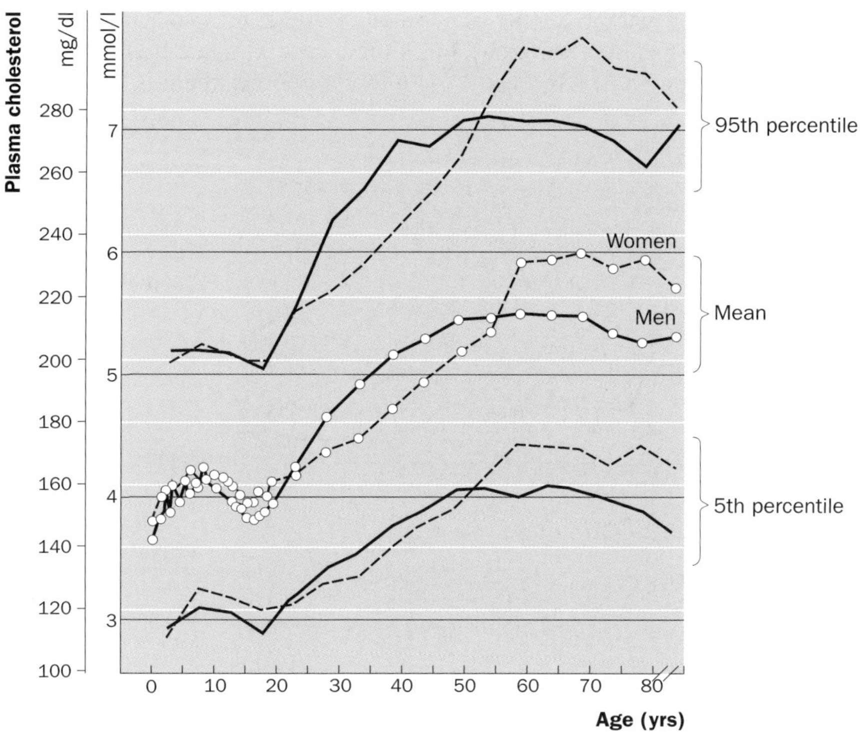

Fig. 16.2 Plasma cholesterol in men and women as a function of age. Redrawn with permission from Durrington PN. *Hyperlipidaemia. Diagnosis and Management,* 2nd edn. Butterworth Heinemann, 1995.

Gender differences in serum triglycerides with age show a third pattern. As with cholesterol, triglycerides rise in men from the time of puberty to levels which on average exceed those of women throughout young adulthood and middle-age. In men reaching their fifties, they decline to a plateau. In women, the upward drift of triglycerides is much slower and does not reach levels similar to those of men until around 70 years of age.

The striking gender difference in CHD risk has been the subject of much interest because understanding it might lead to a means of CHD prevention. The gender difference has most frequently been construed as indicating some protective influence present in women which is lacking in men. This interpretation almost certainly has its origin in some of the ideas about gender and survival prevalent in the 1920s following the discovery of the concept of X-linked recessive inheritance. In this, an additional set of genes, present as a consequence of having two X chromosomes in women, provides protection against disease or developmental abnormality |**9**|. The protective influence in women was widely considered to be oestrogen. However, this does not fit with the facts so far enumerated or at least not

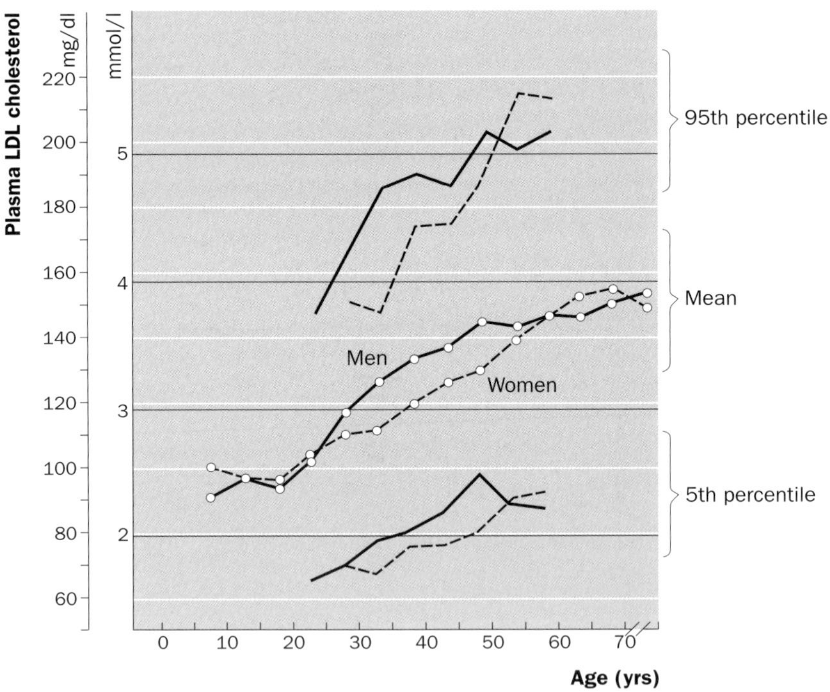

Fig. 16.3 Plasma low-density lipoprotein (LDL) cholesterol in men and women as a function of age. Redrawn with permission from Durrington PN. *Hyperlipidaemia. Diagnosis and Management,* 2nd edn. Butterworth Heinemann, 1995.

all of them. Certainly the higher levels of oestrogen from puberty in women could explain, say the lower levels of cholesterol, which after the menopause rise in women. However, it does not explain why women subsequently have higher levels than men. It also does not explain the drop in HDL cholesterol in boys at puberty with the maintenance of childhood levels in women. Nor does it explain why these do not fall in the typical spontaneous female menopause whereas they rise in men at about the time of the male climacteric.

Perhaps therefore we should regard men as having some characteristic which is positively harmful in terms of CHD, which is lacking in women, being coded for only on the Y chromosome. This actually fits better with biological principles than the alternative hypothesis. In zoology, characteristics which may be important for the survival of the species, but not for the individual tend to be concentrated in the male. The female is thus generally better adapted to survival. Fewer males are required to ensure successful reproduction and they can generally be sacrificed

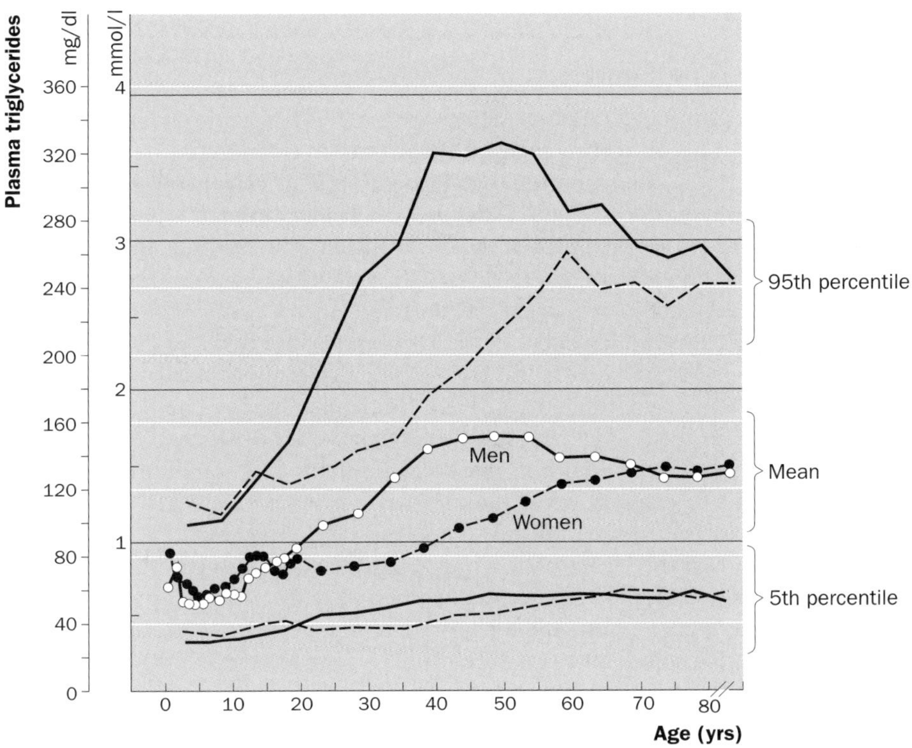

Fig. 16.4 Plasma fasting triglycerides in men and women as a function of age. Redrawn with permission from Durrington PN. *Hyperlipidaemia. Diagnosis and Management,* 2nd edn. Butterworth Heinemann, 1995.

more readily without compromising the survival of the species. The lack of camouflage of the males of many species is but one example. So also is the greater aggressive, devil-may-care, combative tendency of the male and this is not confined to non-human species. In the human, young men almost regardless of intelligence or culture, seem compelled to acts of reckless behaviour which, wrongly directed, are disastrously maladaptive for them as individuals. Indeed war is only possible because of such impulses in young men. As WB Yeats wrote of the fighter pilot in his poem *An Irish Airman Foresees his Death*, 'A lonely impulse of desire drove to this tumult in the clouds'. Similar 'lonely impulses' may colour male behaviour with regard to healthy lifestyle options, such as smoking, diet and alcohol. The female body form too is more suited for survival in many ways, not least her greater ability to survive cold, starvation and water immersion. Clearly, the behavioural and many of the anatomical aspects of the greater susceptibility to premature death in men are due to androgens. Without androgenic effects the human form is female, the most obvious example of which is testicular feminization. These arguments could be further developed with many other examples, but I do not wish to belabour the basic point which is that we should view the gender differences in CHD, as in many other aspects of human existence, not as men lacking some protective feature, but rather that they have a genetic element not possessed by females which leads to their greater susceptibility.

The clinical trials of treating men with oestrogen in an attempt to prevent CHD were a failure |**10**|. Now it seems too that the observational evidence that women are protected against CHD after the menopause by hormone replacement therapy (HRT) is perhaps more the result of the lower risk in women who either choose to receive HRT or who are advised to do so. The first of the randomized controlled trials, the Heart and Estrogen/progestin Replacement Study (HERS), has not shown that HRT diminishes CHD risk |**11**|. It must be admitted that there are questions still to be addressed before this can be fully understood, particularly the role of the progestogenic components of HRT. Immediately after the publication of HERS it was still considered that HRT might prove more successful in primary rather than secondary prevention, but then the first of the randomized placebo controlled trials in primary prevention showed an increase in CHD events with HRT |**12**|. It was also suggested that in the longer term there might be benefits, but in the follow-up of the HERS patients on active HRT this was not the case either |**13**|. There is, however, the evidence that women who have either a spontaneous or a surgical or radiotherapeutic menopause do appear to be at increased CHD risk |**14–16**|. It is perhaps difficult to explain this other than in terms of the loss of some protective influence of oestrogen. Nonetheless, the evidence that the greater risk of CHD in men is related to androgenic effects has much to commend it.

Serum lipoprotein (a) (Lp(a)) does appear to be decreased by HRT |**17**| and this may be because such therapy counteracts the tendency for it to rise after the menopause. Lp(a) has not proved to be strongly related to CHD risk in prospective (as opposed to case–control) studies, although most of these studies have involved men |**18**|. Furthermore there is no randomized placebo-controlled clinical events trial of a therapy which will specifically decrease Lp(a) so its status as a CHD risk factor is uncertain. Nonetheless in HERS, in which HRT was clearly associated with a decrease in Lp(a) levels, a *post hoc* subgroup analysis did show that there was a decrease in CHD events in those women in the highest quartile for Lp(a) [19]. It must be admitted that there was an increased risk of CHD in the lowest quartile of Lp(a) with HRT so it is not certain whether this can be interpreted as indicating a decreased adverse effect of HRT on CHD risk in those with higher Lp(a) levels.

The effect of androgens is to decrease HDL and serum triglycerides and to increase LDL cholesterol, whereas the effect of oestrogen is to raise HDL and triglycerides and decrease LDL cholesterol [20]. The changes in lipoproteins with age in men and women seem to fit with this to the extent that the decrease in HDL and the rise in LDL could at puberty be androgen-induced. The increase in LDL in women at the menopause would be best explained by the decrease in oestrogen, but this does not easily explain why the levels rise to such an extent that they are higher than in men. Serum triglycerides too are not easily explicable. From the action of oestrogens and androgens it might be anticipated that they would follow the reverse of what they actually do, so that levels would be higher in women than in men. It is possible that triglyceride effects are driven more by nutritional preferences and by the effects of increasing obesity and insulin resistance with age. Indeed even the gender differences in HDL cholesterol can be influenced by nutritional and environmental factors [21].

The effect of progestogens depends greatly on their androgenicity and thus their ability to oppose oestrogen effects [1]. The rise in cholesterol during pregnancy may be aided and abetted by other hormones in addition to progesterone. Experience with the combined oral contraceptive (COC) pill raises issues about progestogens. Use of these agents has been associated with a small, but finite increase in risk of acute myocardial infarction [22]. This has never been definitely shown to involve the formation of an atherosclerotic plaque and there is no associated matching increase in angina pectoris as might be expected if atherosclerosis were its basis. The possibility has been raised that thrombosis in non-atherosclerotic coronary arteries (perhaps associated with a migraine-like spasm) could be its basis. Nonetheless, the notion exists that the decrease in HDL cholesterol due to the androgenicity of the progestogenic component of COCs, such as levonorgestrel, might be involved in the increased risk of acute myocardial infarction. The development of COCs containing the less androgenic progestogens, desogestrel and gestodene, was largely driven by this concept. These agents do not lower HDL cholesterol, but because acute myocardial infarction is so rare, studies have thus far been underpowered to establish that these agents do definitely reduce the risk of acute myocardial infarction. Furthermore, the use of these agents has now become controversial because of suggestions that they may be associated with higher rates of venous thromboembolism than earlier agents with similar doses of oestrogen, but more androgenic progestogens [22]. If true, this might hypothetically be because of a weaker opposition by the less androgenic progestogens to the procoagulant effects of oestrogen. Again, uncertainty persists.

There are observations which would suggest that the endocrine effects of male puberty are not simply due to androgens. For example, exogenous testosterone increases HDL in hypogonadal men (see |**23**| for review) and most studies show a positive correlation between plasma testosterone and HDL in adult men |**24**|. Furthermore, higher oestrogen levels in men have been reported in association with myocardial infarction |**25**|. In women, on the other hand, with polycystic ovary syndrome, increased free androgen levels are associated with decreased HDL cholesterol |**26**|. The responses to gonadal steroids may therefore be different in men and women and perhaps vary according to their stage of sexual development. This may depend on the tissue expression of their respective androgen and oestrogen receptors. For example, evidence has recently been summarized that the monocyte-derived macrophages from men, but not women, may accumulate cholesteryl ester in response to androgen |**23**|. This may indicate different androgen receptor regulation of scavenger receptor expression in men and women. It is also naïve to regard such events at puberty and the climacteric as simply reflecting changing levels of androgens in man and of oestrogen in women. Changes in the levels of oestrogen also occur in men and of androgens in women.

Women with insulin resistance, whether or not it is associated with diabetes, have low levels of HDL. They are at increased risk of CHD and diabetic women have a CHD risk which, although less, tends to be much closer to that of their male counterparts than that of non-diabetic women |**3**|. The HDL cholesterol is also decreased in diabetic men and in both men and women the decrease in HDL may precede the onset of glycaemia of diabetic proportions by many years, at a stage when such people would be labelled as having insulin resistance or impaired fasting glucose |**27**|. In women, insulin resistance may contribute to the low HDL cholesterol by a mechanism involving androgenization. In states of insulin resistance, sex hormone binding globulin (SHBG) levels frequently decline. This leads to an increase in free sex steroid levels, particularly androgens. In the most extreme form this causes the masculinization seen in polycystic ovary syndrome and Dunnigan–Kobberling syndrome |**28**|. The mechanism for decreased SHBG in insulin resistance may be that the high insulin levels, occurring as a response to the insulin resistance, suppress hepatic SHBG production |**29**|. In the more common type 2 diabetes and insulin resistance, a tendency to a male pattern of adipose tissue distribution |**6**| may be the manifestation of increased circulating free androgens. Androgenization therefore appears capable of rendering adult women coronary-prone |**6**|.

Thus, in conclusion, it is important that the gender difference in CHD risk is not simply regarded as due to the 'protective' effect of oestrogen. The strongest evidence for that view is perhaps the increased risk in women who have undergone premature menopause and the rise in LDL cholesterol post-menopausally. It is not, however, supported by the changes in HDL throughout life nor by those clinical trials involving oestrogen, which have so far been undertaken. There is thus much about the effect of sex steroids, their binding proteins and receptor expression and other hormones, including insulin, insulin-like growth factor and pregnancy hormones on lipoprotein metabolism, which, were it understood, could clarify some of the present uncertainties.

References

1. Godsland IF, Wynn V, Crook D, Miller NE. Sex, plasma lipoproteins and atherosclerosis. Prevailing assumption and outstanding questions. *Am Heart J* 1987; **114**: 1467–503.
2. Slack J. Risk of ischaemic heart disease in familial hyperlipoproteinaemic states. *Lancet* 1969; **ii**: 1380–2.
3. Barrett-Connor EL, Cohn BA, Wingaard DL, Edelstein SL. Why is diabetes mellitus a stronger risk factor for fatal ischaemic heart disease in women than in men? The Rancho Benardo Study. *JAMA* 1991; **265**: 627–31.
4. Thompson GR, Miller JP, Breslow JL. Improved survival of patients with homozygous familial hypercholesterolaemia treated with plasma exchange. *Br Med J* 1985; **295**: 1671–3.
5. Heiberg A, Slack J. Family similarities in the age at coronary death in familial hypercholesterolaemia. Long-term follow-up of 96 patients. *Arteriosclerosis* 1988; **8**: 163–7.
6. Barrett-Connor E. Sex differences in coronary heart disease. Why are women so superior? The 1995 Ancel Keys Lecture. *Circulation* 1997; **95**: 252–64.
7. Rifkind BM, Segal P. Lipid Research Clinics Program reference values of hyperlipidaemia and hypolipidaemia. *JAMA* 1983; **250**: 1869–72.
8. Christensen B, Glueck C, Kwiterovich P, *et al.* Plasma cholesterol and triglyceride distributions in 13 665 children and adolescents: the prevalence study of the Lipid Research Clinics Program. *Pediat Res* 1980; **14**: 194–202.
9. Huxley J. Why do more boy babies die than girls? *Essays in Popular Science.* London: Penguin Book Ltd, 1926: 55–9.
10. Coronary Drug Project Research Group. The Coronary Drug Project. Findings leading to discontinuation of the 2.5 mg/day estrogen group. *JAMA* 1973; **226**: 652–7.
11. Hulley S, Grady D, Bush T, Furberg C, Henrington D, Riggs B, Vittinghoff E, for the Heart and Estrogen/progestin Replacement Study (HERS) Research Group. Randomized trial of estrogen plus progestin for secondary prevention of coronary heart disease in post-menopausal women. *JAMA* 1998; **280**: 605–13.
12. Writing Group for the Women's Health Initiative Investigators. Risks and benefits of estrogen plus progestin in healthy post-menopausal women. Principal results from the Women's Health Initiative Randomized Controlled Trial. *JAMA* 2002; **288**: 321–33.
13. Grady D, Herrington D, Bittner V, Blumenthal R, Davidson M, Hlatsky M, *et al.*, for the HERS Research Group. Cardiovascular disease outcomes during 6.8 years of hormone therapy. Heart and estrogen/progestin replacement study follow-up (HERS II). *JAMA* 2002; **288**: 49–57.

14. Sznajderman M, Oliver MF. Spontaneous premature menopause ischaemic heart disease and serum lipids. *Lancet* 1963; **i**: 962–5.

15. Robinson RW, Hagano N, Cohen WD. Increased incidence of coronary heart disease in women castrated prior to menopause. *Arch Intern Med* 1959; **104**: 908–13.

16. Parrish HM, Carr CA, Hall DG, King TM. Time interval from castration in premenopausal women to development of excessive coronary atherosclerosis. *Am J Obstet Gynecol* 1967; **99**: 155–62.

17. Espeland MA, Marcovina SM, Miller V, Wood PD, Wasilauskas C, Sherwin R, *et al.*, for the PEPI Investigators. Effect of post-menopausal hormone therapy on lipoprotein (a) concentration. *Circulation* 1998; **97**: 979–86.

18. Sweetnam PM, Bolton CH, Downs LG, Durrington PN, Mackness MI, Elwood PC, Yarnell JWG. Apolipoprotein A-I, A-II and B, lipoprotein (a) and the risk of ischaemic heart disease: the Caerphilly Study. *Eur J Clin Invest* 2000; **30**: 947–56.

19. Shlipak MG, Simon JA, Vittinghoff E, Lin F, Barrett-Connor E, Knopp RH, Levy RI, Hulley SB. Estrogen and progestin, lipoprotein (a) and the risk of recurrent coronary heart disease events after menopause. *JAMA* 2000; **283**: 1845–52.

20. Furman RH, Alaupovic P, Howard RP. Hormones and lipoproteins. *Progr Biochem Pharmacol* 1967; **2**: 215–58.

21. Davis C, Williams DH, Organov RG, Tao S-C, Rywik SL, Stein Y, Little JA. Sex differences in high density lipoprotein cholesterol in six countries. *Am J Epidemiol* 1996; **143**: 1100–6.

22. Report of a WHO Scientific Group. Cardiovascular disease and steroid hormone contraception. *WHO Technical Report Series* 877. Geneva: World Health Organisation, 1998.

23. Ng MKC, Jessup W, Celermajer DS. Sex-related differences in the regulation of macrophage cholesterol metabolism. *Curr Opin Lipidol* 2001; **12**: 505–10.

24. Khaw KT, Barrett-Connor E. Endogenous sex hormones, high density lipoprotein cholesterol and other lipoprotein fractions in men. *Arterioscler Thromb* 1991; **11**: 489–94.

25. Luria MH, Johnson MW, Pego R, *et al.* Relationship between sex hormones, myocardial infarction and occlusive coronary disease. *Ann Intern Med* 1982; **142**: 42–4.

26. Conway GS, Argawal R, Betteridge DJ, Jacobs HS. Risk factors for coronary artery disease in lean and obese women with polycystic ovary syndrome. *Clin Endocrinol* 1992; **37**: 119–25.

27. Haffner SM, Stern MP, Hazuda HP, Mitchell BD, Patterson JK. Cardiovascular risk factors in confirmed pre-diabetic individuals: does the clock for coronary disease start ticking before the onset of clinical diabetes? *JAMA* 1990; **263**: 2893–8.

28. Kumar S, Durrington PN, O'Rahilly S, Laing I, Humphries PJ, Olukoga AO, Bhatnagar D, Mackness MI, Davis JRE, Boulton AJM. Severe insulin resistance, diabetes mellitus, hypertriglyceridaemia and pseudoacromegaly. *J Clin End Metab* 1996; **81**: 3465–8.

29. Plymate SR, Jones RE, Matej RJ, Friedl KE. Inhibition of sex hormone binding globulin production in the human hepatoma (Hep-G2) cell line by insulin and prolactin. *J Clin Endocrinol Metab* 1988; **67**: 460–4.

List of Abbreviations

ABC	ATP-binding cassette
ABL	abetalipoproteinaemia
ACAT	acyl-CoA cholesterol:acyltransferase
ACE	acetylcholine esterase/angiotensin converting enzyme
Ach	acetylcholine
ACS	acute coronary syndrome
Act	Actinomycin
ADA	American Diabetes Association
ADMA	asymmetric dimethylarginine
ADVERT	Atorvastatin Versus Revascularization Treatment
AFCAPS/ TexCAPS	Air Force/Texas Coronary Atherosclerosis Prevention Study
AHA	American Heart Association
AIR	acute insulin secretory response
ALERT	Assessment of Lescol in Renal Transplantation
ALP	atherogenic lipoprotein phenotype
ALT	alanine aminotransferase
AMBP	ambulatory BP
AMI	acute myocardial infarction
ANOVA	analysis of variance
Apo	apolipoprotein
ApoB	apolipoprotein B100 gene
ARH	autosomal recessive hypercholesterolaemia
ASPEN	Atorvastatin as Prevention of coronary heart diseases in patients with type 2 diabetes mellitus
ASCVD	atherosclerotic cardiovascular disease
AST	aspartate aminotransferase
AT	Achilles tendon
ATP	adenosine triphosphate
ATP	Adult Treatment Panel
AUC	area under the curve
azPC	azelaoyl phosphatidylcholine
BH4	tetrahydrobiopteryn
BIP	Bezafibrate Infarction Prevention
BMD	bone mineral density
BMI	body mass index
CABG	coronary artery bypass grafting
CAD	coronary artery disease
CARDIA	Coronary Artery Risk Development in Young Adults study
CARDS	Collaborative Atorvastatin Diabetes study
CARE	Cholesterol and Recurrent Events study
CDK	cyclin dependent kinase
CE	cholesteryl ester
CETP	cholesteryl ester transfer protein
CGN	chronic glomerulonephritis
CH	combined hyperlipidaemia
CHAOS	Cambridge Heart Antioxidant study
CHD	coronary heart disease
CI	confidence interval
CK	creatine kinase
CKMB	creatine kinase MB
COC	combined oral contraceptive
CPK	creatinine phosphokinase

CRD	chylomicron retention disease
CRP	c-reactive protein
CVD	cardiovascular disease
CVE	cardiovascular event
CyA	cyclosporin-A
CYP-450	cytochrome P450
DAIS	Diabetes Atherosclerosis Intervention Study
DBP	diastolic blood pressure
DESIR	data from an epidemiological study on the insulin resistance syndrome study
DGGE	denaturing gradient gel electrophoresis
DHA	docosahexaenoic acid
DIGAMI	diabetes mellitus insulin-glucose infusion in acute myocardial infarction
DM	diabetes mellitus
DMPC	dimyristoyl phosphatidylcholine
ELISA	enzyme linked immunosorbent assay
ENOS	endothelial nitric oxide synthase
EPA	eicosapentaenoic acid
EUROASPIRE	European Action on Secondary Prevention by Intervention to Reduce Events
EXCEL	Expanded Clinical Evaluation of Lovastatin
F	farnesol
FA	fatty acid
FC	free cholesterol
FCH	familial combined hyperlipidaemia
FCR	fractional catabolic rate
FDB	familial defective apolipoprotein B-100
FH	familial hypercholesterolaemia
FHBL	familial hypobetalipoproteinaemia
FHTG	familial hypertriglyceridaemia
FIELD	Fenofibrate Intervention and Event Lowering in Diabetes
Flu	fluvastatin
FMD	flow-mediated dilatation
FOS	Framingham Offspring Study
FPG	fasting plasma glucose
FTR	fractional transfer rate
FXR	farnesoid X-activated receptor
GG	geranylgeraniol
GISSI	Gruppo Italiano per lo Studio della Sopravvivenza nell'Infarto Miocardico
GR	good responder
GOSH	Great Ormond Street Hospital for Children
GRAS	Global Risk Assessment Score
GREACE	Greek Atorvastatin and Coronary-heart-disease Evaluation Trial
GTN	glyceryl trinitrate
GVD	graft vessel disease
HAART	highly active antiretroviral therapy
HAEC	human aortic endothelial cells
HbA_{1c}	glycosylated haemoglobin
HBL	hypobetalipoproteinaemia
HCV	hepatitis C virus
HDL	high-density lipoprotein
HDL-C	HDL cholesterol
HELP	heparin-induced extracorporeal LDL precipitation
HERITAGE	Health, Risk Factors, Exercise Training and Genetics
HERS	Heart and Estrogen/progestin Replacement Study
HF	high flavonoid
hFH	heterozygous familial hypercholesterolaemia
HG	high glucose

HIV	human immunodeficiency virus
HMG-CoA	3-hydroxy-3-methylglutaryl coenzyme A
HoFH	homozygous familial hypercholesterolaemia
HOMA-IR	homeostasis model assessment for insulin resistance
HOPE	Heart Outcomes Prevention Evaluation
HPS	Heart Protection Study
HR	hazard ratio
HRT	hormone replacement therapy
hs-CRP	high-sensitivity C-reactive protein
HTG	hypertriglyceridaemia
HUVEC	human umbilical vein endothelial cells
IAUC	incremental area under the curve
IDL	intermediate density lipoprotein
IFG	impaired fasting glucose
IGT	impaired glucose tolerance
IL	interleukin
ILSA	Italian Longitudinal Study on Aging
IMT	intima-media complex thickness
IRS	insulin resistance syndrome
LBW	low birth weight
LCAS	Lipoprotein Coronary Atherosclerosis Study
LCAT	lecithin:cholesterol acyltransferase
LDL	low-density lipoprotein
LDL-C	low-density lipoprotein cholesterol
LDLR	low-density lipoprotein receptor
LDS	Lipids in Diabetes Study
LF	low flavonoid
LIPID	Long-term Intervention with Pravastatin in Ischaemic Disease
LOD	logarithm of the odds
Lp(a)	lipoprotein a
LPL	lipoprotein lipase
LPP	lipid peroxide products
LRP	low-density lipoprotein receptor-related protein
lyso-PC	lysophosphatidylcholine
M	insulin action
MACE	major adverse cardiac event
MC	mesangial cell
MCP-1	monocyte chemoattractant protein 1
MDA	malondialdehyde
Mev	mevalonate
MI	myocardial infarction
MIRACL	Myocardial Ischaemia Reduction with Aggressive Cholesterol Lowering study
MKP-1	mitogen-activated protein kinase phosphatase 1
MM-LDL	minimally modified LDL
MMP	matrix metalloproteinase
MNC	mononuclear cell
MRFIT	Multiple Risk Factor Intervention Trial
mRNA	messenger ribonucleic acid
MTHFR	methylenetetrahydrofolate reductase
MTP	microsomal TG transfer protein
MVA	mevalonic acid
N	niacin
NASH	non-alcoholic steatohepatitis
NCEP	National Cholesterol Education Program
NGT	normal glucose tolerance
NHANES	National Health and Nutrition Examination Survey
NIH	National Institute of Health
NL	normolipidaemic
NMR	nuclear magnetic resonance
NNT	number needed to treat

NO	nitric oxide
NPL	non-parametric linkage
NS	not significant
NYHA	New York Heart Association
OR	odds ratio
OSTPRE	Osteoporosis Risk Factor and Prevention study
ox-LDL	oxidized low-density lipoprotein
ox-PAPC	oxidized-l-alpha-1-palmitoyl-2-arachidonoyl-sn-glycero-3-phosphocholine
PAFAH	platelet-activating factor acetylhydrolase
PAI-1	plasminogen activator inhibitor-1
PBC	primary biliary cirrhosis
PCI	percutaneous coronary intervention
PCR	polymerase chain reaction
PDC	proportion of days covered
PE	pulmonary embolism
PIB	partial ileal bypass
PKC	protein kinase C
PLTP	phospholipid transfer protein
PMN	polymorphonuclear leukocyte
PON1	paraoxonase-1
POSCH	Program on the Surgical Control of the Hyperlipidaemias
PP	pulse pressure
PPAR-γ	peroxisome proliferator-activated receptor-γ
PPRE	PPAR response element
PR	poor responder
pre-β-HDL	pre-β-migrating HDL
PROCAM	Prospective Cardiovascular Munster study
PROSPER	Prospective Study of Pravastatin in the Elderly at Risk
PSC	primary sclerosing cholangitis
PTB	phosphotyrosine binding
PTCA	percutaneous transluminal coronary angioplasty
PUFA	polyunsaturated fatty acid
PVD	peripheral vascular disease
RANTES	regulated on activation, normal T expressed and secreted
RAU	relative arbitrary units
RCT	randomized controlled trials
recHDL	reconstituted discoidal HDL
rHDL	reconstituted HDL
RLP	remnant-like particle
RLP-C	remnant-like particle-cholesterol
RLP-C	remnant lipoprotein cholesterol
RLP-TG	remnant-like particles triglyceride
RP	retinyl palmitate
RR	relative risk
RT-PCR	reverse transcription-PCR
4S	Simvastatin Scandinavian Survival Study
SBP	systolic blood pressure
SEARCH	Study of Effectiveness of Additional Reductions in Cholesterol and Homocysteine
SENDCAP	St. Mary's, Ealing, Northwick Park Diabetes Cardiovascular Disease Prevention study
Sf	Svedberg flotation rate
SHBG	sex-hormone binding globulin
S(I)	insulin sensitivity
SMC	smooth muscle cell
SNA	simvastatin and niacin (therapy)
SNP	sodium nitroprusside
Sq	squalene
SREBP	sterol regulatory element binding protein
SSCP	single strand conformational polymorphism

sVACM-1	soluble vascular adhesion molecule-1
SWITCH	Swiss Intervention Trial for lowering Cholesterol
SYMPHONY	sibrafiban *vs* aspirin to yield maximun protection from ischemic heart events post-acute coronary syndromes
T2D	type 2 diabetes
TC	total cholesterol
TD	Tangier disease
TG	triglyceride
TGRL	triglyceride-rich lipoproteins
TIMP	tissue inhibitor of metalloproteinase
TLC	therapeutic lifestyle changes
TNF-alpha	tumour necrosis factor-α
TNT	treat to new targets
TPE	therapeutic plasma exchange
TRL	triglyceride-rich lipoprotein
TZD	tiazolidinedione
UDCA	ursodeoxycholic acid
UKPDS	UK Prospective Diabetes Study
VA-HIT	Veterans Affairs Cooperative HDL Cholesterol Intervention Trial
VCAM	vascular cell adhesion molecule
VLDL	very low-density lipoprotein
WBC	white blood cell count
WHHL	Watanabe heritable hyperlipidaemic
WHO	World Health Organization
WHO MSVDD	WHO Multinational Study of Vascular Disease in Diabetes
WOSCOPS	West of Scotland Coronary Prevention Study

Index of Papers Reviewed

General index

Q

R

U

V

W

X

KEEPING UP TO DATE IN ONE VOLUME

Subject matters dealt with in previous volume

The Year in Dyslipidaemia 2002

Lipid disorders

Familial hypercholesterolaemia and related disorders
Hypertriglyceridaemia and mixed hyperlipidaemia
Other lipoprotein disorders
Alcohol, lipids and lipoproteins

Lipid-lowering trials

Trials with clinical end-points
Trials of therapy to modify serum lipids and lipoproteins
Clinical trials in the context of epidemiology and clinical practice
Effects of lipid-lowering drugs on non-lipid risk factors for coronary heart disease and other diseases
Coronary heart disease risk factors and primary prevention

Diabetic dyslipidaemia

Diabetic dyslipidaemia and coronary heart disease
Lipid intervention
Hypertension in patients with diabetes

Lipids and atherosclerosis

Non-lipid lowering effects of vastatins
High-density lipoprotein
Antioxidants and the Mediterranean diet

Current and future trends

Current and future trends in lipid-regulating therapy

Clinical Publishing Services Ltd
Oxford Centre for Innovation
Mill Street
Oxford OX2 0JX, UK

T: +44 1865 811116
F: +44 1865 251550
E: info@clinicalpublishing.co.uk